1910 1915 1920 1925

1909
Charles David Herrold's station broadcasts from San Jose.

1910
Wireless Ship Act of 1910.

1912
Wireless gains publicity by aiding rescue efforts from the *Titanic.*

1912
Radio Act of 1912.

1914
Amateur radio operators form the American Radio Relay League (A.R.R.L.).

1914–1918
Wireless used extensively in World War I.

1915
David Sarnoff writes famous memo on the future of wireless.

1917
Alexanderson's alternator takes on increased importance for international communication.

1919
Owen T. Young negotiates the formation of RCA.

1919
9XM at the University of Wisconsin, Madison, signs on the air. Becomes WHA in 1922.

1920
WWJ in Detroit begins intermittent broadcasting schedules in August.

1920
KDKA in Pittsburgh begins regular programming in November.

1921
Philo Farnsworth outlines to his science teacher the concept of electronic television.

1922–1925
National Radio Conferences. (Four held before new legislation)

1922–1923
National Association of Broadcasters (NAB) is formed.

1922
Toll broadcasting begins at WEAF.

1923
V.K. Zworykin patents the iconoscope pickup tube for television.

1924
International Business Machines Corporation (IBM) is formed.

1926
Case of *U.S. v. Zenith Radio Corp.*

1926
RCA forms subsidiary NBC to operate Red and Blue networks.

1927–1934
Radio Act of 1927 withstands challenges in court.

1927
Radio Act of 1927 is passed. Forms five-member Federal Radio Commission (FRC).

1928
CBS begins when interests are purchased by Wm. S. Paley and Congress Cigar Company.

1929
First NAB "Code of Ethics" is passed.

1929
First broadcast rating by Crosley Radio Company.

1930
Philo Farnsworth applies for permission to experiment with 300-line TV system.

1930
Zworykin visits Farnsworth labs in California to examine 300-line TV scanning system.

1931
Zworykin and RCA officials visit Farnsworth labs in California. RCA later enters royalty agreement with Farnsworth.

1931
KFKB (*Brinkley*) case is decided.

1932
Closed Circuit ETV begins at State University of Iowa.

1932
Shuler case is decided.

1933
President Roosevelt uses radio for "fireside chats" with the public.

1933
Edwin Armstrong demonstrates FM broadcasting for RCA.

1933
Press-Radio War ends with Biltmore agreement.

1934
Communications Act of 1934 is passed. Forms seven-member Federal Communications Commission (FCC). Independent regulatory body.

1934
Mutual network begins as four-station cooperative.

1910 1915 1920 1925 1930 1935

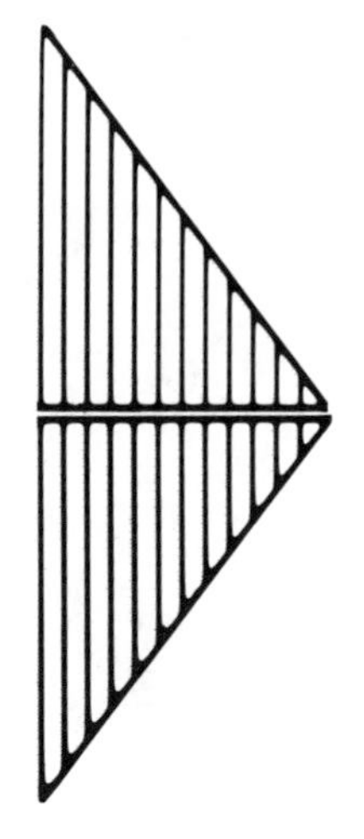

third edition

BROADCASTING and TELECOMMUNICATION an introduction

John R. Bittner

The University of North Carolina at Chapel Hill

PRENTICE HALL, Englewood Cliffs, New Jersey 07632

Library of Congress Cataloging-in-Publication Data

Bittner, John R.,
Broadcasting and telecommunication : an introduction / John R.
Bittner. -- 3rd ed.
p. cm.
Includes bibliographical references and index.
ISBN 0-13-083239-1 :
1. Broadcasting. I. Title.
PN1990.8.B5 1991
384.54--dc20 90-44783
CIP

Editorial/production supervision
and interior design: ***Joanne E. Jimenez***
Cover design: ***Ben Santora***
Manufacturing buyers: ***Debbie Kesar*** and ***Marianne Gloriande***

Research and development are key to network improvements at NYNEX Corporation. Pictured on the cover, from the NYNEX Science & Technology labs, are electrons bombarding an experimental conducting material.

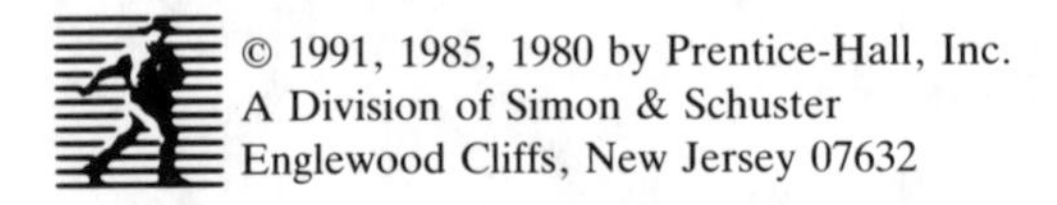

Printed in the United States of America

10 9 8 7 6 5 4 3 2 1

ISBN 0-13-083239-1

PRENTICE-HALL INTERNATIONAL (UK) LIMITED, London
PRENTICE-HALL OF AUSTRALIA PTY. LIMITED, Sydney
PRENTICE-HALL CANADA INC., Toronto
PRENTICE-HALL HISPANOAMERICANA, S.A., Mexico
PRENTICE-HALL OF INDIA PRIVATE LIMITED, New Delhi
PRENTICE-HALL OF JAPAN, INC., Tokyo
SIMON & SCHUSTER ASIA PTE. LTD., Singapore
EDITORA PRENTICE-HALL DO BRASIL, LTDA., Rio de Janeiro

for

FRANK UTLEY FLETCHER

Broadcast Communications Attorney

Friend

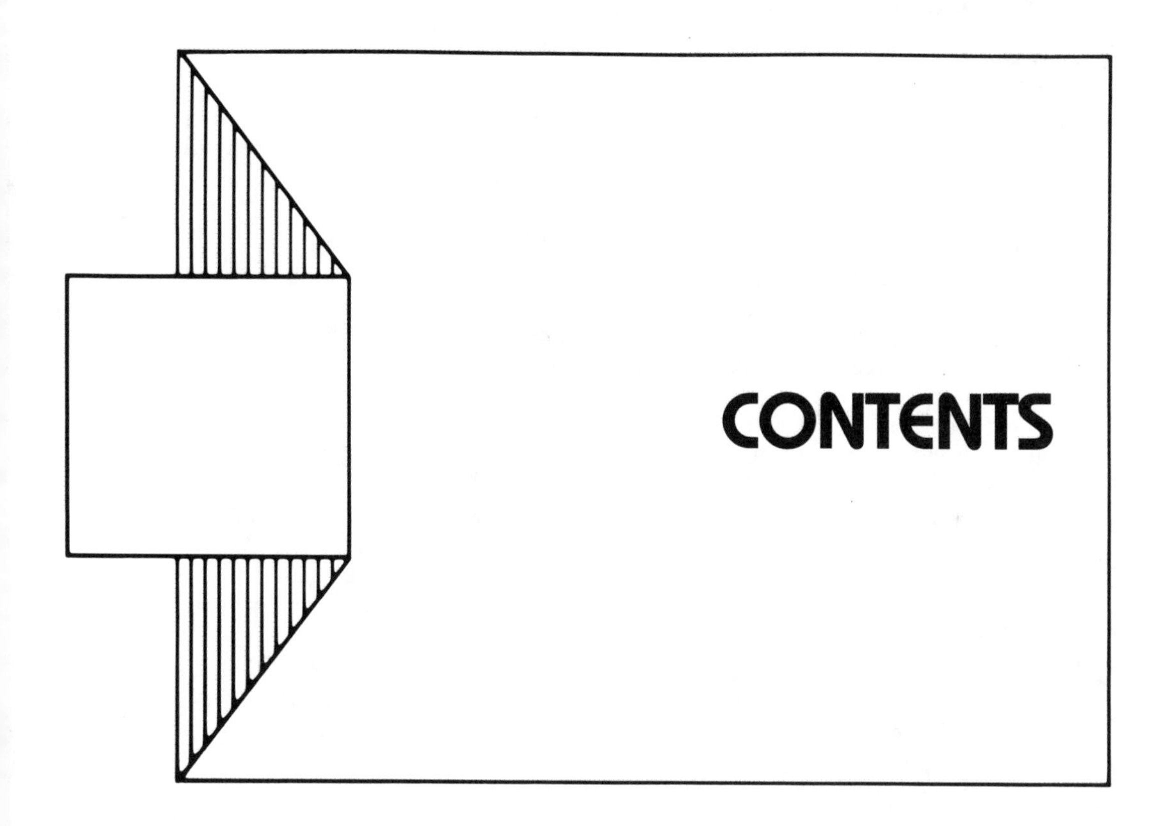

CONTENTS

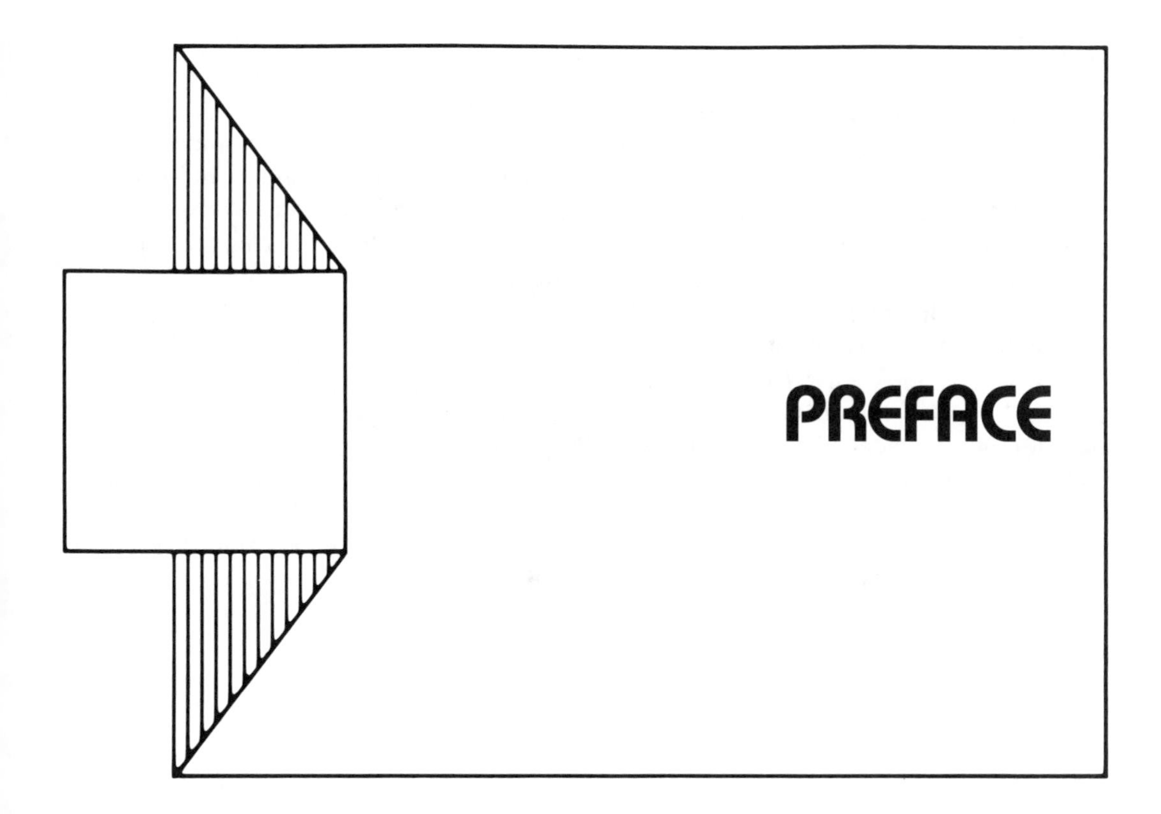

PREFACE

As with the previous editions, this third edition of the text utilizes a total approach to the study of broadcasting and telecommunication. The text incorporates practical applications and examples drawn from the industry as well as the more conceptual frameworks within which broadcasting and telecommunication operate.

The text is designed not only for students aspiring to be practicing professionals but also for students and instructors seeking a more liberal arts perspective, which also prepares students to be responsible consumers and users of broadcasting and telecommunication.

In addition to the strengths that instructors wanted retained from previous editions, this new edition of the text includes:

- **A new expanded emphasis on international broadcasting and telecommunication, including chapters dealing with international domestic, and international overseas systems. Developments in Eastern Europe, the Soviet Union, the two Chinas, Japan, the Middle East, and elsewhere are included, as are international illustrations that highlight other sections of the text.**
- **A new chapter on "Criticism and Ethics" with numerous "Questions for Critical Thinking" to sensitize students to the issues facing working professionals and consumers in our information society.**
- **A new chapter on "Telephone Systems and Fiber Optics" detailing the impact and explaining the operation and application of these technologies on our society.**

- **A new chapter on "High-Definition Television (HDTV) and Consumer Technologies" examining the operation and future of HDTV as well as such technologies as videotex, teletext, and others.**
- **A chapter devoted to "Programming" and the decisions program directors make in a competitive marketplace.**
- **A new chapter on "The Production Process" sketches the artistic decisions behind the development of radio and television programming.**
- **The latest developments in audience and effects research.**
- **Chapters treating such contemporary material as satellites and telecommunication in space; business and industrial television; personal computers; and other aspects of our information society.**
- **A comprehensive chapter on cable television.**

The text includes the pertinent historical material necessary for a responsible understanding of the discipline. A comprehensive *Instructor's Manual* also accompanies the text and information about other instructional aids can be obtained by writing: College Marketing, Prentice Hall, Englewood Cliffs, NJ 07632.

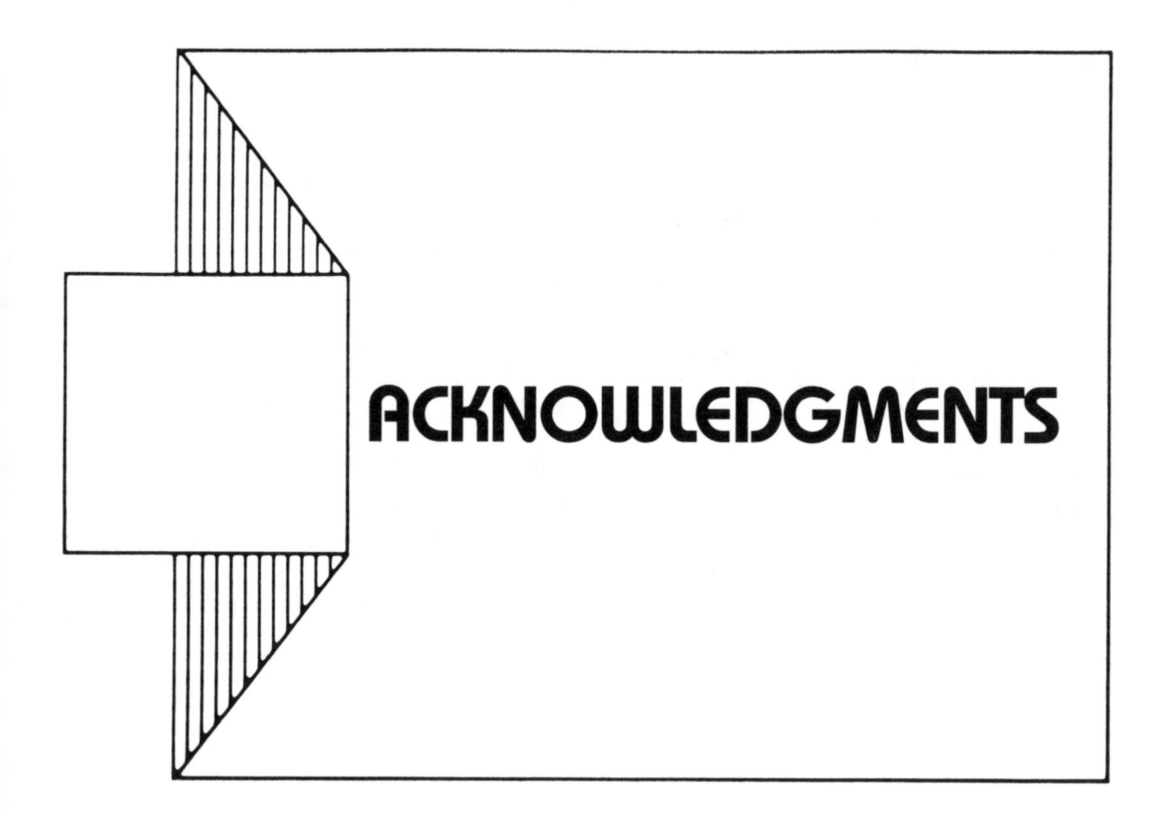

ACKNOWLEDGMENTS

The third edition of the text is being completed while doing research in Europe, made possible partly because of the supportive efforts of colleagues who tend to other duties in my absence.

At The University of North Carolina at Chapel Hill, special friends such as Seth Finn and Anne Johnston are bright spots in the usual daily activities of the university.

Loy Singleton, now serving as chair at the University of Alabama, is thoroughly missed in Chapel Hill, as are the other members of his fine family. Many times his insights became a spark that brought new light to an important concept.

Hal White, a superb communications attorney, author, expert in space law, and Special Deputy Attorney General of North Carolina, added insights to this text and to others that the author has written.

As always, I am indebted to students in my classes at The University of North Carolina at Chapel Hill. Faculty in the Department of Radio, Television and Motion Pictures, the School of Journalism, and the Department of Speech Communication assisted me through their expertise and friendships.

John E. Turtle of the BBC provided important help with illustrations, which will benefit future works.

Many talks, lunches, and the continued friendship of Frank U. Fletcher, special counsel of Capitol Broadcasting Co. and the "gentleman in residence" of the A. J. Fletcher Foundation, were always delightful breaks from the daily routine. Other friends at Capitol Broadcasting, Jefferson Pilot Communications, and the Village Companies are important to my various writing endeavors.

Susan M. Hill at the NAB's Library and Information Center, and Catherine Heinz at the Broadcast Pioneers Library helped through their literature and materials.

Having access to major research libraries at Duke University and elsewhere in the Research Triangle is a major asset for any researcher.

Special thanks go to Professor Edward Funkhouser of North Carolina State University, Professor Larry Z. Leslie of the University of South Florida, and Professor James L. Hoyt of the University of Wisconsin for reviewing this text.

Many people at Prentice Hall continue to help. On this text Joanne Jimenez serves as production editor. Within the rest of the Prentice Hall and Simon & Schuster organizations are so many friends that I must avoid forgetting anyone by offering a blanket "thank you" for all your support.

While this book was being completed, John Charles Bittner and Donald Warrick Bittner were graduated from college.

For so many who keep asking, Stormy is alive and well, but sleeps most of the time.

No one deserves more credit than Denise.

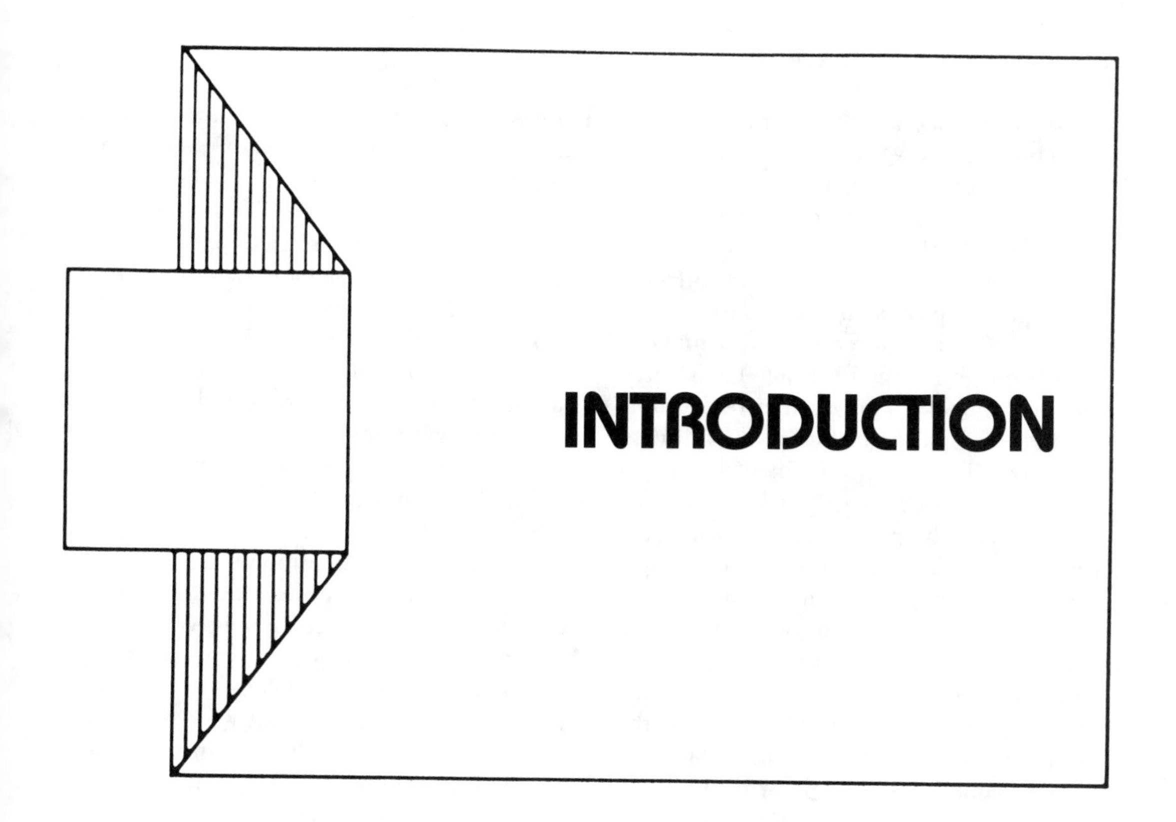

INTRODUCTION

The alarm clock, set earlier in winter since the ferries run only a staggered schedule, goes off at 5:00 AM. By the time the car's headlights slice through the fog over the Neuse River three hours later, the cup of hot chocolate on the dashboard tastes like a weak milk shake. The weather report on the car radio has changed from North Carolina inland crop reports to coastal tide levels. Now and then the waters of Pamlico Sound peek through the fog, and the lights of businesses greeting the dawn break through to join the sun's first rays.

Ahead lies Cedar Island, a corner of detached land lopped off the end of a peninsula that carries a roadway to its end at the ferry dock. There waiting for its first run of the day is the Cedar Island ferry. A crew member wearing a fluorescent orange vest asks for your ferry reservation number, motions you into line, and, if you timed your drive just right and didn't meet the drawbridge at Morehead City, directs you onto the ferry that will take you on the two-hour ride to Ocracoke Island.

If you are observant, you will find the ride to Ocracoke Island one of stark contrast. As the other passengers begin to emerge from their cars to go topside for a better view, the crew casts off the heavy lines, just as crews have been casting off heavy lines on ships for thousands of years. The big vessel grinds and inches forward, the pilings scraping and squeaking against the steel plates welded to the port and starboard. Gradually the speed increases and the bow begins to produce the first splash as it cuts through the calm waters of the channel.

As you look beyond the marsh grass of the harbor, beyond the waterfowl poised for their first morsel of morning seafood, you see the markers showing where crab traps, called "pots," have been lowered to the bottom of the sound by people who make a living from commercial fishing. Their catch will grace the tables of gourmet restaurants and the stick-to-the ribs fare of all-you-can-eat diners. They will put down dozens of crab pots and dozens of markers. To the passengers on the passing ferry it will seem impossible for anyone to locate the traps, even with markers. The small white buoys seem randomly thrown across hundreds of square miles of open water.

Soon the sights of land disappear and the passengers turn to walking around the decks, venturing inside the lounge, and leisurely breaking out a picnic brunch. If you are in the pilot house and are one of the crew you will see an array of sophisticated equipment that includes the green-glowing scope of the depth finders, the frequency markers on the marine radio, and the readout of the radar antenna circling a few feet above. In the distance you may occasionally catch a glimpse of a marine research vessel with its dishlike satellite antenna transmitting and receiving data from satellites 22,000 miles above the earth. The satellite navigation system can tell the location of the research vessel to the nearest foot, a job in years past delegated to the mariner's sextant.

On board the conversations pick up. People who have never met begin talking with each other. The inhibitions and barriers to conversing with strangers begin to fall. Everyone has something in common—they are all on the same vessel at the same time going to the same destination. Most are dressed casually. The conversations will be about where they live, where this trip is eventually taking them, and where is the best place to eat and stay on Ocracoke Island. Some will be retirees enjoying the freedom to travel to places they have never been. Others have people to see on Ocracoke. Still others are couples looking forward to a romantic weekend on the unspoiled beaches of this remote island.

About two hours from Cedar Island the outline of Ocracoke Island begins to appear on the horizon. First the water tower, then the line of trees, then the old lighthouse, then the telephone company's microwave tower. With all the sophisticated navigation equipment aboard oceangoing vessels, the lighthouse still shines as a beacon to craft of all sizes. It is Ocracoke's most famous landmark, but certainly not what the island is most famous for. That distinction is reserved for Blackbeard, the pirate who used Ocracoke's protected harbor as a refuge from the Crown governor.

As the ferry gets closer to Ocracoke the Coast Guard station begins to appear, and in a short time the ferry bears around the channel marker and heads into the harbor. Most of the cars venture toward motels or the beaches, or to the end of the island, where another ferry will take them north to Cape Hatteras. A telephone-company engineer will spend the afternoon on Ocracoke working at the phone company's microwave relay station. The high-technology substation is critical to the island's communication links, for it provides telephone service for the islanders. For some of the residents it links personal computer terminals with data banks thousands of miles away. For the island's Health Center it is a vital link with the mainland.

The day will go quickly for everyone who made the trip. The day on the beach is always too short, the things to see too many. Soon evening will send shadows of twisted cedar trees across sandy sidewalks and shell-laced roadways.

The sun is beginning to set on Ocracoke. Before it rises tomorrow, the fishermen will have found their crab-pot markers, the radio stations on the mainland will be broadcasting the tide levels, the passengers on the early ferry will be chatting and exchanging greetings, and the microwave tower will be receiving and transmitting data at the speed of light. It is an island of contrasts.

An understanding of these contrasts is necessary to an understanding of this book. Broadcasting and telecommunication are part of this contrast. In addition, broadcasting and telecommunication are part of rapid technological change—change that is just as far-reaching as the differences between the hand-held sextant of a ship and the satellite-navigation equipment of today's oceangoing vessel. To understand and appreciate these changes fully it is necessary to first understand the process of communication and then explore how broadcasting and telecommunication fit into the process. We will begin not with the radio station or the satellite-navigation system but with crab pots and people who fish the open water off Ocracoke Island

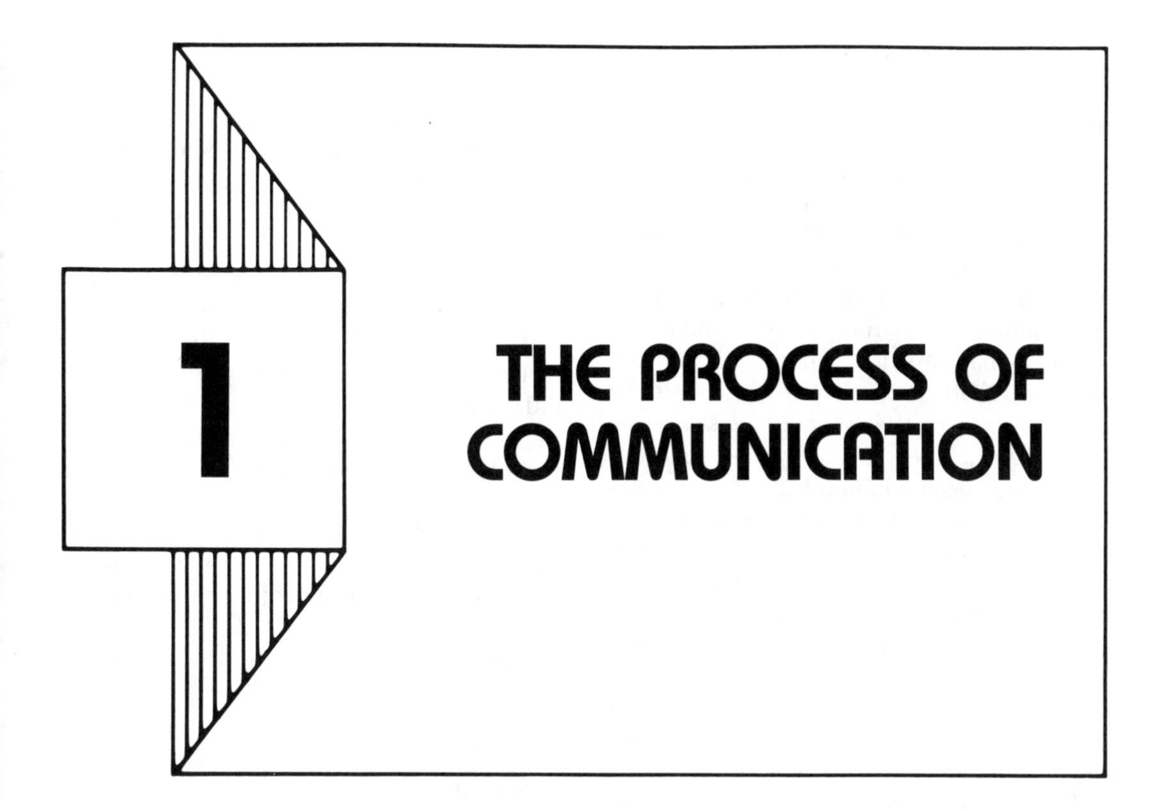

For the fishing boats leaving Ocracoke Island to plant crab pots, communication is an important part of their captain's day. Although they may check a local marine radio station for the weather forecast, their most important asset is their ability to communicate with themselves. That may sound somewhat strange when we consider we are studying broadcasting and telecommunication, but intrapersonal communication, communication within ourselves, is the foundation of all other types of communication.

The people who are scattering the crab pots across open water must go back and find those same pots without the aid of anything but their own instinct and knowledge of the water. The local radio stations do not tell them where their crab pots are located. Nor do their friends. Through years of accumulating bits of information too small to notice, they are able to navigate the waters off Ocracoke.

To understand this process of communication better we will begin by examining three terms: *transmit*, *transfer*, and *transact*.

UNDERSTANDING THE PROCESS OF COMMUNICATION

How would you describe the process of communication? If at first it seems difficult, do not be too disappointed. People who spend their lives researching the subject continue to argue about the process.

Transmit

Whenever we begin discussing communication the term *transmit* pops up. Transmit means to send information.[1] Yet if we transmit something, are we communicating? Consider the person who stands on a hilltop and shouts across the valley to hear the echo. Is that person communicating? Consider the football coach who comes off the sidelines to yell at a referee. Certainly the football coach is transmitting information. But is the coach communicating? Consider the student who tells her roommate to clean up their room. She has transmitted information, but two days later the room remains a mess. What if a television anchorperson asks viewers to write to the station about a community issue but only one viewer replies? Did communication take place? What about the disc jockey who finds she had the smallest number of listeners in the station's coverage area and is told by her boss to begin looking for another job? Was the disc jockey communicating with the listeners?

In all of our examples information was transmitted, but in each case we must ask if the information was received. If it was not, did communication take place?

Transfer

Another term that frequently pops up when we discuss communication is *transfer*. Transfer means to send and receive information. Stop and consider the examples we used. Is the television anchorperson who asks viewers to write the station transferring information? Is the person who stands on the hilltop shouting across the valley to hear the echo transferring information? Certainly that person is transmitting, but does transfer take place if no one hears the shouting? What occurs if someone on the other side of the hilltop shouts back? Does communication take place? Now let's consider the roommate. What if she heard the request but was too busy to clean up the room and did not respond? Was information transferred? Did communication take place? What about the football coach? What if the referee refuses to change the call after the coach yells from the sidelines? A transfer of information took place, but did communication take place? What about the disc jockey who had the smallest number of listeners? Transmission took place, but did transfer occur? Suppose the boss who tells the disc jockey to find work elsewhere leaves but the disc jockey is so shocked by the ratings that she blocks out of her mind her boss's words. The boss has transmitted information, but did she transfer information?

Transaction

A third term that frequently crops up in discussions of communication is *transact*. What happens during transaction? Transaction means information is sent and received and feedback occurs. For example, if the anchorperson's request for mail results in a flood of letters, then information has been transmitted and received and more information sent back to the station through viewers' letters. We might suggest that if the person yelling across the valley hears a reply and then decides to yell back, transaction has taken place. If the roommate at least acknowledges she heard the request to clean up the room, even though she doesn't clean it, transaction has occurred. And what if the referee refuses to change the call but nevertheless has a healthy argument with the football coach? Transaction has occurred. If the disc jockey replies to her boss, "I guess you're right—I'll start looking tomorrow," has communication taken place?

The Dictionary Examines Communication

To go one step further in better understanding what lies behind the process of communication we can consult the dictionary. Our dictionary definition of communicate uses the phrase "to make known; impart; transmit." All of our examples would agree with this phrase—the television anchorperson, the person shouting from the hilltop, the roommate, the football coach, and the disc jockey and her boss. The dictionary also defines communication as "having an interchange of thoughts or ideas, of sharing." This phrase is closest to our term transaction, but certainly includes transfer as well. If we are talking to our instructor about grades on an examination, we are having an interchange of thoughts and ideas. If we are participating in a class discussion, we are having an interchange of thoughts or ideas.

Now let's consider further the dictionary definition of communication and see how it might apply even more directly to the major emphasis of this text. According to our dictionary, communication can also be defined as "a system of sending and receiving messages, as by telephone, television, or computer." We will find later in the text that all three of these technologies play an important role in our understanding of broadcasting and telecommunication. For example, telephone cables may be used to carry television signals between cities. These same cables may provide two-way communication between points hundreds of miles away. This two-way data link is, as we will learn, more characteristic of telecommunication than broadcasting. As the operator interacts with the computer, electronic transaction is taking place. The operator sends information and the computer responds with information that appears on the operator's display terminal. Thus, whereas the examples we first used—the television anchorperson, the football coach, the roommate, the person shouting from the hilltop, and the disc jockey and her boss—were primarily examples of human communication, broadcasting and telecommunication are forms of electronic communication.

We can see how difficult it is to arrive at a single definition of communication. Yet all of the definitions we have considered are correct. It's a dynamic process. It involves information that is transmitted and transferred, and information that becomes part of a communicative transaction. Keeping in mind that there are hundreds of definitions of the term, we will settle on a specific definition for the purposes of this book: *communication is the movement of messages between senders and receivers.*

DISTINGUISHING AMONG TYPES OF COMMUNICATION

Not only are there characteristics that differentiate types of communication such as broadcasting and telecommunication but there are more general distinctions among intrapersonal, interpersonal, and mass communication. As we have seen, *intrapersonal communication* is communication within ourselves, *interpersonal communication* is communication between two or more people in a face-to-face situation.[2]

INTRAPERSONAL COMMUNICATION

As we have learned, the people who spread traps over open water to catch crabs use intrapersonal communication to process the information that tells them where the trap markers can be found. For the people who make their livelihood in this way, a degree of what they might call pure instinct comes into play.

Our senses, our nervous system, and our brain are the main physiological components of the communication process. For example, if we are watching an instructional television program about basic mathematics, our eyes and ears respond to what is on the screen. These two senses of sight and hearing send electrochemical impulses through our nervous system to our brain. After receiving the impulses, our brain feeds back other impulses to our motor nerves, the nerves that influence movement and enable us to pick up a pencil and paper and work the math problem. Different components of the communication process have come into play: the *sender* (eyes and ears), *message* (electrochemical impulses), *medium* (nervous system), *receiver* (brain and central nervous system), and *feedback* (electrochemical impulses).[3] Another component, *noise*, can interfere with the communication process. Your head may ache to the point where you cannot think. A sickness or injury may damage your nervous system, either interrupting the passage of electrochemical impulses or interfering with your ability to respond to commands given your motor nerves by your brain. All of these are examples of one type of noise, *physical noise*.

Intrapersonal communication is the foundation for adapting interpersonally to others. We process all kinds of stimuli when we communicate with other people. Using that intrapersonal processing to communicate interpersonally is one of the most important functions of our internal processing systems.

INTERPERSONAL COMMUNICATION

On the Ocracoke ferry, despite the whir of the radar and the flashing digital navigation devices, interpersonal communication dominates. People talk with other people more freely than they would elsewhere. As we have mentioned, they all have something in common—they are on the same vessel going to the same destination.

As we have learned, interpersonal communication is communication in a face-to-face situation between at least two persons, and often many more, such as a group discussion or a speech to an audience. In interpersonal communication the names of the components of communication are the same as in intrapersonal communication, but the components themselves are different. To continue with our previous example, imagine that instead of watching an instructional television program about mathematics you are attending the instructor's class in person. Now the instructor becomes the sender of communication; the messages become the words spoken by the instructor; the medium is the human voice; and you are the receiver of communication. If you do not understand something the instructor is saying, you can immediately raise your hand to ask her a question. Your hand being raised is a form of feedback to the instructor.

Using our example of the lecture, we can begin to see the reason communication can be referred to as a sharing process.[4] If constructive communication is to take place, we must share certain things with the instructor. One way of examining the process of sharing is to understand that each individual has a field of experience—the accumulation of knowledge, experiences, values, beliefs, and other qualities that constitutes one's self. For effective communication to take place, these *fields of experience must overlap*, a condition we have come to call *homophily*, as seen in Figure 1–1. We must share certain things with another individual. When homophily exists, and our fields of experience overlap, we are more likely to communicate with each other and understand each other.[5] In our example, we must

FIGURE 1–1 The presence of a mass medium, limited sensory channels, a gatekeeper, and delayed feedback are characteristics that distinguish mass communication from intrapersonal and interpersonal communication. Senders, gatekeepers, and receivers all have fields of experience that can overlap and aid in understanding messages. Gatekeepers can be one individual, or an entire organization. Messages disseminated through mass communication both influence and represent the cultures within which the messages are seen and heard.

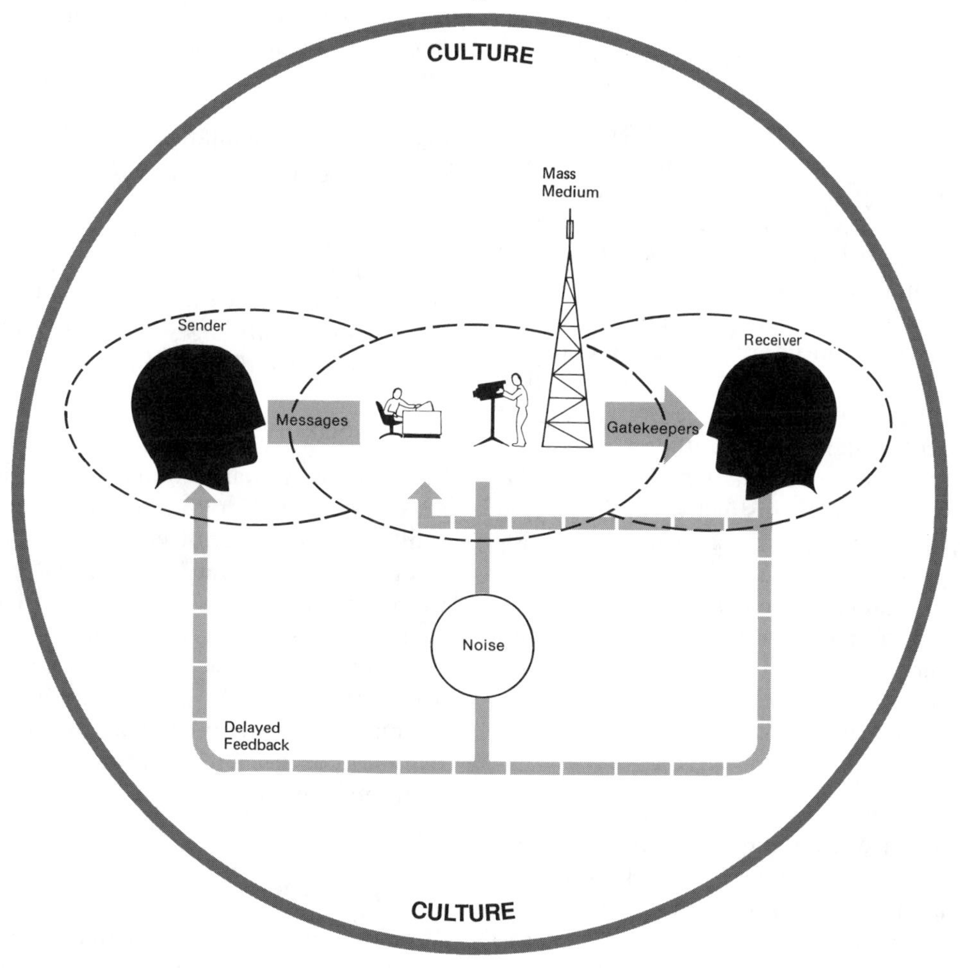

first understand the language being used, both written and oral. Second, we must know something about the subject of mathematics; otherwise, the lecture would have little value for us and we could not begin to work the problems. We may also respect the instructor's ability to teach, perceiving her as having a genuine interest in mathematics whether or not we are able to comprehend the subject. Messages both influence, and represent, the culture within which they occur.

MASS COMMUNICATION

Now that we have a basic understanding of the processes of intrapersonal and interpersonal communication, we need to understand the process of *mass communication* and, specifically, where broadcasting and telecommunication fit into the process. Mass communication is different from intrapersonal and interpersonal communication, but all three types play an important part in our lives. For the people fishing off Ocracoke Island, the weather report from the radio station on the mainland is just as important as their instinctive ability to find their crab-pot markers. And when they bring their catch in, their ability to use interpersonal communication to bargain with the wholesalers directly affects their livelihood.

Defining Mass

First, as the word *mass* suggests, mass communication can reach a large number of people through a mass medium. The number of people who could attend the lecture on mathematics was determined by the size of the classroom. However, a televised lecture could be made available to many thousands, perhaps millions, of people.

The Medium

To make the lecture available to all those people, it is necessary for us to alter our concept of *medium*. No longer is the medium just the human voice or the nervous system; we add a mass medium such as television, the radio, books, or newspapers, depending upon the applicability of the medium to our task. It may be somewhat difficult, although certainly not impossible, to teach our mathematics section by radio. We may even produce a series of articles for the newspaper. If we want to teach music appreciation, radio might be just as effective as television and considerably cheaper. On the other hand, if we want to teach surgical techniques, television would be far superior. In every case, in order to transcend the limitations of interpersonal communication, we would need a mass medium to reach our audience. For our purposes, therefore, we will define mass communication as *messages directed toward a group of people through a mass medium*.

Limited Sensory Channels

A mass medium also limits the number of sensory channels operating between the sender and receiver of communication. In interpersonal communication, all of our senses participate in the process of communication—our sight, hearing, smell, even touch. In mass communication these senses are limited. With radio, for example, we may be able only to hear someone deliver a speech. With television we could hear and see the person but not shake hands.

The Gatekeeper

Besides the presence of a mass medium, another factor traditionally differentiating mass communication from intrapersonal and interpersonal communication is the presence of a *gatekeeper* (Figure 1–2). With the addition of these two concepts, our basic model of communication now represents the process of mass communication.

The term *gatekeeper* was first applied to the study of communication by Austrian psychologist Kurt Lewin, who defined it as "a person or groups of persons governing the travels of news items in the communication channel."[6] Today, the term applies not only to groups of persons but to entire institutions. Within these institutions are both people and technology, all interacting to "govern the travels" of information between senders and receivers. That information is much more than news, as Lewin suggested. It may be strictly informative, such as an evening television news program compiled and produced by hundreds of reporters, camera-operators, editors, engineers, specialists in audio and video recording, researchers, writers, and many others. Or the message may be entertaining and involve producers, directors, costumers, scene designers, musicians, and countless more. The gatekeeper now becomes not only a person or group of persons but people and technology through which the message must pass and be acted upon, and sometimes altered, before it reaches the consuming public.

FIGURE 1–2 Correspondent Bernard Shaw of the Cable News Network (CNN). Gatekeepers are those individuals who "govern the travels" of information between senders and receivers during the process of mass communication. Reporters, as well as camera operators, directors, producers, and others who control the flow of information, are all gatekeepers.

Functions of the Gatekeeper

The function of the gatekeeper is to alter, expand, and limit what we receive from the mass media. Assume one morning that a television assignment editor dispatches a news crew to cover a music festival. When they arrive, the crew finds the festival spread out over a city block. In addition to violinists, pianists, and guitar players, there are groups of musicians playing everything from bagpipes to kazoos.

Upon seeing the television crew arrive, all of the musicians begin to play, each trying to gain attention. The reporter in charge of the story decides to focus on the bagpipe players. She bases this decision on a number of things. For one, the colorful costumes of the musicians will look good on color television. The bagpipes are also something the average viewer does not have the opportunity to see very often. In addition, the leader of the group is from Scotland and has a distinct Scottish accent. His voice alone will help hold the viewers' attention. That night our bagpipe players appear on the evening news.

Now let's examine how gatekeepers—in this case, the news crew—affected the information we received. First, they expanded our informational environment by offering us information we otherwise would not have received. The music festival may have been in an outlying community, and we either may not have had the time or may not have wanted to go to the trouble of driving all that way to attend it in person. On the other hand, the crew also limited the information we received. For instance, many more performers were at the music festival than just those who played bagpipes. However, because the news crew chose to focus upon that one group, we were not exposed to any of the other performers. Had we been present at the music festival, we probably would have seen everyone perform. But because we watched a report of it on the evening news, we were greatly restricted in the amount of information we received.

In summary, gatekeepers serve three functions: (1) they can alter the information to which we are exposed; (2) they can expand our information by making us privy to facts of which we would not normally have been aware; and (3) they can limit the information we receive by making us aware of only a small amount of information compared with the total amount we would have been exposed to if we had been present at an event.

Delayed Feedback

Another distinction between mass communication and other types of communication is *delayed feedback* (Figure 1–3). Remember when you were sitting in the classroom listening to the mathematics lecture? There, as we noted, you could give instant feedback to the instructor. You could raise your hand, ask a question, and probably have your question immediately answered. However, when you were watching the mathematics lecture on television this immediacy vanished. If you did not understand something and wanted to ask a question, you could only telephone the station, if the program were live, or write a letter to the professor conducting the televised lesson. Either of these alternatives is feedback, but this time it is delayed feedback.

New Technology: Altering Delayed Feedback

New developments in technology have altered the concept of delayed feedback in mass communication. New two-way media

FIGURE 1–3 Broadcast ratings are one example of delayed feedback in mass communication. Advertisers and the broadcasting industry are paying increased attention to feedback that provides information about specialized audiences. This *Arbitron Radio Report* analyzes and estimates radio listeners among the Hispanic population in a given geographic area.

permit instant feedback under some circumstances. For instance, the instructor teaching the mathematics course via television may have two different television monitors in front of the lectern, which permit her to view students in two different classrooms hundreds of miles away. In turn, all of the students can see and hear the instructor on the television monitors located in each classroom. A two-way voice connection permits the instructor to hear any questions the students may ask and to answer them immediately. Although messages are being directed toward a large number of people through a mass medium, instant feedback is possible.

Altering the Definition of Mass

At first glance, it may seem as if the appropriate wording of our definition of mass should be messages directed toward a *mass audience*, or large number of people, through a mass medium. Although this traditional definition has merit and in some ways is correct, it has been altered by new applications of mass media, such as the use of radio and television for internal corporate communication. We now find television connecting the boards of directors of two corporations located on different sides of the continent, or even oceans away, for executive conferences. Meetings whose participants are scattered hundreds of miles apart take place regularly in this way. Television is also used to disseminate messages to rather small audiences that cannot communicate face to face. A state-police commander may give a training lecture in front of a television camera. The videotape of the lecture is then played back at regional command centers throughout the state at which groups of ten or twelve troopers view the lecture. In each case the audience is relatively small, far from what we would normally consider a mass audience.

When we consider computers in our definition of mass, we must again alter the way we traditionally perceive the mass audience. For example, we might publish a magazine electronically by placing its contents in a data bank accessed via computer. Let us assume that our magazine is a highly specialized mass medium that reaches a small audience, such as ranchers living in Montana. In addition to obtaining this visual display, our audience has access to an index listing each article in the magazine. A rancher may need only to read an article dealing with beef pricing and disregard the other information contained in the publication. Thus, although the magazine is a mass medium, it is published only in an electronic edition and reaches only a highly specialized audience. Compared with a national television audience the readership of the electronic edition of our magazine is very small—so small we might fail to recognize that it, too, is a mass audience, though not a large one.

The use of new technology such as interactive media is continuing to alter the traditional definition of mass communication. The important thing to remember is that it is not necessarily how many people are exposed to a message, but how many people have access to the message and how it is delivered that helps distinguish mass communication from intrapersonal and interpersonal communication.

Communicative Noise

Noise can exist in mass communication just as it can in intrapersonal and interpersonal communication. Noise can appear in the processing of information through the gatekeeper. Keep in mind that the network of gatekeepers can consist of many different persons or groups of people, all of whom are part of the processing of information. When information is passed from one gatekeeper to another it can become distorted. One example of noise in the communicative process occurred when a group of reporters covered an incident along an interstate highway in the Midwest. A truck carrying two canisters of phosgene gas stopped at a truck stop. The driver of the truck smelled a peculiar odor and decided that one of the canisters was leaking. He became sick and was taken to a local hospital. When state police learned from the invoice what the truck was carrying, they notified authorities at a local army depot. The state police then blocked off an exit on the interstate highway almost twenty miles away. It was the logical place to divert traffic since it was next to a main feeder highway, which made an excellent detour in case the highway immediately adjacent to the truck stop had to be blocked off.

When all of this information was processed into the news media, all under the pressure of deadlines and semicrisis conditions, it was distorted considerably. First, news reports left the impression that the truck was loaded with phosgene gas, and not merely two canisters of it. Obviously, a leak in a tank of gas the size of a gasoline tanker would be much more serious than a leak in a single canister about five feet high and less than two feet in diameter, strapped to the back of a flatbed truck. Second, because phosgene gas had been used in World War I, the wire services began to refer to the canisters as containing "war gas." Added to this was the news of the roadblock twenty miles away, which left the impression that everyone in a twenty-mile radius of the truck stop was in danger of inhaling war gas.

The network of gatekeepers that covered the story included a group of reporters from three radio stations, at least two newspapers, two wire services, and two television stations, and the local and military authorities, who also were dispensing information. The "institution" of gatekeepers was substantial, and much information was processed and eventually distorted.

Reducing Communicative Noise

Just as new technology has altered the concept of delayed feedback, it has also altered noise, primarily by reducing it. In 1950 it would have been almost impossible to carry live pictures and sound from one continent to another. Back then, the speech of a European leader would have been reported first by a correspondent and then fed to a wire-service editor in the United States. The wire-service editor would have then rewritten the correspondent's report before sending it over the teletype to subscribers. This entire process was subject to much distortion and noise because of the number of gatekeepers involved.

Today, although that process still takes place, it is now possible for a videotape of a speech to be sent by satellite into the homes of viewers thousands of miles away. On the evening news the viewer watches the picture and listens to the voice of the political leader in place of the correspondent's interpretations; this reduces the possibility of noise. Even the newspaper reporter can carry a small recorder, almost as inconspicuous as a note pad, and reduce the chance of misquoting a source. Still, few systems of processing information are perfect. Remember that although broadcast technology can reduce noise, the human factor is always present to return some noise to the system.

THE CULTURAL CONTEXT OF MASS COMMUNICATION

Our discussion thus far has concerned messages being sent, processed, and received. Although we have seen how gatekeepers act upon those messages, we should also realize that cultural forces act upon senders, gatekeepers, and receivers, influencing how they react to and process messages (Figure 1–4).

Consider the computer. Information is fed (sent) into the computer, where it is processed and then presented, usually in the form of a printout. You might feed the computer a series of numbers, of which the computer will add and print out the answer. If you fed the same set of numbers into the computer each time, the computer's answer would be the same each time. Such is not the case with messages sent, processed, and received by means of mass communication. People are not computers, and we do not live in a vacuum. Messages causing one reaction at one time may cause an entirely different reaction another time. A politician's speech that attracted one gatekeeper's attention might not attract another's. Let us examine this in more detail.

FIGURE 1–4 The cast of the ABC network television program "thirtysomething." Within the cultural context of a society that places an emphasis on professional achievement, the characters portray the conflicts that occur between their jobs and their families. Messages, such as TV dramatic and entertainment programs distributed through mass communication, can both reflect and reinforce the culture within which the viewing audience lives.

Cultural Environment of Senders

Assume that you have decided to run for a political office and it is time to begin the long, arduous trail to election day. In writing the speech that will kick off your campaign, you want to convey to your audience those qualities you feel will truthfully express your character, your position on the issues, your background, and your intentions. As you approach the podium in a small rural community you think about the times you have seen scenes like this before. The serenity of your childhood, the familiar faces of people you do not know but really do know, the soft, mellow breeze—everything is there, including two gatekeepers, a reporter from each of the two local radio stations.

You begin your speech. You talk about things and individuals that have influenced your life. You talk about farm prices, having grown up on a farm, and you know what you are talking about. You relate your experiences of meeting expenses during the harvest season and borrowing money to buy tractors. You also talk about the plight of those in small business, for after the farm failed your family opened a clothing store. All of these forces had a direct effect on your campaign speech. Now how did your speech affect the two gatekeepers?

Cultural Environment of the Gatekeeper

When you listen to the newscasts of the two radio stations later that afternoon, you are surprised to find that each reporter covered a different part of your speech. One reported your comments on farm prices and only briefly mentioned statements about small businesses. The other station detailed your statements about small businesses but skimmed your comments about farm prices.

Although you considered both reports objective, you wondered why they focused upon different subjects. You discover later that the reporter who cited your comments on farm prices not only grew up on a farm but also owned one. The other reporter grew up in the suburbs; his father had a small business, and he had no love whatsoever for farming. Each reporter had interpreted your speech in accordance with his own background. Unlike a computer programmed to select and process certain information, the two reporters were as different as the forces influencing them.

In research, these phenomena have been called *selective perception* and *selective retention*. Selective perception means we perceive only certain things, such as those that are most familiar to us or that agree with our preconceived ideas. The reporters' backgrounds and resulting selective perception created two different interpretations of the speech. Selective retention means that we tend to remember things that are familiar to us or that we perceive as corresponding to our preconceived ideas. Research implies that what reporters selectively perceive and retain can become even more prominent when they cover controversial issues.[7]

Another influence on the story might be the reporters' peers. The reporters may belong to a professional association and adhere to a code of ethics. This code could in turn directly affect the stories processed by these gatekeepers and consequently received by the public. What if the music festival we discussed earlier had charged a ten-dollar admission fee? And what if the assignment editor, as part of his professional ethics, had prohibited any of the staff from accepting free tickets to any event while assigned to cover that event? Admission to the festival for the news crew would have come to thirty dollars. But what if the manager of the station had refused to pay the thirty-dollar admission fee for "something as unimportant as a music festival"? The editor might have decided finally not to assign a news crew to the festival. Do you agree with that decision?

Social Relationships of Receivers: Opinion Leaders

Just as gatekeepers do not operate in a vacuum, neither do receivers of mass communication. Our family, co-workers, peer groups, and organizations all affect how we receive and how we react to messages from the mass media. In this social realm, interpersonal communication is also very important. For instance, upon hearing the report of your campaign speech over one of the radio stations, one local listener thinks your speech has some strong merits. Yet her friend has an entirely different opinion. Since the listener respects her friend's opinion, she in turn changes her opinion of your speech. In this case, the friend acted as an opinion leader, a person upon whom we rely to interpret messages originally disseminated through the mass media.[8]

Consider another example. Suppose you are watching television and see a commercial about a new headache remedy. The remedy claims to be better than aspirin, to cause fewer side effects, and to work much faster. You have been having trouble with headaches, but instead of running out to buy the new remedy you call your friend, a nurse whose opinion you respect. The nurse recommends the new remedy, and the following day you purchase it and take two pills. It works. Notice, however, that it was not the commercial that convinced you to purchase the medicine. Although the commercial helped, your friend ultimately convinced you. She served as an opinion leader. Had she not recommended the remedy, chances are you might not have bought it then.

Interrelationships of Senders, Gatekeepers, and Receivers

In reviewing our examples of what occurs when information is processed through the mass media, you should begin to see many relationships among senders, gatekeepers, and receivers. For example, it was homophily—the perceived sharing or overlap of experiences between you and the two radio reporters—that caused each reporter to stress a different part of your campaign speech to listeners. Similarly, the radio listeners interpreted your speech in certain ways, also because of this sharing or perceived sharing of experiences, attitudes, and other things. In fact, listeners may even have selected one radio station over the other because of similarities they perceived between themselves and the reporter.

Selecting one radio station over another is an example of *selective exposure*, whereby we expose ourselves to information that we perceive to support our beliefs or ideas. By studying the functional uses of mass media we can examine how we selectively expose ourselves to certain media because those media fulfill a particular need. For example, people waiting out a storm to fish near Ocracoke Island will exhibit selective exposure. They will turn to the radio stations that provide the most accurate weather information, selectively exposing themselves to that station over others. Or they may own a special weather radio locked on a frequency that broadcasts weather reports twenty-four hours a day.

BROADCASTING AS MASS COMMUNICATION

In its most basic sense, *broadcast* can mean "scattered over a wide area" or "in a scattered manner; far and wide." The dictionary also includes such definitions as "to make known over a wide area: broadcast rumors." Certainly a disgruntled loser of an election would agree with that definition. Or consider the definition "to participate in a radio or television program." The guest home economist on an afternoon radio program for consumers would agree with that definition. The farmer in the 1800s, who had never heard of radio or television, would have agreed with the dictionary's definition that broadcast means "to sow [seed] over a wide area, especially by hand." So would the people scattering crab pots in the open water off Ocracoke Island.

Consulting a thesaurus, we find that words similar in meaning to broadcast include "disperse," "cultivate," "publish," and "telecommunication." We would not have to travel far to encounter people who would agree with all of those meanings. The advertising executive would disperse knowledge about a client's product through broadcasting commercials. The supporter of noncommercial public broadcasting would argue that quality programming cultivates an interest in culture and the arts

(Figure 1–5). The broadcast journalist subpoenaed before a grand jury and asked to divulge the source of an investigative report would argue that under the First Amendment to the U.S. Constitution broadcast means the same as publish, and that the journalist's rights to protect the confidentiality of one's news sources are the same as those of newspaper reporters. To the corporate executive, broadcast might be associated more closely with telecommunication. For example, the image of two executives sitting in a corporate boardroom can be reproduced on television monitors a continent away. There, other corporate executives talk back to the boardroom executives via a two-way television system. For our part, we will define *broadcast* as *signals sent via radio or television.*

By now you should have begun to see how broadcasting enters into the process of mass communication. Notice that between the senders and receivers of broadcast communication are the broadcasting stations. These, along with supporting and allied organizations, directly affect the messages sent through this medium of mass communication. Broadcasting stations consist of standard-broadcast radio and television stations as well as cable television—

FIGURE 1–5 Television has the ability to reach mass audiences with high-quality programming. For example, public broadcasting has achieved recognition for cultivating an interest in the arts and making the arts available to the public through programs such as the "Dance in America" series produced by WNET (Channel 13) in New York.

commonly called *community-antenna television (CATV)*—and *closed-circuit television (CCTV)*.

SUPPORT STRUCTURES OF ELECTRONIC MEDIA

Both broadcasting and other electronic media as forms of mass communication are affected by numerous support structures. These range from the committees of Congress that hammer out legislation affecting the industry to small-town municipalities debating a cable-television ordinance; from creative minds at a metropolitan production center to the local merchant preparing a drugstore commercial. We will divide these support structures (Figure 1–6) into program suppliers, supporting industries, professional organizations, control mechanisms, technical services, audience-measurement services, and management services.[9]

Program Suppliers

Program suppliers provide stations, cable operators, and others with programming ranging from Hollywood game shows to

FIGURE 1–6 The institutions of electronic media include many support systems. For example, the programs we view on television can be affected by the program source; by regulatory and social controls; by technical services; by feedback from the audience; and by management, advertising, and employee services such as professional societies and labor unions.

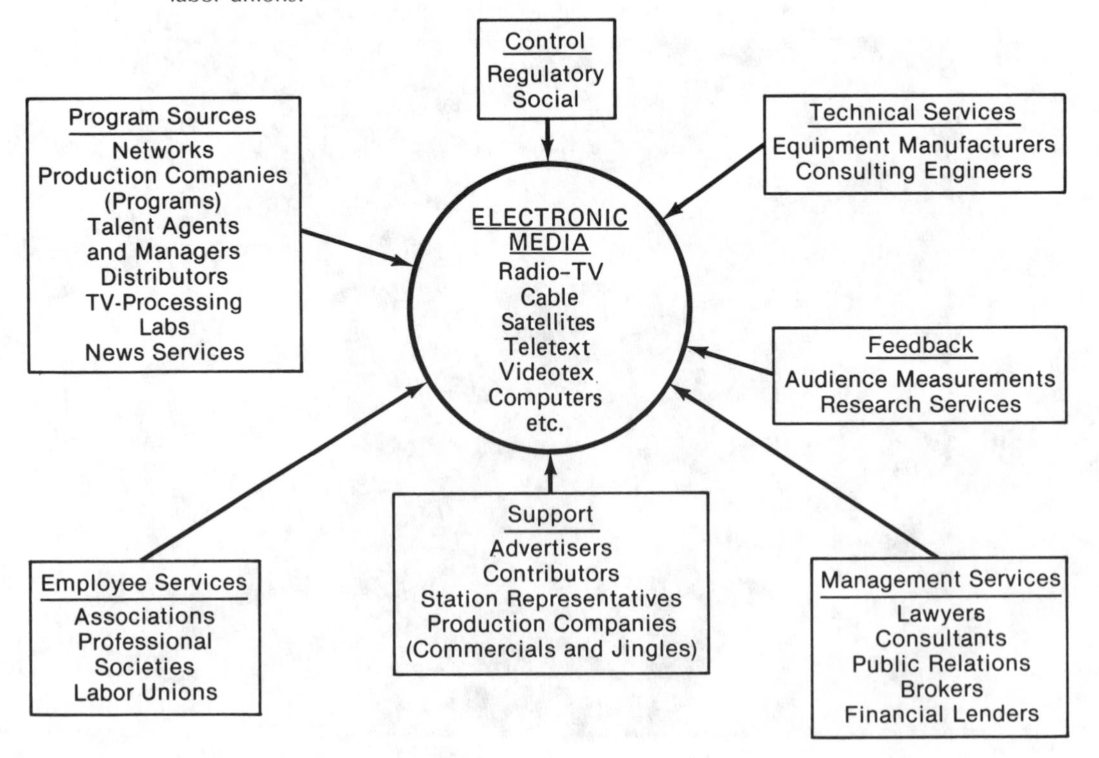

spectaculars. Many of these suppliers are already familiar to us. They include such major television networks as *CBC* in Canada; *BBC* in Great Britain; *NHK* in Japan; and *ABC*, *CBS*, *NBC*, and *PBS* in the United States. Television production houses, such as MTM Enterprises, are other program sources. Their programs are either sold directly to the networks or distributed through major distribution companies, such as Viacom. Not all program sources deal with entertainment. News program sources have become increasingly important as communication links with satellites continue to shrink the world and whet our interest for international events. Two widely used radio news program sources are *United Press International Audio* and *Associated Press Radio*.

Supporting Industries

These consist of advertising agencies, which place commercials on stations, and station representatives, who act as national salespersons for a station, group of stations, or cable systems and other forms of electronic media.

Professional Organizations

Within any industry or profession are services that link employees together for a variety of reasons, from professional to purely social. For example, broadcasting's version of such a service is the *National Association of Broadcasters (NAB)*. More narrowly defined professional organizations include the *Radio-Television News Directors Association (RTNDA)* and *American Women in Radio and Television (AWRT)*. There are over a hundred other broadcast-employee services in the United States alone. Labor unions constitute a large share of the broadcast-employee membership, especially in metropolitan stations and the networks. Major unions having a foothold in broadcasting include the *International Brotherhood of Electrical Workers (IBEW)* and the *Communication Workers of America (CWA)*.

Control Mechanisms

Control of electronic media ranges from governmental to social. At the national level, governmental control is represented by the *Federal Communications Commission (FCC)* and the *National Telecommunications and Information Administration (NTIA)*. In the former, control takes the form of specific laws and regulations. In the latter, it is oriented more toward policy issues. State and local governments may also control broadcasting, cable in particular.

In the social-control arena, public-interest groups, such as *Action for Children's Television (ACT)*, lobby both legislators and the stations themselves. Hearings on television violence held by another group, the National Congress of Parents and Teachers (PTA), culminated in a report to the industry and pressure to reduce television violence.

Advertisers and stockholders also exercise control over broadcasting. In fact, a small-market radio station may fear the loss of its biggest advertiser just as much as a visit from an FCC inspector. Why? Because advertisers, especially in smaller communities, can often "influence" the content of broadcast programming. If the local car dealer spends a huge sum of advertising money on a station, his drunk-driving charge may conveniently be absent from the morning news, all on the strong suggestion of the station manager. Or sponsors may refuse to air their ads during violence-filled programs.

Technical Services

The hardware components of electronic media have spawned a giant industry consisting of everything from the production of television and radio receivers to engineering consulting. General Electric, Zenith, SONY, Panasonic, RCA, Motorola, and others all vie for this lucrative broadcasting market. In addition, companies and governments actively produce and service satellite and microwave systems that span the globe. The industry also fosters its own technical service—the consulting engineer. When an antenna on a 2,000-foot tower needs fixing, it is hardly the job for the local TV repair shop.

Audience-Measurement Services

An audience is the lifeblood of any mass medium. Measuring this audience requires the talent of a host of survey companies. Other such companies specialize in customized surveys, such as measuring the effectiveness of a station promotion, undertaking a station-image survey, or initiating a personality-recognition survey among viewers.

Management Services

With the increasing complexity of electronic media, few broadcast managers have the skills necessary for handling all functions. They must therefore rely on management consultants. Among the most important of these are attorneys hired to help them process the mountain of governmental forms they now must file, and to give advice on complicated legal matters. Most of the major communication law firms are in Washington, D.C., close to the heart of government.

Promotion services and brokers are two other management services that are important to the industry. Media are becoming highly specialized where more competition evolves every day. Sophisticated advertising and promotion campaigns are necessary if a station is to thrive in the marketplace. Professional promotion consultants are available who handle such things as the station's public relations or special advertising campaigns. Brokers are the real-estate professionals of the industry. If we want to buy or sell a station or media property, we will probably use a broadcast broker.

Although we have discussed each of these allied organizations and services separately, keep in mind that they are interrelated. The production company is just as concerned about the FCC's stand on obscenity as is the broadcaster. The attorney's advice is just as valuable to the advertising agency producing a broadcast commercial as it is to the station manager. The organizations and interrelationships constitute the interactive process of broadcasting in our society.

DEFINING TELECOMMUNICATION

Now that we have examined definitions and processes of communication and looked at how mass communication and broadcasting fit into these definitions, we want to understand the other term that appears in the title of this book— *telecommunication*.

Telecommunication is not a new term, but its use is somewhat more recent among teachers and researchers of broadcasting. Although we did not define it at the time, we have already discussed telecommunication. The microwave relay tower on Ocracoke Island sends and receives information that includes telephone conversations and computer data. Both telephone and computer have traditionally played an important role in defining telecommunication (Figure 1–7). If we return to the dictionary and examine the definition of telecommunication, we find such words and phrases as "electronic communication," "transmission of impulses," "telegraphy," "telephone," "cable," "radio," "computer," "television," and "messages communicated electronically." We know from examining the Greek term *tele* that it means "at a distance" or "far off." Thus, we can see in the juncture of *tele* and *communication* a meaning that includes "distant communication."

At the same time, telephone is also derived from *tele*, and in its common usage telecommunication incorporates as much a sense of communication by telephone as it does the meaning of long-distance communication. Until the recent development of two-way cable-television systems, which we will discuss later, computer data and video communication traveled primarily through telephone lines or through microwave-relay systems that in many cases were owned by the telephone company.

The emergence of two-way interactive media, such as cable television systems that permit viewers to talk back electronically to their television sets and select information from central data banks, has enabled us to see how the differences between technologies are being diminished. Telecommunication has become a broad term that centers more and more in electronic communication, of which the computer, radio, television, cable, telegraph, and telephone are all a part. It encompasses broadcasting in its more traditional sense of a radio or television station sending signals to the masses as well, and includes the electronic magazine in Montana accessed via a home computer. It encompasses the radio stations that broadcast weather reports to the fishing vessels off Ocracoke Island and the radar signals emanating from the antenna on top of the ferry leaving Cedar Island. Electronic signals may travel through the air and be broadcast to a wide region. The radio sta-

FIGURE 1–7 The traditional definition of mass communication is being altered by interactive communication technology. This interactive work station includes a telephone, on-line directory, personal computer, and facsimile machine. It can automatically distribute a document to as many as 100 different locations by dialing preprogrammed numbers stored in its computer. Privacy can be guaranteed by code words available to users of other units who enter the correct code word to receive their document. The computer can be connected to other computers, thus creating a "network" of users who engage in interactive communication.

tion on the mainland broadcasting weather reports uses radio waves, which we will later learn are part of the lower end of an electronic yardstick we call the *electromagnetic spectrum*. These waves are not relayed via telephone lines or other facilities but travel directly to the listeners tuned to the station. Such stations are truly broadcasting in the traditional sense.

At the same time, however, a local radio station airing a newscast that originates in New York must first receive it via telephone lines through a satellite system. In addition, a local cable system may pick up the signal from the radio station and feed it to its subscribers on one of the cable channels. We can see from this that the term *broadcasting* is simply not broad enough to be accurately applied to all of the technologies that are now part of our world of electronic communication. Thus, we have adopted the broader term *telecommunication*. We will define telecommunication as *electronic communication involving both wired and unwired, one-way and two-way communications systems*. We can see that this definition includes broadcasting.

CONTEMPORARY APPLICATIONS OF TELECOMMUNICATION: WHERE THIS BOOK WILL TAKE US

The present chapter has helped us define key terms and understand some examples of them, but we have only scratched the surface of telecommunication. In the chapters that follow we will learn more about broadcasting and other fields in this important realm of technology.

The Development of Telecommunication

Although this is not a history book, Chapters 2 through 4 will examine some of the historical foundations of telecommunication (Figure 1–8). We will begin by examining the first technologies that could be called the ancestors of modern telecommunication, the telegraph and telephone. From the wires that stretched across Europe and the pony-express routes of the Great Plains to the first sound that emanated from Alexander Graham Bell's telephone, the telegraph and telephone are part of our technological heritage. Today these two technologies stretch beyond the confines of any geographic region to satellites traveling thousands of miles in space beaming telephone and telegraph signals across continents.

In later chapters we will also learn how the computer gradually integrated itself into these technologies and brought about a new frontier of communication. We will also be introduced to and learn more about such terms as *microprocessor, random-access memory (RAM)*, and *interactive video*.

Broadcast and Information Technologies

In Chapters 5 through 8 we will look more closely at some of the technologies of broadcasting and telecommunication. As we learn about how radio waves bring us the morning weather and our favorite programs, we will examine the electronic yardstick, or electromagnetic spectrum, that we referred to earlier. We will also learn about *microwaves*, which appear higher on our electronic yardstick and help carry telephone

FIGURE 1–8 The history of broadcasting and telecommunication is filled with a rich heritage of experiences, achievements, and developments in technology, which today permit us to enjoy the programs and services in our information society. This photo, by Alonzo Pond for Wisconsin Public Radio, captured the thrill of hearing sounds and voices in the early days of radio.

and data communication from microwave towers such as the one on Ocracoke Island. Microwaves can travel thousands of miles into space, bouncing back to earth thousands of miles from the point of origin and bringing us everything from our evening television programs to our long-distance telephone calls.

We will examine satellite communication (Figure 1–9), which has helped advance the technology and applications of telecommunication. From the navigation antenna on board a marine-research vessel to the rooftop antenna of a remote Alaskan village, satellites have challenged the boundaries of our minds and the boundaries of cultures.

We will discuss cable communication, which began as experimental antennas on mountaintops in the late 1940s and today is a billion-dollar industry that "wires" cities and greatly expands the number of television channels and other services we can receive. Two-way cable systems are capable of providing interactive video, whereby a small home terminal can activate services such as home banking, shopping, theater purchases, and airline reservations.

FIGURE 1–9 Satellite communication permits instantaneous transmission and reception of audio, video, and data signals anywhere in the world where ground stations are located. International and domestic satellite systems will continue to revolutionize worldwide telecommunication in the 1990s.

Of all the applications of new technology that are available to the public, teletext and videotex have perhaps received the most attention. Teletext is primarily a one-way system that operates much like a television signal but consists of textual information that may also be electronically illustrated. Videotex is a wired, two-way interactive textual system carrying information and electronic illustrations. Our example of the electronic magazine for the Montana rancher is an application of videotex. With a home terminal an individual can access a computer data bank. A "menu" of the information in this bank can be called up on a television screen, and the subscriber can then select from it.

Other communication technologies are also being developed. New cellular mobile radio systems permit many more mobile telephones to operate than ever before. Because mobile telephones, the kind we could use in our car, employ radio waves to transmit and receive messages, the number of these phones in each city, for example, used to be limited by necessity to prevent interference. But by dividing each city up into "cells" and using different frequencies for different cells, we have made it possible for more telephones to be licensed to the same geographic area.

The same satellites that carry data and other information into space carry the pictures and voices of businesspeople conducting meetings via a process known as *teleconferencing*. Using video and voice hookups between distant locations, a group of executives in, for example, Columbus, Ohio, can talk and see another group of executives in, say, San Francisco, all via television monitors. The expense of a two-way audio-video link is much less than the travel costs and lost time of business executives who need to cross the country for a meeting. Teleconferencing is another of the technologies whose principles and applications we will investigate.

Systems and Programming

In Chapters 9 through 12 we will look at some of the telecommunication systems that are part of broadcasting and other technologies. The major networks and the public broadcasting services—for years, important parts of the distribution system for radio and television—are now being joined by distribution via syndication. Through syndication, programs are sold directly to stations. We will examine both networks and syndication in Part III. We will also look more closely at how telecommunication affects our educational system. What started in the late 1930s as a crude closed-circuit educational television program that ushered in the era of educational television (ETV) has expanded today into educational telecommunication whereby students may sit at their own personal computer terminal and learn such subjects as statistics and accounting. At the same time, a group of managers in a nearby assembly plant may spend part of their lunch hour enrolled in a telecourse, a course taught by television. We will discuss the development of educational telecommunication and contemporary applications of telecommunication in business and industry, as well as broadcast programming. Chapters 13 and 14 focus on international systems (Figure 1–10).

Regulatory Control

As we have seen, control is one of the components of broadcasting's support structure. In Chapters 15 through 18 we will expand our knowledge of the controls that affect broadcasting and telecommunication. We will begin by examining the historical basis for the system of laws and regulations that affect telecommunication. We will then analyze the most prominent regulatory agency affecting telecommunication, the Federal Communications Commission (FCC). Looking more closely at the content of radio and television programming and the operation of broadcasting stations, we will study such regulations as the Fairness Doctrine and Section 315 of the Communications Act of 1934. We will also explore some of the steps one follows when seeking permission from the FCC to construct a new radio or television station. We will observe the regulatory structure affecting common carriers, such as telephones and other interstate communication systems. The provisions of a typical cable-television ordi-

FIGURE 1–10 International broadcasting and telecommunication are discussed in depth in later chapters of the text. This billboard, located in Berne, Switzerland, is promoting an FM radio station.

nance and how local governments deal with such ordinances will give us an insight into this emerging arena of municipal law. The increasing technological capacity to reproduce information has resulted in new issues in copyright law, ranging from cable-television systems to photocopying.

Economics and Evaluation

An inside look at any commercial radio or television station will uncover an economic base necessary to keep the station operating. In Chapters 19 and 20 we will examine some of the financial issues and procedures found in a typical station and study the important contribution of broadcast promotion to a successful operation. We will also take an inside look at broadcast ratings. Ratings in many markets are an indicator of station success, and a station's income is directly related to how well it does in the ratings—how many people are listening to or viewing the station.

From the station advertising director trying out a new promotional campaign to the college professor completing a study on television violence, research in telecommunication is necessary for intelligent decision making by everyone from legislators to station managers. In Chapter 21 we will examine the different types of research in telecommunication and some of the issues surrounding them. Much of this research focuses on the audience and users of telecommunication. Our study of broadcasting and telecommunication will conclude with an examination of the audience of radio and television programming and how it is affected by and reacts to it (Chapter 22) and a discussion of criticism and ethics (Chapter 23) related to the broadcasting industry.

SUMMARY

The basis of the process of communication is intrapersonal communication—communication within ourselves. In intrapersonal communication our senses become the senders of communication, our brain processes the messages sent by our senses, and we react to feedback messages sent to our muscles.

The basic components of the communication process—sender, messages, medium, receiver, feedback, and noise—apply both to intrapersonal and interpersonal, or face-to-face communication. Interpersonal communication encompasses intrapersonal communication. In interpersonal communication the sender of communication is one individual and the receiver another individual. The medium of communication is the human voice and messages are words. Feedback occurs when the receiver reacts to the message of the sender. Both physical and semantic noise may interrupt interpersonal communication as they do intrapersonal communication.

To understand better the process of communication we frequently use a communication model, a diagram that serves as a stop-action picture of the process. For effective interpersonal communication to take place, the sender and receiver must have certain things in common as they communicate. A high degree of homophily—the term for these overlapping fields of experience—can aid interaction.

Mass communication is somewhat different from intrapersonal and interpersonal communication. The term *mass* denotes the presence of a large number of people. Frequently, but not always, mass communication reaches millions of people. Moreover, mass communication involves the presence of a mass medium. Radio, television, and cable are examples of electronic mass media. We define mass communication as messages directed toward a group of people through a mass medium. Not all of our senses participate in the process of mass communication, as they do in interpersonal communication. We cannot touch the other person, we cannot smell the other person, and he or she cannot respond immediately to our sensory feedback.

Mass communication entails the presence of gatekeepers. A gatekeeper governs the flow of information in a communication system. Today, gatekeepers can be individuals or institutions, a single reporter, or a television-network news operation. Because they have access to more information than we do about a given topic, gatekeepers both expand our informational environment by giving us more information and restrict that environment.

In mass communication, feedback is delayed, whereas in interpersonal or intrapersonal communication it is immediate. Writing a letter to a politician we see on television is a form of delayed feedback. New technology is altering the way we present feedback via the mass media. Two-way interactive communication systems permit instant communication. From home shopping to public-opinion polling, today's emerging technologies allow us more opportunity for immediate feedback.

These same technologies are also changing the traditional definitions of the term *mass*. For example, a computer bank may store the contents of an electronic magazine from which a small number of subscribers may access a single article through their home-computer terminals. These smaller, highly specialized magazines reach a smaller, more specialized audience.

As with intrapersonal and interpersonal communication, noise can enter the process of mass communication. Physical noise ranging from interruptions in the living room of a viewer to static on the television screen can affect messages between sender and receiver. The inexactness of language increases the chances of semantic noise also being present.

Broadcasting—messages sent via radio or television—is a form of mass communication. Telecommunication—electronic communication involving wired and unwired one-way and two-way communication systems—is a much broader concept that has come into use in the broadcasting industry as it begins to consider many new technologies. These new technologies range from teletext and videotex systems to home computers.

The support systems surrounding broadcasting include such areas as program suppliers, supporting industries, professional organizations, control mechanisms, technical services, audience-measurement services, and management services.

OPPORTUNITIES FOR FURTHER LEARNING

BARNOUW, E., ed. *International Encyclopedia of Communications*, 4 vols. New York: Oxford University Press, 1989.

COMPAINE, B. M. *Political and Social Implications of New Information Technology*. Norwood, NJ: Ablex, 1988.

CREEDON, P.J., ed. *Women in Mass Communication*. Newbury Park, CA: Sage, 1989.

DENNIS, E.V. *Reshaping the Media*. Newbury Park, CA: Sage, 1989.

MCQUAIL, D. *Mass Communication Theory: An Introduction*. Newbury Park, CA: Sage, 1987.

O'KEEFE, D. D. *Persuasion: Theory and Practice*. Newbury Park, CA: Sage, 1990.

SCHEMENT, J. R., and LIEVROUW, L. *Competing Visions, Complex Realities: Social Aspects of the Information Society*. Norwood, NJ: Ablex, 1988.

SINGLETON, L. A. *Global Impact: The New Telecommunication Technologies*. New York: Harper & Row, 1989

THOMAS, S., ed. *Studies in Communication, Vol. 1: Studies in Mass Communication and Technology*. Norwood, NJ: Ablex, 1984.

WENNER, L.A., ed. *Media, Sports, and Society*. Newbury Park, CA: Sage, 1989.

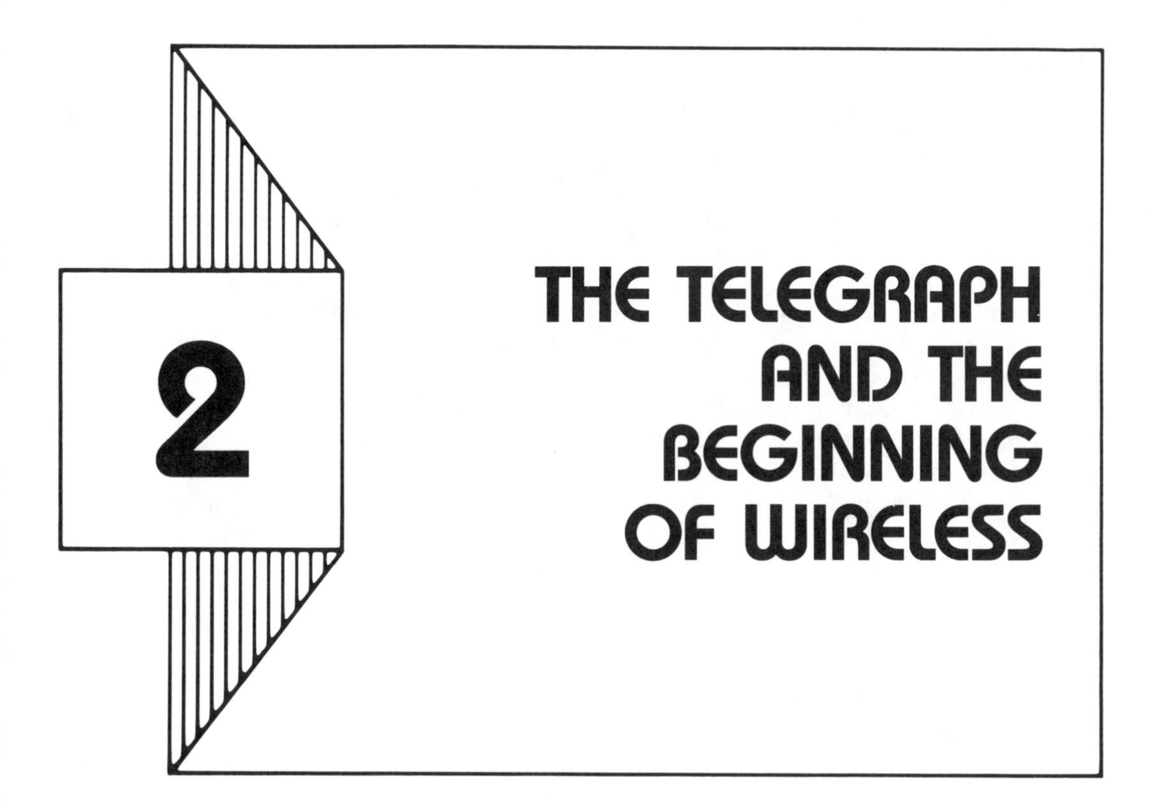

2 THE TELEGRAPH AND THE BEGINNING OF WIRELESS

When we turn the dial on our radio or switch channels on our television, it is hard for us to imagine the hundreds of years of theory building and applied technology that paved the way for the modern era of electronic communication. Our dream of capturing electricity and applying it to the communicative process dates back centuries. As we gradually learned about electricity and began to apply its power, the era of the telegraph and then wireless, later to be called "radio," began to emerge. Both were responsible for whetting the appetites of the inventors who would bring electronic mass communication into our homes and eventually creating an electronic link with the other side of the world.

APPLYING THEORY TO PRACTICE

In 1791 Luigi Galvani, an Italian physician and professor of anatomy at the University of Bologna, published the results of his research on the nervous system of frogs.[1] Galvani sent an electrical current into the nerve of a dead frog and watched as the frog's leg contracted. Galvani discovered he could achieve a similar reaction by touching the nerve with different metals, such as copper and iron. Probably because of his background in anatomy, Galvani attributed the movement of the leg to the presence of "animal electricity" in the frog.

Greatly skeptical of Galvani's research, Alessandro Volta told the Royal Society in

London in 1800 that Galvani's "electricity" wasn't to be credited to the frog but to the different metals, and that he, Volta, had proved the theory by constructing what was to become known as the *voltaic pile*. Volta first placed a zinc disk on top of a silver disk and then placed a cardboard or leather disk soaked in brine on top of the metals. On top of this he placed metal and then cardboard or leather disks in a series until he had formed a small pile of disks. What Volta had invented was the first practical energy cell. Now scientists had at their disposal a continuous source of electricity.

Hans Christian Oersted and André Marie Ampère

Research into the uses of Volta's battery continued, but not until twenty years later did a professor of physics in Copenhagen, Hans Christian Oersted, discover that an electrical current could cause a nearby compass needle to rotate. Oersted's discovery accomplished two things: (1) it provided proof of the relationship between electricity and magnetism, and (2) it joined the scholarly disciplines of electricity and magnetism. The same year, French physicist André Marie Ampère refined Oersted's discovery by applying mathematical formulas to electromagnetism.

Michael Faraday and Joseph Henry

Scientists had yet, however, to actually observe phenomena that would verify Oersted's and Ampère's theories. That task was left to the English chemist and physicist Michael Faraday and the American physicist Joseph Henry. Henry did not manage to publish the results of his research until after Faraday's work had achieved world recognition.

The son of a blacksmith, Faraday left school at thirteen and while working for a bookbinder read an article on electricity in one of the volumes he was stitching. He landed a job as an apprentice at the Royal Institution and eventually became one of the most respected scientists of his day, later to head the institution.

Faraday's work climaxed late in 1831. In his experiments in the late summer and fall of that year Faraday was attempting to discover whether magnetism could produce electricity. The discovery came on November 4, when he moved a copper wire near the poles of a large horseshoe magnet and produced a measurable electric current. Faraday said the phenomenon was caused by "lines of force," as can be illustrated by placing a magnet near small iron filings. Faraday continued research on lines of force for the next twenty years, and in a research paper prepared in 1852 he alluded to lines of force radiating into the atmosphere, thereby generating electricity.

James Clerk Maxwell

The next step in the development of Faraday's theory came shortly after his death in 1867. Industrial leaders were calling for an updating and modernization of science instruction in English universities. In 1874 the James Henry Cavendish Laboratory at Cambridge was established. James Clerk Maxwell, a respected Scottish physicist and mathematician who had been appointed to an endowed chair of physics in 1871, became director of the laboratory. Having the luxury of devoting all his time to scientific research with no pressure for results, Maxwell could work at his own speed, building theory upon theory. As a trained mathematician, he extended Faraday's theories into mathematical predictions. His Dynamical Theory of the Electromagnetic Field stated that electromagnetic action travels through the atmosphere in waves, and that the atmosphere has the capacity to carry these waves at the speed of light.

By the late nineteenth century, the scientific community in Europe was experiencing support and growth. Whereas England's scientific movement was supported by lobbying in Parliament and endowments to universities, in Germany science was supported as a business. What had once been an agricultural region was now beginning to experience the profits of industrial growth. Raw materials and the technology to transform them into industrial products signaled changes in the economy and the labor force. Technology and industrial growth necessitated the support of scientific inquiry, and statesmen began to place their firm support behind research and instruction in the sciences at the universities. Moreover, government subsidies and national research organizations created an atmosphere that nurtured new knowledge and fostered experimentation with existing theories. German scientists found themselves treated with more respect than any other scientists in Europe, and the universities geared themselves to train and employ not only the brilliant but also the many people of average intelligence who had the necessary persistence and fortitude.

Heinrich Hertz

It was at the beginning of this era, in 1857, that Heinrich Rudolph Hertz was born to a middle-class family in Hamburg, Germany. Taught an hour a day by tutors and obtaining the rest of his learning in his spare time, young Hertz developed a keen interest in science and outfitted himself with his own home laboratory. Engineering first whetted his appetite, but after a year of study at the University of Munich he moved to Berlin to study pure science under the well-known German scientist Hermann Ludwig Ferdinand von Helmholtz. It was there, under the lure of the Berlin Prize of 1879, that von Helmholtz encouraged his twenty-two-year-old apprentice to further inquiry into electromagnetic forces. Their early experiments were not very fruitful, and for a while Hertz occupied his time with other experiments. However, he never ceased to be fascinated by the potential of proving the Faraday-Maxwell theory of the electromagnetic energy through space.

One day while lecturing, Hertz noticed that when a spark gap was introduced into a wire coil it produced a current in an adjacent wire coil. What Hertz had stumbled across were very high-frequency electromagnetic waves generated by the spark. From there the investigations proceeded systematically, and from 1886 through 1889 Hertz, using both transmitting and receiving spark gaps at high frequencies, was able to prove the hypotheses of Faraday and the predictions of Maxwell. Hertz had discovered electromagnetic waves—today we also call them radio waves—which catapult radio, television, and other communication around the world and into outer space.

Hertz's work carried him beyond radio waves and into light waves. He learned about both the penetrating qualities of electromagnetic waves and their reflective qualities at ultrahigh frequencies—frequencies approaching those of light. This electromagnetic theory of light gained as much attention from future researchers as Hertz's discovery of electromagnetic waves. Both von Helmholtz and Hertz died in 1894.

THE EARLY TELEGRAPH

Until now we have dealt with scientists who were concerned with inquiry for the sake of new knowledge and who did not necessarily apply that knowledge to some commercial principle. Pure scientific inquiry has an established place in history and continues to enjoy great esteem. Without new knowledge, the inventive minds of scientists may never have created new technology capable of grasping the attention of everyone from world leaders to the common people. We now examine how inventors applied new scientific knowledge to the field of communication.

Prior to the telegraph, many devices had been developed for signaling over long distances. The most familiar to us are the smoke signals used by Indians to signal the approach of warriors or the success of a hunt. The cannon volley from an early troop vessel sailing off Ocracoke Island would signal to Blackbeard and his pirates that they were about to encounter a fleet from the Crown governor. Flags on a mast and semaphores on the railroad were both early forms of telegraphic communication, and some are retained even today.

Early French Signaling

About 1790 Frenchman Claude Chappe developed a series of semaphores (mechanical flags) that relayed messages across land in France. A similar network later crossed southern England. Chappe based his system on the work of Englishman Robert Hooke, who more than a century earlier had considered the use of a signaling system employing the newly invented telescope. The Chappe system consisted of a series of towers, each with a person standing on top with a movable wooden beam whose different configurations represented the different letters of the alphabet. Each person standing watch would read the signals from one tower and repeat them to the next tower. The system stretched more than 140 miles outside Paris, but it was slow, cumbersome, and in bad weather subject to serious limitations. About the same time, the Spanish physicist Francisco Salva theorized that single-wire telegraph lines could be insulated and laid across the ocean, enabling water to act as the "return wire."

The Telegraph in Europe

The discovery of Volta's battery led the German scientist Samuel Thomas von Soemmerring to apply a steady current to sending and receiving units that were joined by a complex array of thirty-five wires. He tested his apparatus over relatively short distances of a few hundred feet and reported his findings on August 29, 1809, to the Munich Academy of Sciences. He incorporated an alarm whereby a spoon would fill with liquid and fall on a bell arrangement; this would alert the receiving operator that a signal was about to come down the lines. Soemmerring's telegraph was improved by his friend and colleague, Paul von Schilling-Cannstadt (also spelled Shilling).

Another early application of electricity to telegraphic communication was made by Sir Francis Ronalds in 1816. To generate electricity, Ronalds used a friction machine that was much like an electric generator turned by a hand crank. The sending and receiving apparatus consisted of two round revolving disks, each with a small opening near its outer rim. Positioned behind the disk were the letters of the alphabet and the numbers 0 through 10. Power from the friction machine would make the disks rotate in sequence, and the position of the receiving disk would be the same as that of the sending disk. By reading the different positions of the receiving disk in sequence one could understand the message. The system lacked two important elements: (1) a steady power supply that was available in the form of Volta's pile but was not used by Ronalds, and (2) speed of transmission. The slow rotation of the disks did not even approach the rapid transmission of later systems. Ronalds tested the system by constructing two wooden frames twenty yards apart and stringing eight miles of wire between them. By connecting the sending and receiving apparatus to the two ends of the wire and placing the ends next to each other, he could watch how one disk reacted immediately to the other.

THE MODERN TELEGRAPH

Developments in England

The modern telegraph developed in England through the resourceful efforts of William F. Cooke. While traveling in Heidelberg, Germany, Cooke learned about the telegraph of Schilling-Cannstadt. Cooke knew that if the machine could be further developed, it would have practical application in England. He immediately began work on his own telegraph, copying the designs of Schilling but using magnetic needles that would point to different characters of the alphabet. He later refined Ronalds' telegraph by powering the rotating disk with a battery instead of Ronalds' friction machine. In consultation with Faraday, Cooke further refined the telegraph's power supply. He also made the acquaintance of Professor Charles Wheatstone of Kings College. Working together, the two men continued to develop the telegraph, increasing it to a four- and finally a five-needle system. Cooke and Wheatstone formed a legal partnership, and on June 12, 1837, a patent for their telegraph was issued.

The Cooke-Wheatstone telegraph had a keyboard with five keys, one for each of the needles on the telegraph. Each key would engage current into the circuit and thereby cause the corresponding needle to turn and point to a letter of the alphabet. The Cooke-Wheatstone telegraph became the major long-distance communication medium in England—so much so that Samuel F. B. Morse's first attempts to introduce his telegraph in England were unfruitful. The Cooke-Wheatstone system received its first commercial test when it was installed on the Great Western Railway. Although their partnership was strained at times, Cooke and Wheatstone continued to develop their telegraph, and became more prosperous because of the railroad's use of it. The Cooke-Wheatstone system is important in that it differed from Morse's telegraph and was patented first.

DEVELOPMENTS IN AMERICA

The ship *Sully* was journeying home from England to the United States in the fall of 1832. On board, two gentlemen talked about the use of electricity for telegraphy in England and Europe. One of the men, Samuel F. B. Morse, was so enthralled listening to his companion, Dr. Charles T. Jackson of Boston, discuss electricity, that upon reaching America he began to work arduously on his own version of the telegraph. An artist by day, Morse spent his nights building the telegraph system that would simplify the transmission of messages over long distances. After considerable refinement, he demonstrated the Morse telegraph in 1837 and patented the system on June 20, 1840.

Experimenting with the Morse Telegraph

The system differed significantly from the Cooke-Wheatstone model in that Morse used a thin paper tape on which indentations were made as it slid across a wooden bar (Figure 2–1). Signals consisted of short and long marks, which became known as the *Morse code* although they were said to be the brainchild of Morse's partner, Alfred Vail, the son of a manufacturer who had invested money in the new device. Similar aid was given Morse by Leonard D. Gale, a chemistry professor at the University of the City of New York, where Morse had been working as an art teacher. Morse, though blessed with an inquisitive mind and ingenuity, didn't know much about science. Art was his vocation.

Morse continued to work on the telegraph, and he mounted public displays in an attempt to garner support for widespread development of the device. An early demonstration came in the fall of 1842, when Morse tried to span a river with a telegraph cable only to have a ship hook the underwater wire and cut it. He then went abroad for financial backing, but was rebuffed. His chance arrived in 1843 when Congress appropriated $30,000 to build an experimental telegraph line between Washington, D.C., and Baltimore. In 1844 the system was completed, and using a greatly improved transmitting device, Morse conducted a successful test whose famous message "What hath God wrought" signaled the telegraph's full-scale arrival in the United States (Figure 2–2). The Morse telegraph spread throughout the nation, linking western boom towns with eastern ports. Although he had to fight infringements upon his patent, Morse continued to develop the telegraph in America and later in Europe, where he was both honored and well compensated for the use of his system.

FIGURE 2–1 Samuel F. B. Morse's telegraph, which used a paper tape to record the code. A more efficient way of sending and receiving messages by Morse code became the dot-dash "clicks" of the telegraph key.

FIGURE 2–2 This artist's depiction of Morse shows the inventor sending the famous message "What Hath God Wrought" over the telegraph line from the Capitol in Washington, D.C., to Baltimore, Maryland, on May 24, 1844. Early work in telegraphic communication led other inventors, such as Marconi, to use the theory of electromagnetic energy in other ways, thus resulting in the development of the wireless.

The Telegraph Expands: Western Union and the Atlantic Cable

It was in this setting that a Rochester, New York, businessman named Hiram Sibley established a telegraph line in 1851 from Buffalo to St. Louis. With other investors, Sibley formed the New York & Mississippi Valley Printing Telegraph Company. In 1856 the company changed its name to the Western Union Telegraph Company. The telegraph business, closely tied to the development of the railroads, began to expand. Telegrams were also expensive: Twenty dollars, a substantial sum in those days, was not an uncommon fee. Through a variety of lease options, the company secured the rights to the Morse telegraph west of Buffalo. It then began immediately to buy up other smaller telegraph companies and established one large telegraph system.

By now Congress realized the need for a wireless connection of the West and East coasts. After all, gold had been discovered in California, ships were making regular passages from New York to California around the tip of South America, and the nation needed a communication link between eastern and western commerce. On September 20, 1860, Western Union was awarded a $40,000 contract to build a telegraph line connecting the eastern and western lines. Sibley hired Edward Creighton to survey the route. Creighton, who later helped establish Creighton University in Omaha, faced a great expanse of plains, rugged mountains, and unfriendly Indians. The Overland Telegraph Company, based in San Francisco and backed by California telegraph interests, and the Nebraska-based Pacific Telegraph Company began work on the line, one starting from each end on July 4, 1861. The eastern end of the line ran from Omaha to Salt Lake City and was supervised by Creighton. The western end connected Sacramento and San Francisco with Salt Lake City and was supervised by James Gamble.

Construction was not uneventful. It wasn't until permission from the Shoshone Indians was obtained and Mormon leader Brigham Young gave his blessing to the project that the line could be completed. On October 24, 1861, three months and twenty days after ground had been broken and slightly less than ten years earlier than the experts predicted, the lines met. Shortly thereafter, work began on another telegraph line through the Pacific Northwest, Canada and across Alaska to Russia. The project was suspended in 1867, however, upon completion of the Atlantic Cable, which bridged the communication gap between Europe and North America.

Now the telegraph sped news across the Atlantic. Stock-market quotations, shipping news, and economic fare dominated the transatlantic news flow. Yet despite all the popularity of the telegraph, it still could not operate without wires. And although the railroads were important to commerce, they were no less important than cargo ships, which were without communication once they left sight of land. The wires simply couldn't follow them. So for the next thirty years, the telegraph would continue to function as it had originally been conceived.

The work of the theorists would not be confined to their scientific curiosity. It would not be confined to land-based communication systems or to cables running under oceans for thousands of miles. The lines of force that James Clerk Maxwell witnessed, the spark that Hertz saw, would become steps upon which another inventor would climb.

THE WIRELESS IS BORN: MARCONI THE INVENTOR

The telegraph had captivated America and Europe. On April 25, 1874, two years after the death of Samuel Morse, the second son of Giuseppe and Anna Marconi was born. By late-nineteenth-century standards, Guglielmo Marconi's parents were quite well to do.[2] But the young, restless Guglielmo was not like the rest of his family, comfortable with gracious Italian living. Often he irritated his father by interrupting the quiet conversation at an evening meal with persistent, unrelated questions. There was no improvement when, after reading a scientific magazine, Guglielmo developed a keen interest in the work of Heinrich Hertz. Finally, having experienced his father's rancor and his mother's reinforcement, Guglielmo Marconi began to experiment in the top floor of the home. With crude tables, boards, hanging wires, and other paraphernalia he set about duplicating the experiments of Hertz.

Early Experiments in Italy

To the Marconi family, the work of the young son in his upstairs laboratory was intriguing but of questionable value. Guglielmo's father felt that his son was wasting the best years of his life, but he became more interested when Guglielmo asked him for money to advance his work beyond the experimental stage.[3] A stern and practical businessman, the senior Marconi first wanted a demonstration. This was followed by a long discussion as to how he would get a return on his money. Little did he realize that the boy's corporate empire would eventually gross billions. Finally, the two agreed to an initial investment, and Guglielmo began building his first transmitting device. Then, using a reflector sheet strung between two poles he first managed to receive a signal across the room (Figure 2–3). His receiver utilized a *coherer*—a small glass tube filled with metal filings and with wires in each end. The filings would collect between the two wires whenever electricity was applied.

Marconi, already familiar with the work of Samuel Morse, immediately realized the potential of his own device for long-distance communication.[4] He also had a sense of urgency, because to him the principle of his invention was extremely simple. Why had someone not thought of it before or, more important, applied it? His experiments became more and more frequent and the range of his signals more and more distant. On top of a hill twenty minutes from home, the experiments reached a threshold. Could the signal go beyond the hilltop? If the invention were to be a success, it would have to be able to leap over hills, mountains, buildings, and oceans.

On the day of the crucial test, Marconi's brother and two helpers carried the receiver and antenna over the hilltop out of sight of the family's villa. Guglielmo's brother also carried a gun with instructions to fire it into the air to confirm the signal. No sooner had Guglielmo fed current to the transmitter than the shot rang out. Now the funds that his father had provided had to be increased

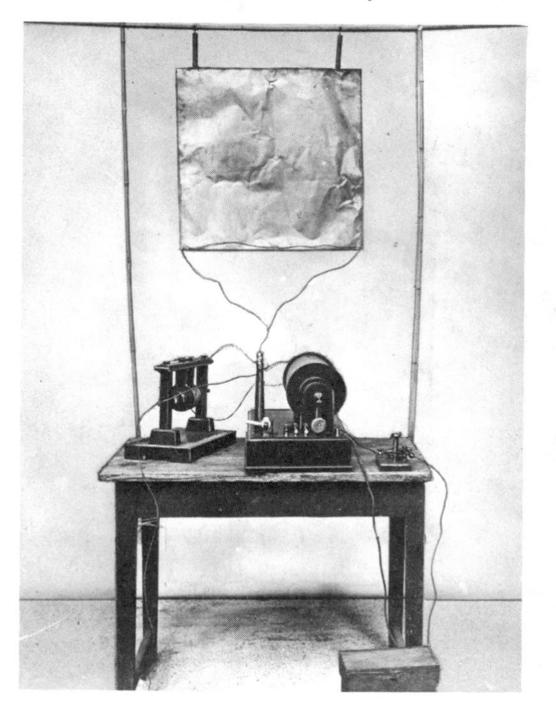

FIGURE 2–3 Marconi's first transmitter used in his early experiments in 1895. The large piece of tin suspended above the table served as the antenna.

before the experiments could progress. A letter was sent to the Italian Post Office Department in an attempt to obtain government backing for Guglielmo. The reply was negative. But if Italy were to say no, perhaps the great naval power of the day would say yes. Accompanied by his mother's encouragement, Marconi was off to England.

Experiments in England

The first stop was customs. Here the journey hit one of its low points, as ignorant customs inspectors ripped at the equipment until it was all but destroyed. Marconi managed to reconstruct the broken pieces, which had been crated so carefully in Italy. The next step was to be sure no one else captured the idea. For four months Marconi and his mother slaved over the papers that were to be presented to the London Patent Office.[5] The first specifications were filed on June 2, 1896. The complete diagrams and detailed specifications were filed on March 2, 1897, under the title "Improvements in Transmitting Electrical Impulses and Signals, and an Apparatus Therefor." On July 2, 1897, patent number 12,039 was granted to the twenty-three-year-old Italian inventor. The experiments could now be resumed, but it still was necessary to get from the government the funds with which to develop the invention to its full potential.

The help Marconi needed came first from the chief engineer of the British Post Office, William Henry Preece, who took a liking to the young inventor. With Preece's support Marconi began his experiments in England, first a transmission between two buildings and then a major demonstration across the Bristol Channel, a distance of about three miles (Figure 2–4). The press noticed Marconi's wireless and published the news to the world. More attention was bestowed on

FIGURE 2–4 Three officials of the British Post Office Department examine the equipment Marconi used to test the first successful wireless across the Bristol Channel in 1897. The British Post Office Department provided both encouragement and financial support for Marconi's early work.

the device than the young inventor had ever dreamed of. Along with offers to buy the rights to his invention came offers of marriage from women who said Marconi's waves made their feet tickle.[6] The distance of his experiments increased from three to thirty-four miles. Publicity abounded again when Marconi was commissioned to install a *wireless* on a tugboat to report the sailing races at the Kingston regatta. He secured other patents. One of the most important, a patent for a selective tuning device, was granted in 1901.

Wireless Across the Atlantic

The year 1901 was also the year of the most convincing experiment of the power of wireless communication. Still to be hurdled was the vast expanse of the Atlantic Ocean. Marconi left England for America in February of that year and headed for Cape Cod, the point he felt was best suited to the test of his wireless. But as with any stretch of New England coastline, harsh winter winds on Cape Cod can play havoc with any structure not built for permanency. The same is true of the English coast. For Marconi, 1901 held a double disaster. News arrived that storms had toppled the antenna at his installation at Poldhu, England. Within weeks, the same fate befell the Cape Cod station. Marconi now decided to transfer operations to Newfoundland, then a British colony. Using a bit of intrigue, he told local officials he was attempting to communicate with ships at sea; he made no mention of the real purpose, transatlantic communication. Instead of antenna towers, he planned to use balloons, and packed six kites as a backup.[7]

The experiments in Newfoundland started on December 9, 1901. First, a balloon was tested, but a line broke and the balloon headed for open sea. The next decision was to try one of the large kites. Marconi's assistants, George Kemp and P. W. Paget, sent the kite soaring hundreds of feet up, stringing behind it the antenna wire connected to the essential receiving equipment on top of nearby Signal Hill. Serious monitoring started on December 12. There were no results in the morning; nothing was heard from Poldhu. Spirits were low as the men continued to listen for the tapping signal that would indicate that England was calling. At 12:30 P.M., Guglielmo Marconi listened intently as the tapping sound of three dots, signaling the letter *S*, crackled through the earphone (Figure 2–5). Marconi handed the earphone to Kemp, and the assistant verified the signal.

Reaction to Transatlantic Wireless

The world would spend the rest of December reading about it. The *New York Times* called it "the most wonderful scientific development of recent times" and headlined the story WIRELESS SIGNALS ACROSS THE ATLANTIC. Across the ocean, the *Times* of London headlined WIRELESS TELEGRAPHY ACROSS THE ATLANTIC.[8] The London paper described how Marconi had authorized Sir Cavendish Boyle, the governor of Newfoundland, to "apprise the British Cabinet of the discovery, the importance of which is impossible to overvalue." Not forgetting his beloved Italy, Marconi informed the Italian government himself. Magazines were equally enthusiastic about the feat. *Century* magazine called Newfoundland "the theatre of this unequaled scientific development." *World's Work* labeled the transatlantic transmission "a red letter day in electrical history." *McClure's* magazine mused, "Think for a moment of sitting here on the edge of North

FIGURE 2–5 Guglielmo Marconi seated at the receiving set at St. Johns, Newfoundland, from where the historic wireless link across the Atlantic occurred. On December 12, 1901, at 12:30 P.M., Marconi and his assistants verified contact with the wireless station at Poldhu, England. The *New York Times* called the wireless contact "the most wonderful scientific development of recent times." For Marconi, it meant the start of major international business ventures.

America and listening to communications sent through space across nearly 2,000 miles of ocean from the edge of Europe!"[9] Not everyone, however, was happy in Newfoundland. The apparent threat of competition between wireless and the cable telegraph surfaced immediately. Cable stocks declined shortly after the announcement of the transatlantic broadcast.[10] The Anglo-American Telegraph Company, which had a monopoly on telegraph communication in Newfoundland, was quick to threaten reprisals if Marconi did not stop the experiments. A few days later, the inventor received a letter from the company stating:

> Unless we receive an intimation from you during the day that you will not proceed any further with the work you are engaged in and remove the appliances erected for the purpose of telegraphic communication, legal proceedings will be instituted to restrain you from further prosecution of your work and for any damages which our clients may sustain or have sustained: and we further give you notice that our clients will hold you responsible for any loss or damage sustained by reason of your trespass upon their rights.[11]

The Canadian government, however, obviously seeing the chance to emulate its neighbor, immediately offered Marconi its full cooperation. Public sentiment toward the action taken by the telegraph company was unfavorable on both sides of the Atlantic. The *New York Times* criticized the action, and letters to the editor of the London *Times* expressed similar sentiments. All of this soon became history as the world began to utilize the results of the December 1901 experiments.

WIRELESS EXPANDS: THE MARCONI COMPANIES

Marconi respected those who pursued pure science, but he was much more interested in applying results and harvesting financial rewards. Thus, it was only a short time after his patent had been issued in England that he began formulating a world corporate empire that would stretch over all the continents and involve millions of dollars in capital.[12] The company that had the most direct effect on wireless development was the Marconi Wireless Telegraph Company, Limited, formed on July 20, 1897, as the Wireless Telegraph and Signal Company, Limited.[13] It was Marconi's father who insisted that the family name be attached to the venture. The beginning capital amounted to 100,000 English pounds, of which £15,000 went to Marconi for his patents. It was from this £15,000 that he paid the cost of organizing the company. He also received 60,000 of the 100,000 initial shares, valued at £1 each. The remaining 40,000 shares went on the open market.

England: The Marconi Wireless Telegraph Company, Ltd.

A year after the company's formation, its operating capital increased by another £100,000. Although wireless had captured the imagination of the British, there were warnings for unwary investors. *Investors World* remarked in 1898 that "from all we can gather, the public will be well advised to keep clear of this concern. . . . Marconi's ingenious ideas do not seem to have made much headway, and it would be interesting to learn what the government officials reported about them."[14] The warning had little effect, and although for years to come investments did not show much success in terms of dividends, the public was always ready to buy up new shares whenever they were placed on the market.

In March 1912 a contract between Marconi's company and the British government became tainted with rumors of corruption. One rumor suggested that Marconi was treated favorably because of his close friends in Parliament. Some government officials had made a huge profit by selling their Marconi stock when it peaked after the news of the contract was signed. Another set of charges was of manipulation of stock by the American Marconi Company. A committee was appointed by Parliament to investigate the matter. After due deliberation, the committee came out strongly in favor of Marconi, but the matter was not over. Another committee investigated the role of middlemen and stockbrokers, and still another the role of the House of Lords. Libel actions were taken and the company stock tumbled. Because of the publicity from the scandal, the company enjoyed only briefly the prosperity for which Marconi had long hoped. The future development of Marconi in England would have to wait until the end of World War I (Figure 2–6). In North America, the story was much the same.

FIGURE 2–6 Much of the importance of wireless was attributable to its use in ship-to-shore communication. As wireless was installed on more and more vessels, international agreements were signed to foster cooperation among nations. Trained shipboard telegraph operators, capable of sending and receiving high-speed Morse code, were an important part of naval and private shipping interests. Also see Figure 2–8.

Marconi's Interests in Canada

Marconi's corporate interest in Canada dated from his experience with the cable-telegraph authorities in Newfoundland. He erected a station at Glace Bay, Nova Scotia, and began major attempts to achieve reliable transatlantic wireless communication. The first transatlantic service opened on the night of December 15, 1902, when the *Times* of London correspondent at Glace Bay cabled a newspaper report across the Atlantic. Two nights later it was arranged that the American station at Cape Cod would send a message from the United States to the King of England. The signal would be relayed to Glace Bay and from there to Poldhu. As it turned out, the atmospheric conditions were so good that the station in England picked up the signal directly from America.

The *Times* was so infatuated with the prospects of transatlantic service that it convinced Marconi to open the station again so that its correspondent could send news flashes to England. But a little more than a week later an ice storm sent the Glace Bay antennas crashing to the ground. The station was later reconstructed, this time with a large umbrella antenna.

The Marconi Wireless Telegraph Company

When Marconi came to the United States in 1899 to report the America's Cup races by wireless, he also began an American subsidiary of his English company so that he could utilize his patents in America. The

Marconi Wireless Telegraph Company was incorporated in the state of New Jersey in the fall of 1899 (Figure 2–7). The first equipment installed by the company was on the Nantucket Light Ship and its shore station on the eastern shore of Nantucket Island, and was used to warn ships of bad weather and coastal conditions.

That same year, the American company ran into trouble over a proposed United States Navy contract for the installation of Marconi wireless on Navy ships. After a series of tests, the Navy recommended buying the Marconi equipment. But when it asked Marconi the cost, it received word that the company would not sell the equipment; the Navy would have to rent it. At that point the Navy backed out. Captain L. S. Howeth, writing later about the negotiations, said: "In light of future events, the Marconi leases and stipulations have proved a blessing in disguise. The foresight of the authorities in not permitting themselves to be shackled with its restrictions, which would have persisted for more than a decade, allowed the Navy a free hand in guiding and assisting in the development of radio in this country."[15]

Despite the loss of the Navy contract, the Marconi Wireless Telegraph Company received a boost in its assets in 1912 when it won a patent suit against the United Wireless Company. Using a case in England as a legal precedent, the American company charged United Wireless with infringing upon the patent for the Marconi tuner that could select different signals from a single aerial. United Wireless pleaded no defense,

FIGURE 2–7 The stock certificate of the Marconi Wireless Telegraph Company of America. Incorporated under the laws of the state of New Jersey in 1899, the company installed its first equipment on the Nantucket Light Ship and at its shore station on the eastern bank of Nantucket Island.

SHARES $5.00 EACH

CERTIFICATE FOR LESS THAN 100 SHARES

067323

MARCONI WIRELESS TELEGRAPH COMPANY

OF AMERICA

INCORPORATED UNDER THE LAWS OF THE STATE OF NEW JERSEY.

This is to Certify that Mary C. Bennett, Adm
is the owner of ONE shares of the par value of Five Dollars ($5.00) each, fully paid, and non-assessable in the Capital Stock of Marconi Wireless Telegraph Company of America transferable on the books of the Company in person, or by duly authorized attorney, on surrender of this Certificate properly endorsed. This Certificate shall not be valid until countersigned by the Transfer Agent and the Registrar.
Witness the seal of the Company and the signatures of its duly authorized officers this day of 19

TREASURER

VICE-PRESIDENT

and Marconi assumed control of the company and all of its assets and contracts. In the end, the U. S. Navy ended up using Marconi equipment when World War I began, since it had either taken over or closed all commercial and amateur wireless stations, many of which used Marconi equipment (Figure 2–8).

A young boy named David Sarnoff had been hired by the American company in September of 1906. Shortly after World War I ended, American Marconi was purchased by the newly formed Radio Corporation of America (RCA). Sarnoff became part of RCA management and later headed the company. We shall learn the reasons for the sale and discuss RCA's early development in Chapter 3.

IMPROVEMENTS IN WIRELESS RECEPTION

Marconi's success in transmitting signals across the Atlantic and developing a world corporate empire was greatly aided by subsequent developments in wireless communication. One of the most important needs was for a device that would more efficiently detect and receive electromagnetic waves. As it was, the receiving and sending antennas were the size of football fields. Yet the current that entered a radio antenna and received an electromagnetic wave was minute. The great challenge was how better to detect these tiny, almost indistinguishable currents of energy hitting the football-field size antenna of a radio receiver. For

FIGURE 2–8 The wireless room aboard the *Lusitania*. The cumbersome antenna rigging was gradually eliminated as equipment became more advanced and more sensitive receivers were manufactured. Many of today's oceangoing vessels are equipped with sophisticated telecommunications gear, including satellite navigation instruments and worldwide telephone communication.

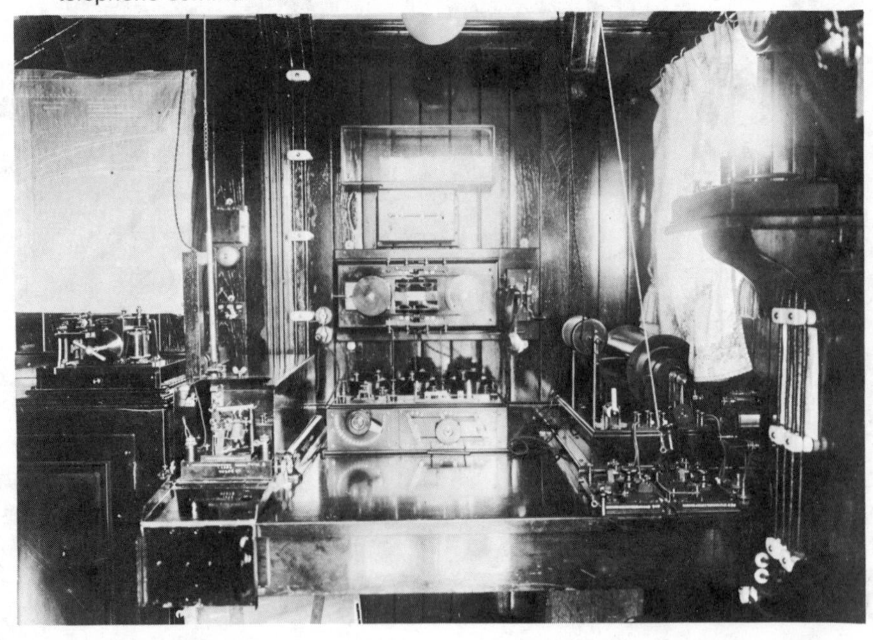

radio to become a household appliance, the huge receiving antennas had to be made much smaller.

Edison's Contributions

Some of the first experiments leading to an improved detector came during the study not of radio but of electric light.[16] Thomas Edison, while in the process of inventing the light bulb, had experimented with a two-element bulb but had found it impractical. The bulb consisted of two metallic elements—a plate and a filament—in a vacuum. If a battery were attached to the bulb so that its positive connector attached to the plate and its negative connector to the filament, current would flow through the bulb. If the connectors were reversed, the current would stop. What Edison had invented but discarded was later to be called a *valve*, since it could "shut off" current running in one direction, much like a valve controls steam or water.

The Fleming Valve

One of the keys to unlocking future developments in wireless technology was to find some way to measure electromagnetic waves in order to understand better their behavior and frequencies. J. Ambrose Fleming, an employee of Marconi, determined that the best way to do this would be to invent a means of measuring the waves as they flowed in only one direction. The secret lay in Edison's two-element light bulb. Fleming went to work perfecting the device, which became known as the *oscillation valve*, or *Fleming valve*. He patented it in England in 1904 and, through the Marconi Wireless Telegraph Company, in the United States in 1905. One worked the device by attaching the plate to the antenna, attaching a wire from the filament to the ground, and then hooking a telephone receiver into the completed circuit. The receiver could then detect the presence of the electromagnetic waves. It was not long, however, before Fleming's device was greatly improved by the inventive hand of Lee de Forest.[17]

LEE DE FOREST AND THE AUDION

The work of Lee de Forest ranks close in significance to that of Marconi in the development of radio. Born in Council Bluffs, Iowa, in 1873, de Forest was the son of a Congregational minister who was later to become president of Talladega College in Alabama. After attending Mt. Hermon School in Massachusetts, de Forest entered a mechanical engineering program at the Scheffield Scientific School at Yale University. Having completed a dissertation entitled *Reflection of Hertzian Waves From the Ends of Parallel Wires,* he was granted a Ph.D. degree in 1899. The research done and knowledge gained at Yale and his desire to apply pure science, first to inventions, next to patents, and then to profits, led him to a remarkable career that spanned much of his more than eighty years. He died in 1961 in Hollywood, where he was closest to one of his most beloved works, talking motion pictures. Our emphasis here, however, is on de Forest's invention of the *audion*, a three-element vacuum tube that revolutionized radio.

Adding the Grid to the Vacuum Tube

Lee de Forest discovered that a third element—a tiny grid of iron wires—could be added to Fleming's two-element vacuum tube valve. The result was characterized as follows in an early book on the development of radio:

> This may not seem much to the uninitiated, but that miniature gadget was the truest "little giant" in all history . . . that the brain of man ever created. It set unbelievable powerful currents in motion, magnifications of those which flicked up and down the antenna wire, and thus produced voice amplification which made radio telephony a finished product. By adding another tube and another, the amplification was enormously increased.[18]

The vacuum tube now had a filament, plate, and grid. De Forest first announced the tube—named the *audion* by his assistant, C. D. Babcock—in a paper presented to the October 26, 1906, meeting of the American Institute of Electrical Engineers in New York. After the paper was reproduced in the November 3, 1906, issue of *Electrical World*,[19] it was not surprising that one of the first reactions to his discovery came from Fleming. In a letter to the editor of *Electrical World*, Fleming attempted to diminish some of the importance of de Forest's invention:

> There is a remarkable similarity between the appliance now christened by de Forest as an "Audion" and a wireless telegraphic receiver I called an oscillation valve. . . . Dr. de Forest's method of using this appliance as an electric wave detector appears, so far as I can judge from published accounts, to be a little different from mine, but nevertheless the actual construction of the apparatus is the same. . . . Even if Dr. de Forest has discovered some other way of employing the same device as a receiver, I venture to think that my introduction and use of it should not be ignored, as I believe I was the first to apply this device . . . as a means of detecting electric oscillations and electric waves.[20]

De Forest did not let Fleming's suggestions go unchallenged. He replied with a letter to the same magazine, which was published two weeks later. In it de Forest credited German scientists Johann Elster and F. K. Geitel, not Fleming:

> Prof. Fleming has done me the injustice of expressing an opinion based on an extract only of my paper regarding the "audion." In a more complete abstract of that paper published in the *Electrician of London*, it is seen that I mention not only the device described by Prof. Fleming in 1904, but point out the real genesis of this device by Elster and Geitel in 1882, or eight years prior to its rediscovery by Prof. Fleming in 1890. . . . The difference which Prof. Fleming questions may be tersely stated as that between a few yards and a few hundreds of miles, between a laboratory curiosity and an astonishingly efficient wireless receiver employing the same medium but operating on a principle different in kind.[21]

The Feud with Fleming

The rift between Fleming and de Forest did not end in the pages of *Electrical World*. Lee de Forest went on to patent his audion, but as we have seen, the Fleming valve also had been patented, in both England and the United States. It was the American patent that provoked a lawsuit by the Marconi Wireless Telegraph Company. The case went in favor of the company, which contended that Lee de Forest had read the paper presented by Fleming to the Royal Society of England in 1905 in which Fleming described the oscillation valve, and that de Forest had then used this knowledge to begin the experiments that resulted in the audion.[22] The case was appealed, and again the court ruled in favor of the Marconi Wireless Telegraph Company and the Fleming patent. Two years passed between the lower court's decision and the appeal. In the meantime both de Forest and the Marconi Wireless Telegraph Company continued to manufacture the radio tubes. To make matters more complex, the court held that although de Forest had infringed on the Fleming valve, the Marconi Wireless Telegraph Company had infringed on the audion. The result was that neither company could manufacture the devices without the other's consent.[23] The situation was chaotic until the Fleming patent with Marconi expired in 1922. Incredibly, the U. S. Supreme Court ruled in 1943 that the Fleming patent had never been valid in the first place!

De Forest's account of the conclusion of his dispute with Fleming is worth reading, partly because of its humor. Most important, it captures a rivalry between two men that was typical in its intensity of feuds between companies and inventors during the early development of the radio:

> [Shortly after the Supreme Court decision], Sir John Fleming, still unregenerate at ninety-two, published an amazing article in which he ignored all the earlier work . . . claiming even the discovery of the so-called "Edison effect," but never mentioning Edison's name! For this omission I wrote him in righteous reproach, incidentally calling to his attention the recent Supreme Court decision. Fleming's reply evinced profound disdain for what a mere Yankee court might think of his best-loved child. Having married a young opera singer at 84, he lived to be 95, dying in 1945. He never yielded in his firm conviction that he was radio's true inventor.[24]

De Forest's modesty is not convincing when we remember that he entitled his autobiography *The Father of Radio*. He also had some choice words for a group of radio executives about what radio had become: "The radio was conceived as a potent instrumentality for culture, fine music, the uplifting of America's mass intelligence. You have debased this child, you have sent him out in the streets in rags of ragtime, tatters of jive and boogie-woogie, to collect money from all and sundry."[25]

BREAKING THE VOICE BARRIER: RADIOTELEPHONY

The second great development in the early history of the radio was the advance from the "dit-dahs" of Morse code and the "What hath God wrought" of the telegraph to the "O Holy Night" of Reginald Fessenden's Christmas Eve radio broadcast in 1906. The story of voice transmission starts long before 1906, back when early experimenters examined the capacity of the ground and water to act as a conductor for *wireless telephone* conversations.

The system had been used by telegraph operators in 1838.[26] It applied a process known as *conduction*, in which the ground or water provided the "second wire" in a telegraph hookup. Morse used it in his New York experiments. It was not long before inventors discovered that they did not need any wire at all to communicate between the transmitter and receiver over short distances. Because a current in one antenna would produce a current in another one nearby, a process called *induction*, two antennas close to each other would make the system work. This was a different principle from that of electromagnetic waves traveling through space, which Marconi and others used. Induction created an electrical disturbance in the atmosphere that was detectable only in the immediate vicinity of the transmitter.

NATHAN B. STUBBLEFIELD AND HIS WIRELESS TELEPHONE

Before Marconi mastered the Atlantic and while Lee de Forest was studying at Yale, a farmer and experimenter named Nathan B. Stubblefield developed a way to transmit the voice as much as three miles by means of induction (Figure 2–9).[27] Near his home in Murray, Kentucky, and later on the Potomac River in Washington, D.C., he successfully transmitted the human voice without using wires. It was in Murray that he received his first publicity. Dr. Rainey T. Wells witnessed Stubblefield's experiments:

> He [Stubblefield] had a shack about four feet square near his house from which he took an ordinary telephone receiver, but entirely without wires. Handing me these, he asked me to walk some distance away and listen. I had hardly reached my post, which happened to be an apple orchard, when I heard "Hello, Rainey" come booming out of the receiver. I jumped a foot and said to myself, "This fellow is fooling me. He has wires somewhere." So I moved to the side some 20 feet but all the while he kept talking to me. I talked back and he answered me as plainly as you please. I asked him to patent the thing but he refused, saying he wanted to continue his research and perfect it.[28]

The demonstration was reported to have taken place in 1892. A modified Bell-type transmitting device provided the signal, which emanated from a large, circular metal antenna. Other residents of the small town of Murray witnessed a similar demonstration in 1898. Claims of Stubblefield's accomplishments published in the St. Louis *Post-Dispatch* generated so much interest that he was brought to Washington, D.C., for a public demonstration on March 20,

FIGURE 2–9 Nathan B. Stubblefield (left) and his wireless telephone. As early as 1892, Stubblefield is reported to have sent voice by wireless over short distances at his farm near Murray, Kentucky. His son, Bernard (right), later became an employee of Westinghouse. Stubblefield used a Bell-type transmitting device and formed his own Wireless Telephone Company of America. Although receiving a patent for his invention, commercial development never materialized.

1902. Following the demonstration Stubblefield said, "As to the practicality of my invention—all I can claim for it now is that it is capable of sending simultaneous messages from a central distributing station over a wide territory. . . . Eventually, it will be used for the general transmission of news of every description."[29]

Commercial Exploitation

Commercial exploitation of the invention was not far behind, and in 1903 Stubblefield became director of the Wireless Telephone Company of America. Demonstrations in Philadelphia and Washington, D.C., created more interest in the device. Yet it is here that the rest of Stubblefield's life becomes somewhat obscure. There are various reports of what happened to him. One suggests that he became disillusioned with how the stock for the company was being handled and on one occasion even charged it with fraud.[30] Stubblefield returned to Kentucky and with the help of local citizens obtained a patent for the device on May 12, 1908. Obviously disenchanted with the commercial aspects of his wireless telephone, he went into seclusion and continued research in his workshop shack near Murray.

If you travel through the Kentucky countryside near Murray today you may pass the place where Stubblefield was found dead on March 30, 1928; the cause of death was listed as starvation. Or you might drive by Murray State College, where students perform a dramatization of Stubblefield's life, *The Stubblefield Story*. Then as you go downtown you can tune your car radio to 1340 kHz and hear a blend of rock, "easy listening," and country, and the news "centrally distributed" from WNBS radio. At a certain time the announcer will tell you "You are tuned to WNBS 1340 on your radio dial in Murray, Kentucky, the birthplace of radio."

THE EARLY WORK OF REGINALD FESSENDEN

Although Nathan Stubblefield had created a working wireless telephone, no one had yet mastered the ability to transmit the voice beyond very short distances. Some of the most productive experiments toward this goal were carried out in 1899 by Reginald A. Fessenden at Allegheny, Pennsylvania, at the Western University of Pennsylvania, later to become the University of Pittsburgh. Fessenden, a Canadian by birth and a professor at the University of Pittsburgh, worked to improve both the detection of electromagnetic waves and a means by which a human voice could be placed "piggyback" on electrical oscillations and sent into the atmosphere. Later he developed an improved detector, which would subsequently be called the *heterodyne circuit*. Fessenden applied for patent papers for the improved circuit in 1905. Simultaneously, he continued to improve the transmitting and antenna systems for wireless.

Experiments at Cobb Island, Maryland

Fessenden's early work was completed under government contracts, and was aimed more at improving wireless communication than at seeking out a new method of radiotelephony. An experimental "station" was established by Fessenden on Cobb Island, sixty miles south of Washington, D.C., in the Maryland section of the Potomac River. In a letter dated January 4, 1900, from Willis J. Moore, Chief of the U .S. Weather Bureau, Fessenden was informed of the terms of his agreement:

> You will be employed for one year in the Weather Bureau at a salary of $3,000 per annum. The Bureau will pay your actual expenses while on the road to an amount not exceeding $4.00 per day. You will be allowed to remain in Allegheny and continue your local connections for not longer than 3 months. Two active young men of the Weather Bureau will be assigned to duty as your assistants, and if you are especially desirous of retaining the one you at present employ, he will also be employed in the Weather Bureau for one year, at a salary of $1200. Such apparatus as described in your letter of the 29th ultimo will be purchased at the expense of the Government, one of our own men making the purchases and auditing the accounts: the property to belong to the Government. At the end of the year, if your work is successful, your services may be continued at a salary not less than that paid the first year.[31]

Fessenden accepted the offer and, traveling part of the way by river steamer, transported his equipment to Cobb Island. There, Fessenden's emphasis was not on distance but on the exact measurement of signals, which could verify and expand on some of the theoretical applications of the work that Fessenden had first tested in the 1899 experiments.[32] Two fifty-foot masts were erected for antennas. In a report of the experiments published in *Popular Radio*, Fessenden described the Cobb Island system:

> The exact method of transmission of the waves was experimentally determined by means of ladders placed at varying distances from the antennas. The course of the waves in the air was fully mapped out up to distances of several hundred yards from and to the antennas, and by burying the receivers at different depths in the ground and immersing them in different depths in sea water, the rate of decay below the surface and the strength of the currents flowing in the surface were accurately determined.[33]

Although poor in quality, intelligible human speech was transmitted between the two antennas.

EXPERIMENTS IN NORTH CAROLINA AND VIRGINIA

At the end of the year, both parties were satisfied with the arrangement and decided to renew the contract. This time Fessenden was to erect experimental stations in North Carolina on Roanoke Island and Cape Hatteras, and on Cape Henry in Virginia. These three stations formed a large triangle that enabled Fessenden to test the system over longer distances than were possible at Cobb Island. The new round of experiments were directed not as much toward perfecting speech as toward improving telegraphic communication, and especially the receiving circuit, the key for long-distance communication by voice or telegraph. Fessenden's letters to his patent attorney reveal his early success with the improved circuitry. After overcoming the considerable frustration of equipment parts that did not meet specifications, Fessenden wrote from Roanoke Island:

> I could hear every click of the key at Hatteras, and got every dot and dash as plainly as could be and as fast as they could send.
>
> To do a little figuring. The resonator should increase the effect 10 times. The prolonged oscillations about 5 times. The vacuum about 20 times. Longer waves about 5 times. Salt water instead of insulating water about 5 times. Good coils about 4 times, i.e., the sensitiveness can be increased about 1,000,000 times over this crude apparatus. This would give about 1,000 times the distance or 50,000 miles.
>
> As it is perfectly selective, perfectly positive, i.e., can give no false dots and cannot omit dots or dashes, I think we are at the end of all our troubles.[34]

Later in a letter of April 3, 1902, from Roanoke Island he wrote:

> I have more good news for you. You remember I telephoned about a mile in 1900—but thought it would take too much power to telephone across the Atlantic. Well I can now telephone as far as I can telegraph, which is across the Pacific Ocean if desired. I have sent varying musical notes from Hatteras and received them here with but 3 watts of energy, and they were very loud and plain, i.e., as loud as in an ordinary telephone.—I enclose telegram which was received with less than 1/500 of the energy which it took to work the coherer. The new receiver is a wonder!!![35]

Experiments at Brant Rock, Massachusetts

Despite the success of the experiments, Fessenden's relationship with the Weather Bureau began to deteriorate and he left Roanoke in 1902 after beginning a series of business arrangements. First, Fessenden licensed Queen & Company, Instrument Makers, of Philadelphia to fulfill contracts for his communication system.[36] Then, through an arrangement with his patent attorney, two Pittsburgh financiers, Thomas H. Givens and Hay Walker, Jr., put $2 million behind Fessenden's work and the four of them formed the National Electric Signalling Company.[37] Besides $300 a month in salary, Fessenden also received stock in the new venture. After conducting experiments on the Chesapeake Bay, Fessenden moved in 1905 to Brant Rock, Massachusetts. Here the next chapter in wireless history would be written. Trying continually to improve Marconi's invention, Fessenden constructed a high-power station at Brant Rock and radically altered his antenna design. Instead of the series of umbrella-like wires used in Marconi's experiments, Fessenden constructed an "antenna tower." It stood 420 feet high and consisted of a series of telescopic metal tubes 3 feet in diameter at the bottom, held in place by guy wires and insulated at all points from the ground. The result was a signal that penetrated the atmosphere, which Marconi's station could not reach. Signals were received in Puerto Rico and at a station in Scotland even during the summer months, when static normally interferes with transatlantic broadcasts. These first achievements at Brant Rock were shadowed by excitement as voice broadcasting moved out of the laboratory.

FESSENDEN AND ALEXANDERSON'S ALTERNATOR

The problem that plagued Fessenden was how to increase the number of transmitted oscillations so that the human voice would be audible. A telephone-type receiving apparatus had already proved successful. Marconi had used it to hear the signals from England in his famous Newfoundland experiment, and wireless operators on ships used headphones to listen to messages in Morse code. The problem was to generate a frequency where the voice would travel with the signal and not be drowned out by the sound of the current passing through the headphones. To accomplish this, Fessenden enlisted the help of the General Electric Company in Schenectady, New York.

The General Electric Team

There in the General Electric laboratories, a young Swedish scientist named Ernest Alexanderson was placed in charge of the engineering team assigned to produce the Fessenden alternator. Both trial and error and difficulties in meeting Fessenden's wishes slowed the project. Alexanderson first developed an alternator that utilized a revolving iron core called an *armature* (Figure 2–10). Fessenden, however, demanded a wooden core, and the work started over. Fortunately for Fessenden, his two Pittsburgh financial backers continued to pour money into the project. Fessenden tried another company, the Rivett Lathe Manufacturing Company in Boston, but their device failed when the bearings burned up at the high speeds necessary to produce a frequency of 50,000 cycles, the amount Fessenden felt would be needed for voice transmittals. Finally, in September 1906, Alexanderson and the GE team delivered the wooden-armature alternator.

Voice Broadcasting from Brant Rock

Within a few hundred miles of Brant Rock were ships filled with crews celebrating the mixed merriment and loneliness of Christmas at sea on that December night in 1906. In the wireless rooms the operators were on duty as scheduled, exchanging messages and receiving the food and good cheer of fellow officers, when the splitting sound of "CQ,CQ" came through their headphones. The universal call alerted them that a message would immediately follow. But instead of the dit-dah of Morse code came the sound of a human voice. Officers were called to

FIGURE 2–10 Dr. Ernst F. W. Alexanderson, a General Electric engineer, developed the high-frequency alternator that gave America a big edge in early long-distance voice broadcasting. The alternator, shown here next to Alexanderson, was one of many developed by him between 1905 and 1920. It was used to send transatlantic broadcasts from the RCA station at Rocky Point, Long Island.

the room to witness the phenomenon. The voice was that of Reginald Fessenden. "O Holy Night" rang out through the cabin, followed by the words "Glory to God in the highest, and on earth, peace to men of good will." Voice broadcasting had reached as far away as Norfolk, Virginia, and the West Indies, shouting the world of wireless into a new era.

THE CANADIAN CONTROVERSY: NATIONAL ELECTRIC SIGNALLING COMPANY IS BANKRUPT

Many of the early wireless experimenters managed to amass considerable fortunes from the new medium, and even those who at first had lost money later reaped a profit. For Reginald Fessenden, fate had the opposite in store.[38] With the Brant Rock experiments a success, Fessenden's backers wanted to develop some profit potential for the company, which until now had been devoted to pure research. But Fessenden was at odds with Givens and Walker over a proposal to open a Canadian subsidiary. The Canadian company had evolved from a plan by the three men to give Marconi competition in transatlantic broadcasts. Fessenden went to England and made an agreement with the British Post Office Department: if his station at Brant Rock could communicate with a station in New Orleans, a distance of about eighteen hundred miles, the British Post Office would approve a fifteen-year license for Fessenden's company to establish a reliable communication link between Canada and England.

Fessenden successfully completed the Brant Rock-New Orleans experiments, and then the trouble started. Fessenden, a Canadian by birth and the chief negotiator in the British contract, felt the Canadian subsidiary should be controlled mainly by himself, the Canadians, and the British. Despite providing the capital for the new venture, to say nothing of the millions they had already invested, Givens and Walker were not to serve in any position of authority. Naturally, both men objected strongly, whereupon Fessenden resigned and sued, collecting $460,000. The National Electric Signalling Company declared bankruptcy in 1902, and Marconi and his companies were once again the undisputed leaders in wireless communication.

DE FOREST GAINS PUBLICITY

After inventing the audion, Lee de Forest began to experiment with voice communication at the same time Fessenden was developing his heterodyne circuit and conducting the Brant Rock experiments.

The French Experiments

Using a high-frequency arc to modulate the signal, de Forest succeeded in transmitting a voice across the length of a room during the same year Fessenden gained recognition for his Brant Rock experiments with ocean vessels. De Forest was quick to see the potential of voice broadcasting and felt that good publicity would bring investors to his own company. Although voice broadcasts were well known in the United States, they were unknown in Europe. So in the summer of 1908, de Forest traveled to France and conducted demonstrations of radiotelephony from atop the Eiffel Tower, communicating with stations about twenty-five miles away.

Broadcasts from the Metropolitan Opera

The European experience whetted de Forest's appetite for more publicity at home. Always an opera buff, the inventor contacted the Metropolitan Opera in New York. He arranged to place a transmitter in the attic of the music hall and connect it to the microphones on stage. Although not very clear by modern standards, the microphones were the new Acousticon models manufactured by the National Dictograph Company.[39] On January 13, 1910, Enrico Caruso and Ricardo Martin bellowed *Cavalleria Rusticana* and *Pagliacci* to a small audience listening to receiving sets in New York. A master at gaining publicity, de Forest could rival Buffalo Bill Cody in obtaining press coverage for a show. The opera broadcasts were no exception. "The newspapers had been tipped off in advance and reporters were listening in at the Terminal Building, 103 Park Row, the Metropolitan Tower station, at the Hotel Breslin, on one of the ships downstream, and at our factory in Newark."[40] Although World War I and patent squabbles would slow the growth of modern radio until the late teens, de Forest's publicity helped set the stage and arouse the public's enthusiasm for what would occur in the decades ahead.

WIRELESS GAINS POPULARITY: CRYSTALS AND HAMS

Up to this time, the wireless had remained in the hands of the large companies, such as Marconi, and the major users—the Navy in the United States and the Post Office Department in England.

Dunwoody and Pickard: The Crystal Radio and Detector

All that changed in 1906 with the invention of the crystal radio receiver by General Henry C. Dunwoody. That same year, Greenleaf W. Pickard perfected a silicone-crystal detector. These two devices contributed two important words to the wireless vocabulary: availability and inexpensive (Figure 2–11). Remember, the audion was still being perfected, and vacuum tubes were expensive. As late as 1915, radio receiving equipment ran anywhere from $20 to $125—prices that were beyond the reach of young experimenters attracted to the lure of wireless. But by using the silicone crystal and a long, outside antenna, the general public could listen in on everything from opera to Navy broadcasts.

Amateur Radio

These early experimenters were called *amateur radio operators*, better known today as *hams*.[41] They were primarily of two types: (1) those who were interested in using radio to test new equipment, and (2) those who wanted to use the new medium to communicate with others. In each type the spirit of the other was fostered. It was these early hometown inventors who did much to see radio mature. Although the inventors and the big companies provided capital for international expansion, the ham operators were responsible for many of the early developments and experiments aimed at improving radio. In 1909, the first known amateur radio club was formed in New York City. The group started with five youngsters, and their advisor was Reginald Fessenden. A second organization, the Wire-

FIGURE 2–11 Invented by General Henry C. Dunwoody, the crystal receiving set brought wireless to the general public. The sets were relatively inexpensive and became a favorite of experimenters who wanted to listen to wireless but who couldn't afford the prices charged for radio receivers equipped with vacuum tubes.

less Association of America, was started by Hugo Gernsback, publisher of *Modern Electrics*. The membership roster jumped from 3,200 in 1909 to 10,000 in November 1910. The association published the first wireless *Blue Book*, which listed ninety amateur stations as members. A second *Blue Book* followed a short time later, and by 1911 the circulation of *Modern Electrics* had soared to 52,000. Sensing a lucrative market, the D. Van Nostrand book publishing company put *Wireless Telegraph Construction for Amateurs* on the bookstore shelves. By now, other radio clubs were rapidly forming, including the Radio Club of Salt Lake City, the Wireless Association of Central California, and the Radio Club of Hartford.

Ham radio was also gaining stature because it could be relied upon when other communication systems failed. In March of 1913 a major storm hit the Midwest, knocking out power lines and telephone communication. Ham radio operators, including those at Ohio State University and the University of Michigan, carried on communication and relayed emergency messages for seven days following the storm. This sparked Hiram Percy Maxim, famous as an inventor of an automobile and an engine silencer, to form the American Radio Relay League in 1914, an outgrowth of the Hartford Radio Club of which he was a member.

Ham radio has continued to thrive as a hobby and has developed throughout the field of wireless communication. When radiotelephony replaced wireless, hams began to chat "in person," but the Morse code remains even today a cherished language of those amateur experimenters. They communicate worldwide, using teletype, teleprinters, video-display terminals, and television. In cooperation with NASA, relay satellites have been launched for use by hams in international communication.

SUMMARY

In this chapter we traced the beginnings of electronic communication back to the eighteenth century and the experiments of Luigi Galvani. Subsequently, Alessandro Volta stored electricity in a stack of zinc and silver disks that came to be called the voltaic pile—the first storage cell. Oersted established the link between electricity and magnetism, and in doing so unified research and scholarship in these two areas. Ampère mathematically proved this relationship, and Faraday and Henry observed the phenomena. James Clerk Maxwell hypothesized the presence of electromagnetic energy, and Heinrich Hertz proved its existence by observing the presence of electromagnetic waves created by an electrical spark.

Our desire to communicate over long distances sparked the invention of two devices—the telegraph and telephone—that whetted the appetites of those who would later apply technology to wireless communication. The telegraph found wide acceptance in England and Europe and was developed in the United States by Samuel F. B. Morse. Morse conducted successful telegraph experiments between Washington and Baltimore in 1844, and through Western Union's efforts the telegraph later spanned the United States.

Drawing on Hertz's discoveries, Guglielmo Marconi first transmitted wireless signals over a short distance near his home in Italy. When the Italian government showed little interest in his invention, he traveled to Great Britain, where he received financial support from the British Post Office Department. After tests near the Bristol Channel, in 1901 he succeeded in receiving wireless signals from across the Atlantic.

Marconi continued his experiments while expanding his corporate interests. His companies began to spring up in many countries; they included the Marconi Wireless Telegraph Company, Ltd., in England, and the Marconi Wireless Telegraph Company in the United States. Improvements in the wireless were also made by J. Ambrose Fleming, creator of the Fleming valve, and Lee de Forest, inventor of the audion.

It was not long before people started to transmit the human voice over the airwaves. Using the devices that had been developed for telephone communication, scientists came closer and closer to quality voice transmission. A Kentucky farmer named Nathan B. Stubblefield performed short-distance wireless voice transmission. Then Reginald Fessenden developed the heterodyne circuit. With this improved detector of electromagnetic waves and with the help of a large alternator developed by General Electric and Ernest Alexanderson, Fessenden transmitted the human voice in December 1906. After disagreements with his financial backers, Fessenden was overtaken by de Forest and others in his quest for what was called radiotelephony. At the same time, radiotelephony became practical, as inexpensive receiving sets using silicone crystal detectors were manufactured and sold. The general public was becoming interested in what was now being called radio, and amateur ham operators talked first across city blocks and eventually across continents.

OPPORTUNITIES FOR FURTHER LEARNING

ABRAMSON, A. *The History of Television, 1880 to 1941.* Jefferson, NC: McFarland, 1987.

BROWN, R. D. *Knowledge Is Power: The Diffusion of Information in Early America: 1700–1865.* New York: Oxford University Press, 1989.

DOUGLAS, S. J. *Inventing American Broadcasting, 1899–1922.* Baltimore: The Johns Hopkins University Press, 1987.

INSTITUTION OF ELECTRICAL ENGINEERS. *International Conference on the History of Television: From Early Days to the Present.* London: Author, 1986.

MARVIN, C. *When Old Technologies Were New.* New York: Oxford University Press, 1990.

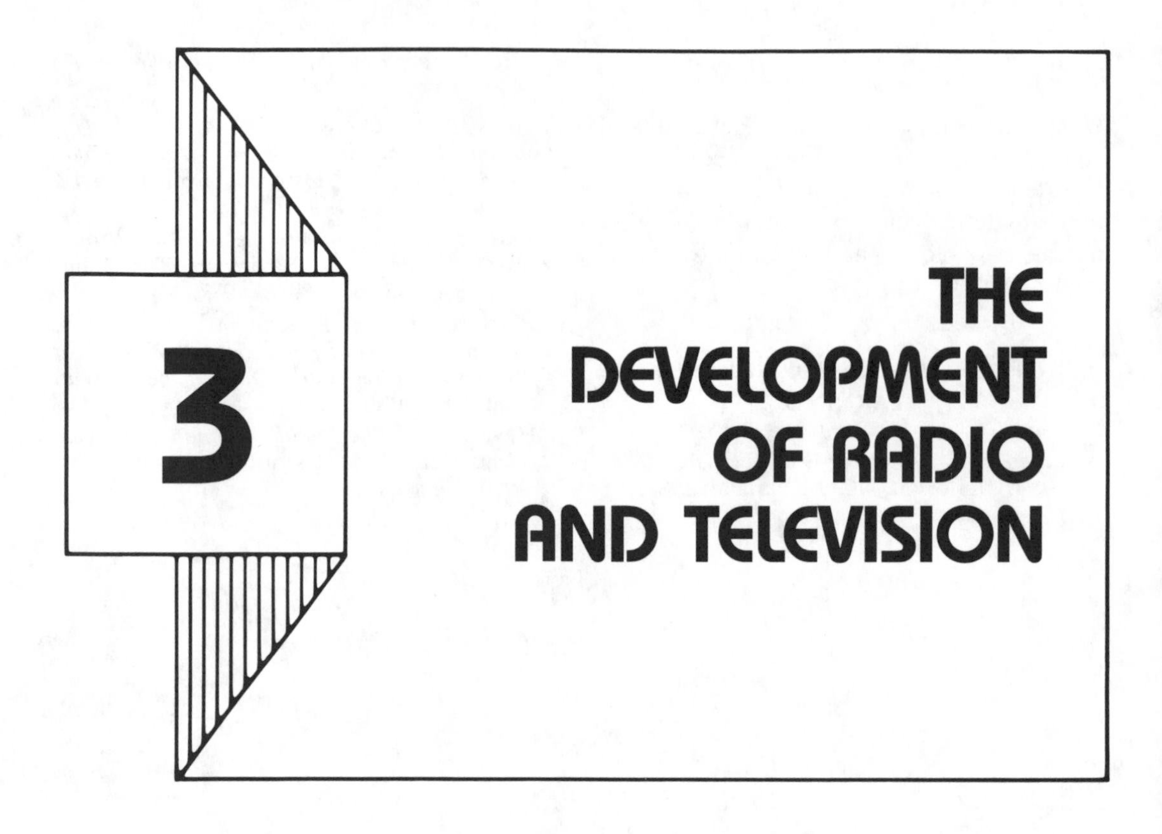

The excitement of the first wireless signals, the thrill of the first voice broadcasts, and the world of the radio amateur—all came from an era of pioneer spirit and experimental technology. Radio was magic, and people welcomed it with open arms. They could set a black box on their kitchen table, stretch a wire into the evening sky, and pick voices and music right out of the air. There was no need to have it delivered by the paper carrier, no need to walk to the country store to get it. The sounds of presidents, operas, big bands, and sporting events were live and immediate (Figure 3–1). Needless to say, people wanted all the radio they could get, and the stations that gave it to them grew in stature and power. Some of the earliest stations are still household words, and by learning about them we can catch some of the spirit of early radio.

THE PIONEER STATIONS

Much like trying to identify the inventor of radio, it is hard to put a label on the town, place, or person responsible for the first broadcasting station.

Basic Criteria of a Broadcasting Station

R. Franklin Smith has established several basic criteria for modern broadcasting stations.[1] First, *a broadcasting station transmits by wireless*. The signals must travel through

FIGURE 3–1 The immediacy and popularity of radio was displayed in this "radio organ," which traveled the streets of London in the early 1920s. The wagon had a four-tube receiving set with loudspeakers that could be heard 200 yards away.

space as *electromagnetic waves*. Smith does not consider ETV a form of broadcasting, nor closed-circuit wired college stations. Second, *a broadcasting station transmits by telephony*. The sounds of the station should be intelligible to the general listener. Third, *a broadcasting station transmits to the public*. It is distinct from other types of communication, such as telephone or telegraph, and from such special services as safety, aviation, and marine use. Fourth, *a broadcasting station transmits a continuous program service*. Programming is interconnected and is recognizable as a program service. Last, *a broadcasting station is licensed by government*. In the United States, the government licensing arm is the Federal Communications Commission.

Although these criteria are too limited for our purposes, they are helpful in outlining the history of broadcasting. Merely finding the station that first met these five criteria is difficult, since definitions of broadcasting were changing even in the early 1920s. Service, license, call letters, and ownership were often short-lived and sporadic.

Still, four stations are considered important to an understanding of the historical development of broadcasting. These are KCBS in San Francisco, which evolved over the years from an experimental station established in San Jose, California, in 1909; noncommercial WHA at the University of Wisconsin in Madison; WWJ in Detroit; and KDKA in Pittsburgh.

Charles David Herrold Begins in San Jose

Professor Charles David Herrold is credited with operating one of the first broadcasting stations in America. Others broke the airwave silence before him, but as early as 1909, residents of San Jose, California, could spend a Wednesday evening with their crystal sets tuned to news and music broadcast by Herrold. A classmate of Herbert Hoover at Stanford, Herrold had gone on to become owner of the School of Radio in San Jose.[2] The radio station was the school's medium of advertising—advertising that was aired more than ten years before KDKA in Pittsburgh and WWJ in Detroit began regular programming. Herrold had constructed a huge umbrella-like antenna in downtown San Jose, and from the Garden City Bank Building the wire structure hung out in all directions for a city block. Although it was a far cry from the eastern giants that could carry football games and political speeches, the little San Jose wireless station became one of the famous firsts in the broadcasting industry.

After 1910, the station handled regularly scheduled programs with operators on regular shifts. Even Herrold's first wife, Sybil M. True, had an air shift, which made her one of the earliest female disc jockeys.[3] She would borrow records from a local store and play them as a form of advertising. When listeners went to the store to purchase the recordings they would register their name and address, thereby giving the station an indication of its extent and influence. The California station gained national recognition at the Panama Pacific Exposition in 1915, and when Lee de Forest spoke in San Francisco in 1940 he called it "the oldest broadcasting station in the entire world."[4]

WHA in Madison, Wisconsin

Radio station WHA traces its inception all the way back to 1904 in the physics laboratory at the University of Wisconsin, where Earle M. Terry was working his way toward a Ph.D.[5] Graduating in 1910, Terry stayed on as an assistant professor, and in 1917, with the help of colleagues and assistants, he began experimental broadcasting of voice and music. The equipment was makeshift, and the three element tubes were not the sturdy successors of the 1920s. Rather, they were a mixture of creative craftsmanship, hand-blown glass, and immense frustration, especially when they burned out.

By 1922, station 9XM had been legitimized by the Department of Commerce with a license and the new call letters WHA. The same year, Professor William H. Lighty became WHA's program director. He developed the station into one of the first "extension" stations, responsible for bringing universities to the public with everything from news to college courses. WHA made other great strides in programming: broadcasts of the University of Wisconsin Glee Club, regular weather and road reports, farm and market reports, symphony broadcasts, and the famous Wisconsin School of the Air.

To aid listeners, Professor Terry taught them how to build their own radio sets. He even distributed some of the raw materials free of charge. The radio rage of the early 1920s caught many of the large equipment manufacturers unprepared. Loudspeakers had not yet replaced the earphone, and Professor Terry first demonstrated amplified radio reception in the Wisconsin Exposition Hall.

Meanwhile, WHA's farm and market reports and weather broadcasts were being picked up by the newspapers, and weather forecasting stations as far away as Chicago were using WHA data to aid prediction. Letters poured in from listeners as far away as Texas and Canada. WHA has since been joined by WHA–FM and WHA–TV. At the University of Wisconsin in Madison a historical marker reads: "The Oldest Station in the Nation . . . the University of Wisconsin station under the calls 9XM and

WHA has been in existence longer than any other."

WWJ and the Detroit News

After leaving the historical marker at the University of Wisconsin, one can travel east around Chicago and the tip of Lake Michigan to another pioneer station still operating—WWJ in Detroit.[6] When broadcasting was still in its infancy, some forward-thinking newspaper publishers realized that it would be better to reap some of its profits rather than always compete against it (Figure 3–2). William E. Scripps of the *Detroit News* had such a vision. He presented the idea to his colleagues, and they responded by appropriating money for construction of a makeshift radiotelephone room on the second floor of the *Detroit News* Building.

At 8:15 P.M. on August 20, 1920, an Edison phonograph played two records into the mouthpiece of the de Forest transmitter; probably no more than a hundred amateur operators heard the signal. There was no advance warning of the trial broadcast; no publicity draped the pages of the *Detroit News*. Everything worked perfectly, and the staff began preparations for the next day's broadcast of a Michigan election. When the election returns began to trickle in, it was the radio, not the newspaper, that first brought them to the public. Like a proud parent doting on a child's accomplishments, the September 1 issue of the *News* reported: "The sending of the election returns by the *Detroit News* Radiotelephone Thursday night was fraught with romance, and must go down in the history of man's conquest of the elements as a gigantic step in his progress."

FIGURE 3–2 WWJ's radio news truck, which operated in conjunction with the *Detroit News*. WWJ, an early pioneer station, claims some of the "firsts" of early broadcasting. It went on the air on August 20, 1920, and programmed two records over a Lee de Forest transmitter. About a hundred radio operators were reported to have heard the signal. The sixteen-piece Detroit News Orchestra was organized to provide music for the station's programming.

The early programming of WWJ, originally licensed under the call letters 8MK, reflects much of the same programming that other early stations experimented with and sometimes nurtured into long-running popular fare. The election returns were supplemented with a sportscast the following day, a preview of the World Series on October 5, and reports of the Brooklyn–Cleveland match-up. Returns from the Harding–Cox Presidential election were heard on November 2, 1920, the same returns that later became KDKA's claim to the "first station" honor.

So important was music that WWJ organized the sixteen-piece *Detroit News* Orchestra expressly for broadcast. It also expanded its studios to auditorium proportions after which they were described as "magnificent," having perfect acoustics, two-tone blue walls, and a white ceiling with a silver border. The latest equipment took WWJ's news microphone onto the road and into the air. A single-engine prop aircraft with NEWS painted in big letters on one wing was equipped for direct broadcast. Its news and photographic team thus became one of the first mobile units, now a common element of radio stations even in small communities.

With all of these early credits to its name, it is not surprising that on the front of an antique microphone illustrating the promotional literature of WWJ is the inscription WWJ RADIO ONE, WHERE IT ALL BEGAN, AUGUST 20, 1920.

KDKA in East Pittsburgh

Radio Station KDKA also established its place in broadcasting history in 1920.[7] The story of KDKA begins with Dr. Frank Conrad (Figure 3–3). Assistant chief engineer at the Westinghouse Electric Plant in East Pittsburgh, Conrad had constructed a transmitter licensed in 1916 as 8XK. After the World War I ban on nonmilitary use of radio was lifted, Conrad began his experimental programming. Through an arrangement with a record store in the nearby community of Wilkinsburg, Pennsylvania, he received records in exchange for mentioning the name of the store. The station's popularity grew so rapidly that Horne's Department Store in Pittsburgh ran an ad for inexpensive receiving sets.

To H. P. Davis, a Westinghouse vice-president, the ad was the inspiration for a license application including the call letters KDKA, granted on October 27, 1920. A month later, the new call letters identified the station as it sent Harding–Cox presidential election returns to listeners with amateur receiving sets and to a crowd gathered around a set at a local club. The crowd called for more news and less music, and KDKA's mail reported reception of the signals even at sea. The success of the broadcast gained widespread publicity, overshadowing WWJ's similar effort. Moreover, the combination of a publicized event and a major effort to get receivers into the hands of the public made the KDKA broadcast a milestone.

Manufacturing receivers was Westinghouse's definition of "commercial" broadcasting. Addressing an audience at the Harvard Business School, H. P. Davis remarked, "A broadcasting station is a rather useless enterprise unless there is someone to listen to it. . . . To meet this situation we had a number of simple receiving outfits manufactured. These we distributed among friends and to several of the officers of the company."[8]

As the popularity of the station grew, so did the staff, and when one day a Westinghouse engineer walked into the transmitting shack, he became the first full-time an-

FIGURE 3–3 Dr. Frank Conrad of KDKA radio, which began in East Pittsburgh, Pennsylvania. Originally licensed in 1916 as an experimental station under the call letters 8XK, KDKA broadcast under its new call letters beginning in 1920. The station carried continuous programming following its sign-on, and it also claims to be the country's "first" station.

nouncer in radio. Harold W. Arlin's broadcasting experiences were quite a change from his duties as an electrical engineer. During his career he introduced to KDKA's listeners such famous names as William Jennings Bryan, Will Rogers, Herbert Hoover, and Babe Ruth.

Today, clear-channel, 50,0000-watt KDKA can be heard over a wide area of the Northern Hemisphere late at night during good atmospheric conditions. If you are traveling in the Pittsburgh area, you might even pass the former home of Dr. Frank Conrad in the suburb of Wilkinsburg, where a plaque reads: "Here radio broadcasting was born. . . . "[9]

RCA IS FORMED

Although it may seem as though the pioneer stations and their owners were to become the corporate giants of broadcasting, by 1920 a new worldwide corporation was already operating with the blessing of the United States government. It would soon become a giant not only in broadcasting but also in other communication areas. The company was the Radio Corporation of America (RCA). Its beginning was full of international intrigue, skilled corporate maneuvering, and presidential politics. Even the United States Navy played a role.

The play begins at the close of World War I, when the United States government still controlled all wireless communication. Turning a major share of the American wireless interests back to Marconi was more than President Woodrow Wilson wanted to do. After all, the Marconi Company was still substantially British in influence if not in stock ownership. Communication and transportation were now recognized as important keys to international power. Great Britain had a network of cable systems in Europe and the United States, and its shipping industry and strategic location gave it an edge in transportation. Although not necessarily a threat, the British were at least to be treated with caution. Moreover, President Wilson was a fan of radio in his own right, having seen the benefits of his famous Fourteen Points spread throughout Europe by an American station using the huge General Electric alternator designed by Ernst Alexanderson.[10]

Government Attempts to Keep the Alternator

In 1918, two bills were introduced in Congress that were indirectly designed to bring wireless under control and to retain American control over Alexanderson's alternator. Seemingly harmless at the time, the proposed bills suggested the use of technical-school radio stations for experiments but failed to mention anything about the ham stations. Although the legislation had the support of President Wilson and the Department of the Navy, neither counted on the lobbying efforts of the amateurs. In Chapter 2 we learned of the mushrooming popularity of radio and the growth of the amateur organizations. When World War I began and the government took control of broadcasting, the hams (amateur radio operators) were silenced and their equipment did little more than collect dust. But now with the war over and with an army of war-trained operators wanting to continue their experiences as hobbyists and to share their exciting tales of war escapades involving radio, the legislation did not have a chance. The scathing attacks on the bills even claimed that they would prohibit the youth of the country from participating in investigation and invention.[11] The proposed legislation was tabled permanently.

Bullard, Young, and Sarnoff

The next scene cast the General Electric Company, President Wilson, and Admiral William H. G. Bullard in leading roles. The war was a period of considerable government support for General Electric (GE), especially for its Alexanderson alternator. When the war ended, the company faced substantial layoffs because of the lack of government contracts.

Although patriotism had taken precedence over trade during the hostilities, an end to the conflict meant GE was free to trade with any company it chose. By coincidence, that trading was about to begin with the British Marconi Company. But President Wilson wanted the new technology of radio to remain in American hands. Although the details of the conversation are unclear, we know President Wilson at least spoke to Admiral Bullard, who was Chief of Naval Operations Service, about keeping the Alexanderson alternator on home ground.[12] Bullard then took it upon himself to speak to General Electric's general counsel, Owen T. Young. He managed to convince Young and GE to take the giant leap of forming a new, all-American company in the wireless business.

A man known for his significant corporate maneuvers, Young managed to coordinate international negotiations that not only formed the Radio Corporation of America but also facilitated the purchase by RCA of the American Marconi Company. GE also bought the American Marconi Company stock owned by the British Marconi Company. The new corporation had American directors and stipulated that no more than 20 percent of its stock could be held by foreign nationals. For American Marconi, becoming part of RCA was a necessity if it was to overcome its "British" image in the face of American patriotism. It also needed the alternator to succeed just as much as GE needed customers. As it turned out, the merger maintained the jobs of American Marconi employees and directors.

One of the more famous American Marconi directors was David Sarnoff (Figure 3–4). As a wireless operator he had "worked" the messages from the ships rescuing the survivors of the *Titanic*. In 1916, in a now-famous letter, Sarnoff wrote his boss, Edward T. Nally, suggesting a commercial application of radio:

> I have in mind a plan of development which would make radio a "household utility" in the same sense as the piano or phonograph. The idea is to bring music into the house by wireless. . . . The receiver can be designed in the form of a simple "Radio Music Box": . . . supplied with amplifying tubes and a loud-speaking telephone, all of which can be neatly mounted in one box.
>
> Aside from the profit to be derived from this proposition the possibilities for advertising for the company are tremendous, for its name would ultimately be brought into the household, and wireless would receive national and universal attention.[13]

FIGURE 3–4 David Sarnoff taught himself Morse code and began his career as a wireless operator with the American Marconi Company on Nantucket Island.

Named commercial manager of RCA when the merger took place, Sarnoff later headed the corporation.[14]

PATENTS, CROSS-LICENSING, AND COMPETITION

KDKA's experiments and its accompanying publicity put the major corporations involved in wireless communication into a small turmoil. Corporate giants RCA, GE, and American Telephone and Telegraph (AT&T) had entrusted their futures to a joint enterprise that would effectively, if not completely, control the development of radio. But the vision of the triumvirate had been marine communication and radiotelephony, not the type of communication KDKA created with its November 1920 demonstration. Now Sarnoff's memo, which had originally gone politely unheeded, took on new significance. Perhaps there was money to be made from using broadcasting for mass appeal. The empire that Owen T. Young had built already had acquired allies in GE and AT&T—each had previously acquired important broadcasting patents, which the three now shared by agreement.

Sharing the Discoveries

Some of the earliest patents belonged to Lee de Forest. The audion, the forerunner of a series of improved vacuum tubes, was the most important link to the future of communication, at least to AT&T. In 1913 AT&T began buying de Forest's patents to the vacuum tube and having its own engineers improve the device. By 1915, using the latest equipment, including German-manufactured vacuum pumps that sucked the air out of the tubes, the company had perfected the first commercially successful vacuum tube.[15] AT&T used it for the first transatlantic telephone call.

As we learned in Chapter 2, the courts ruled that the audion infringed on the vacuum tube invented by Ambrose Fleming and that Fleming's patents belonged to the American Marconi Company. Yet war has its peculiar benefits, and breaking this AT&T–American Marconi conflict was one of them. The United States government stepped in and called for all companies to forge ahead as part of the war effort; thus, all became immune from patent-infringement suits.

The demand for vacuum tubes also involved GE and Westinghouse. Each had the capacity to manufacture light bulbs. The equipment that could suck air from a light bulb also could perform the same task in the manufacture of vacuum tubes.[16] General Electric, as we learned, also had the Alexanderson alternator.

So for the duration of World War I everyone worked in harmony, but each with an important part of the pie that could be reheated after the war ended. When it did, each had something the others needed. Thus for the future of radio it was advantageous for RCA, GE, and AT&T to enter into a complex arrangement of cross-licensing agreements, which permitted each to share in the others' developments but clearly divided the way in which radio would be marketed to the public.

Armstrong's Superheterodyne: Westinghouse Asset

Westinghouse, meanwhile, had been scrambling to compete with the RCA-GE-AT&T alliance. Just a month before KDKA's November 1920 broadcast, Westinghouse shrewdly bought the patents to a new type of circuitry invented by a graduate student at Columbia University, Edwin H. Armstrong. While Armstrong was serving in France in World War I, he became interested in finding a way for antiaircraft guidance systems to hone in on the radio waves emitted by aircraft engines.[17] Although his invention never aided the war effort, it did spark the development of the *superheterodyne circuit*, an improvement on Fessenden's heterodyne circuit. The superheterodyne changed the frequency of incoming radio waves, amplified them, then changed them to an audible signal. Westinghouse also acquired some patents held by Michael Pupin, a Columbia University professor who had worked with Armstrong, permitting him to use his laboratory and financing some of his work.[18]

When KDKA showed its stuff, Westinghouse was invited to become the fourth member of the RCA-GE-AT&T alliance. Still another company, United Fruit, joined because of its patents on crystal detectors. Under agreements among the big four, (1) GE and Westinghouse would manufacture radio parts and receivers; (2) RCA would market and sell them; and (3) AT&T would make, lease, and sell radio transmitters.[19] All of the companies were free to start their own broadcasting stations, and they did. But the agreements were concerned mostly with wireless telephony and telegraphy.[20]

When the stations did get under way they signed on fast and furiously. KDKA was only the beginning. More and more amateurs with number prefixed call signs applied and were granted licenses to operate broadcasting stations in the same fashion as KDKA. Westinghouse did not stop with that Pittsburgh station: before long it had signed on WBZ in Springfield, Massachusetts; WJZ in Newark; and KYW in Chicago (it was later assigned to Philadelphia). WJZ was sold to RCA in 1923.

RCA started its own station, WDY, in 1921 in New York. Although it stayed on the air only three months, the station tried some innovative programming, including a remote broadcast from the New York Electrical Show featuring Metropolitan Opera star Ann Case.

General Electric entered broadcasting by signing on WGY in Schenectady, New York. But of all the stations on the air in the early 1920s, the one to stir the attention of the public and the industry alike was AT&T's WEAF in New York City.

TOLL BROADCASTING: WEAF, THE AT&T STATION

The idea of commercial broadcasting was realized at AT&T on June 1, 1922, with the licensing of station WEAF. The station initiated the concept of *toll broadcasting*. This meant that anyone wishing to use the station could do so by paying a toll. Sponsoring a program meant buying the entire time segment and using it for whatever purpose desired. At first, the idea had few takers. To fill the programming void, the station used AT&T personnel as announcers (Figure 3–5). One of the earliest was Helen Hann, a member of AT&T's Long Lines Department. The first sponsor to try the new toll concept was the Queensboro Corporation of New York, which on August 28, 1922, began a set of five short programs over five days to sell real estate.[21] At a cost of fifty dollars the Queensboro Corporation had begun the era of modern commercial broadcasting.

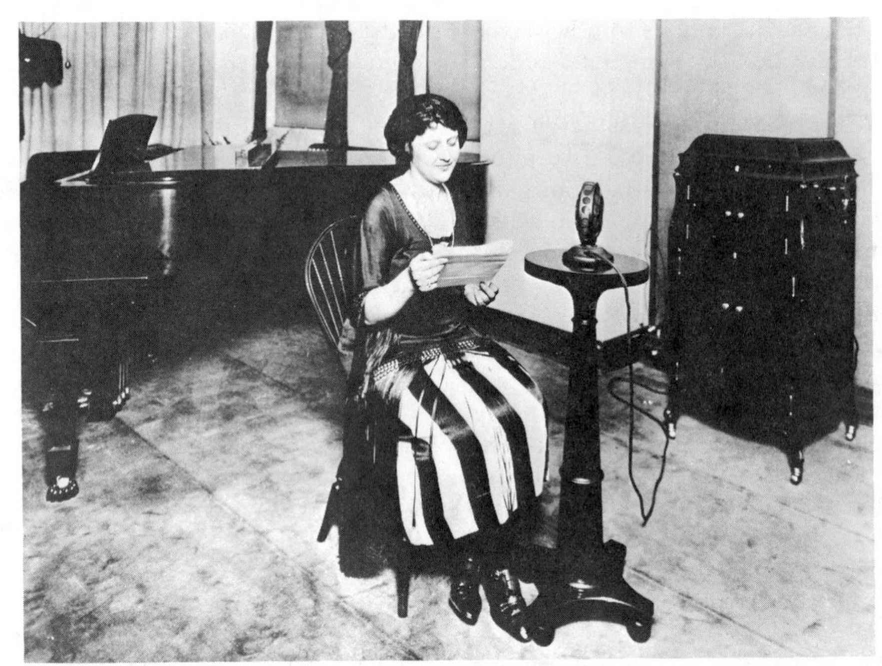

FIGURE 3–5 WEAF's early studio with Helen Hann, the announcer. WEAF is considered the first station to engage in commercial broadcasting, then called "toll" broadcasting. The first sponsor was the Queensboro Corporation of New York, which used the station to sell real estate.

Criticism of Toll Broadcasting

Not everyone liked the idea. Arguments against commercial radio started surfacing in the trade press. The *American Radio Journal* suggested three alternatives: (1) have municipalities undertake programs on a civic-entertainment basis; (2) charge the public and collect revenues from a large number of "radio subscribers"; or (3) tax the manufacturers of radio equipment, the people who distribute it, and the people who sell it.[22] *Printer's Ink*, the trade journal of early advertising, concluded:

> Any attempt to make the radio an advertising medium, in the accepted sense of the term, would, we think, prove positively offensive to great numbers of people. The family circle is not a public place, and advertising has no business intruding there unless it is invited. . . . The man who does not want to read a paint ad in the newspaper can turn the page and read something else. But the man on the end of the radio must listen, or shut off entirely. That is a big distinction that ought not be overlooked.[23]

But despite the skeptical reviews, advertising revenue gradually dribbled in to WEAF.

Through some political maneuvering with the U.S. Department of Commerce, the station managed to secure a more favorable *frequency* and extended hours. Both were important since stations then did not have the protection from interference that they do today. In fact, sometimes three or more stations had to share the same frequency and split up the broadcast day, each vying for the audience when another signed off.

As WEAF attracted more advertisers, AT&T began pouring money into the station, building new studios, and obtaining the finest equipment that its manufacturing arm, Western Electric, could manufacture. The equipment became the envy of the broadcasting industry, and when other stations started to request it, AT&T was reluctant to fill their orders. The short-term profit of transmitter sales was less important to AT&T than the potential of a national advertising medium under its control. When AT&T increased WEAF's remote broadcasts the audience clamored to listen, and when WEAF's competition made remote broadcasts, AT&T responded financially.

Finally, AT&T concluded that it would be in its best interests to block remote hookups on AT&T lines by its old allies, RCA, GE, and Westinghouse. Resentment, fueled by profits, lit a spark that inflamed the industry. While its three competitors were scrambling to use Western Union lines for broadcasts, AT&T was arguing that it alone should be permitted to engage in toll broadcasting, based on nothing less than the 1920 cross-licensing agreements that spelled out the rights to manufacture and distribute radio equipment.

As time went on, the stakes grew higher. While WEAF's income from its toll venture continued to climb, hundreds of smaller companies ate away at the profits of RCA, GE, and Westinghouse by manufacturing radio receivers in defiance and sometimes in ignorance of patent rights. It was clear that the future lay in commercial broadcasting to a mass audience. AT&T even went so far as to collect license fees from some stations before permitting hookups into AT&T *long lines*. The company also strung together a group of stations on which an advertiser could buy time separately or all together. This *chain*, as early network broadcasting was to become known, was a prime example of how toll broadcasting could work.[24] Although AT&T was receiving some severe criticism in the press, it continued its toll concept.

Finally, the accusations of infringement on the 1920 agreements escalated into open confrontation, and an arbiter—Boston lawyer Roland W. Boyden—was called in. The parties agreed to adhere to the verdict he would issue.[25] Simultaneously, the Federal Trade Commission, apparently completely unaware of the arbitration action, issued a sobering report claiming the existence of a monopoly in the radio industry and placing the blame on none other than AT&T, RCA, GE, Westinghouse, and the United Fruit Company.

The Antitrust Issue

Taking his time in this delicate matter,[26] Boyden finally presented a draft opinion that effectively ended AT&T's claim to exclusivity in toll broadcasting. The opinion caused the telephone company to try an end run. First, the AT&T attorneys issued their reaction:

> We believe that the referee's unavoidably incomplete knowledge of the extremely intricate art involved in this arbitration [coupled with] his effort to cooperate in the attempt of the parties to work out this situation have misled him into a radical departure from the contract which the parties actually made, and into conclusions which amount to an attempt to make a new contract for them.[27]

They then got an opinion from none other than John W. Davis, who had helped draft the Clayton Act, that major piece of antitrust legislation passed the same year that the Federal Trade Commission was formed.

Davis argued that if Boyden were correct, then the original cross-licensing agreements of the 1920s were illegal and an infringement of antitrust laws. It was a crafty move on AT&T's part, effectively suggesting that it did not have to agree to the arbitration because the agreement was illegal in the first place.

Despite all the turmoil, AT&T was very conscious of public opinion. Waging open warfare to gain control of broadcasting was not an image it wanted to acquire. Consequently, the next scene would see the power structure of American broadcasting change dramatically.

EARLY NETWORK RADIO

Whether RCA realized AT&T did not want to begin battle, or decided it was time it went into toll broadcasting, is open to speculation. Undoubtedly, both thoughts crossed the mind of David Sarnoff as he and other RCA officers watched AT&T organize its broadcasting interests into a separate corporation in May 1926 and call the new subsidiary the Broadcasting Company of America. At RCA a similar move was afoot: in September 1926 the RCA broadcasting interests were consolidated into a company called the National Broadcasting Company (NBC). Shortly thereafter, RCA bought WEAF for $1 million. WEAF was eventually consolidated into WJZ, which RCA had previously purchased from Westinghouse.

As for AT&T, the future forecast a healthy income from fees paid by broadcasters for the use of long lines for remote and network broadcasting. It also lifted the weight of negative public opinion from AT&T's shoulders. Although it might have won the court battles and the arbitration, and even survived the wrath of the Federal Trade Commission, AT&T felt comfortable with its network of "wires"; it would let NBC shoulder public opinion on the new "national network." In a major display of public pronouncement, NBC advertised its new venture in newspapers, promising "better programs permanently assured by this important action of the Radio Corporation of America."

NBC's Red and Blue

The National Broadcasting Company operated two basic networks as part of its nationwide coverage plan. The Blue network served some stations exclusively, as did the Red network, and a number of the stations had the option of drawing programming from both. Although still consolidated under NBC, the flagship station of the Red network was none other than WEAF. The Blue network chose its old rival WJZ. Not surprisingly, the rivalry continued. In 1932, NBC executives began to consider giving a separate status to the Blue network and having it operate even more competitively with the Red. One of these executives was Mark Woods, later to play a key role in ABC's development. There was no change at the Blue network until 1939, however, when a separate Blue sales department was established, followed by other departments separate from the Red network. Undoubtedly, an impetus for the changes was the Federal Communication Commission's announcement in 1938 that it was planning a full-scale inquiry into network broadcasting.

The FCC's Report on Chain Broadcasting

Out of the inquiry came the FCC's 1941 *Report on Chain Broadcasting*. Among other things, the report was critical of

NBC's interest in talent management. This interest developed early in 1931, when, because of its need for talent, NBC acquired a 50 percent share of the Civic Concert Service, Inc., in order to complement an artist-management division of the company. Increasing its share in the Civic Concert Service until it owned it, the network became the target of conflict-of-interest charges by the FCC. The report stated, "As an agent for artists, NBC is under a fiduciary [hold in trust] duty to procure the best terms possible for the artists. As employer of artists, NBC is interested in securing the best terms possible from the artists. NBC's dual role necessarily prevents arm's-length bargaining and constitutes a serious conflict of interest."[28] Scrutiny of the artists' service was only part of the investigation. The report also examined NBC's growing interest in its transcription business, which included recordings for libraries and other services.

The FCC concluded that stations could not be bound by exclusive network contracts prohibiting them from airing programming from other networks; that network contracts were to be for a period of one year; and that stations were to be the sole determiner of programming, a right not to be delegated to the networks. The most important statement, however, hit at the very heart of NBC's dual-network concept: "No license shall be granted to a standard broadcast station affiliated with a network organization which maintains more than one network."[29] Seasoned veteran David Sarnoff, now president of RCA, set the wheels in motion to protect RCA's investment. He immediately organized the Blue network as a separate corporation. The action was an attempt (1) to pacify the FCC, at least temporarily, and (2) to get an accurate reading of exactly how much the Blue network was worth by creating a separate accounting system. The handwriting was on the wall—Blue had to be sold.

Edward J. Noble Launches ABC

When it became clear that NBC's disposal of the Blue network was inevitable, major industrialists began to consider the jump into broadcasting. They included the Mellons in Pittsburgh, Marshall Field, Paramount Pictures, and Edward J. Noble, a former undersecretary of commerce who had amassed a sizable fortune making and selling Lifesavers candy. In the summer of 1943 Noble posted $1 million of Blue's purchase price and made arrangements to pay RCA the remainder from his own pocket and with loans from three New York banks.[30] The FCC, meanwhile, had delayed enforcing the 1941 *Report on Chain Broadcasting* in order to permit the sale of the Blue network in a calm atmosphere that wouldn't depress the price.

On October 12, 1943, the FCC announced it was approving the sale of the Blue network to Edward J. Noble.[31] Mark Woods was retained as president. In approving the sale, the FCC stated that the transaction "should aid in the fuller use of the radio as a mechanism of free speech. The mechanism of free speech can operate freely only when controls of public access to the means of a dissemination of news and issues are in as many responsible ownerships as possible and each exercises its own independent judgment."[32]

For Edward J. Noble, the challenge to develop the Blue network was sizable. World War II was raging, and American business, although geared up for war production, was in a state of uncertainty. A total of 168 stations and 715 employees were now Noble's responsibility. Already on the climb, however, were Blue's credits as an independent organization.

While still part of NBC, the Blue network showed promising opportunities as an investment. It instituted a special daytime-rate package permitting advertisers to buy at a discount over a series of daytime hours. Another discount package provided savings for advertisers who steadily bought programming time on more and more stations. Institutional advertising permitted companies to sponsor one-time programs publicizing important accomplishments. Typical were the famous "Victory Broadcasts" calling attention to the war effort. Noble also inherited the "strip" broadcasts, which permitted companies to sponsor programming over a strip of four to seven evenings per week. Some of the early takers included Metro-Goldwyn-Mayer, which sponsored the antics of Colonel Lemuel Q. Stoopnagle, heard five nights a week for five minutes a night over fifty-four stations.[33]

Despite all its recent accomplishments, the Blue network still had not made a profit. So Noble pulled together his own team of experts and named Adrian Samish, a New York advertising executive, vice-president of programs. Samish, in his mid-thirties, had worked on the stage, and he realized the Blue network did not have the big-name talent that was pulling audiences to the other networks. He was also faced with a diehard group of female followers who lived for the tensions, intrigues, and love affairs of the soap operas aired on the other networks. To compete with these he instituted a series of game shows, and although they did not set the world on fire, they did provide the Blue network with alternative programming.

Working with Samish was Robert Kinter, a former Washington correspondent. Vice-president in charge of special events, Kinter seemed like a public-relations troubleshooter until he began showing everyone he had a head for management decisions. By the turn of the decade he was serving as executive vice-president, and he would later be named president of the network.

Noble had formed a separate corporation, the American Broadcasting System, Inc., in order to purchase the Blue network. On June 15, 1945, affiliate stations heard announcer James Gibbons say, "This is the American Broadcasting Company." The influence of the war effort and the patriotic mood of the country were reflected in Mark Woods's comments about the new name. The name was chosen, he said, "because 'American' so completely typifies all that we hope, and believe, this company will be and will represent to the people of the world. The tradition of independence and of free enterprise, liberality in social philosophy, belief in free education for all and in public service—all of this and much more is inherent in the name."[34]

Today these words to some might seem overstated, but we must remember that Woods was appealing to the heart of a nation headed toward victory in global conflict. Patriotism was also present later that evening when ABC officially retired the label Blue with an hour-long program. Entitled "Weapon for Tomorrow," it discussed the "importance to a democracy of a freely informed people."[35]

CBS Is Born

When ABC began network broadcasting it had three formidable competitors—the Red network, which later became NBC; CBS; and Mutual. CBS can trace its beginnings to January 27, 1927, when a company called United Independent Broadcasters, Inc., was formed for the dual purpose of selling time to advertisers and furnishing programs for stations. Acting as the sales arm of United was another company and stockholder, the Columbia Phonograph Broadcasting System, Inc. Sixteen stations were included in the original United network. United had devised a plan by which it would pay them $500 per week to furnish it with ten specified hours of broadcasting. But the cost was simply too high, and it was not long before the venture became less than profitable. In the fall of 1927 the Columbia people withdrew from the venture and United bought the stock. United also changed the name of the organization to the Columbia Broadcasting System, Inc. The network revised its rate agreement with the affiliate stations, having suffered losses of over $220,000 in its first nine months of operation.[36] The new agreement cut the losses, but it was not until William S. Paley arrived that things began to brighten.

Paley's father owned the Congress Cigar Company, one of the sponsors on the old United network. When cigar sales jumped from $400,000 to $1 million per day in six months, radio got the credit. Congress's advertising manager, the owner's son, went to New York with an eye on buying the faltering sixteen-station network. Taking control of 50.3 percent of the stock, the Paley family entered the broadcasting business. Ten years after Paley arrived, the network had grown from 16 to 113 affiliates. In its first year of operation it sank more than $1 million into programming and moved its facilities to new quarters.[37]

The Mutual Broadcasting System

The Mutual Broadcasting System started in much the same way as the United network, except for two major differences.[38] First, Mutual did not enter into agreements to pay unmanageable sums of money to affiliate stations. Second, it started small. Mutual began with four stations—WOR in Newark; WXYZ in Detroit; WGN in Chicago; and WLW in Cincinnati. The four stations agreed that Mutual would become the "time broker" and pay them their regular advertising rate, first deducting a 5 percent sales commission and other expenses such as advertising-agency fees and line charges. Mutual expanded in 1936 by adding thirteen stations in California and ten in New England. In 1938 a regional network in Texas added twenty-three more stations to the chain. By 1940 Mutual had 160 outlets. Yet the network operated more like a co-op than a profit-making network like NBC or CBS. A special stock arrangement gave some stations a greater voice in the network's operation as well as in special sales commissions.

FM BROADCASTING

Many stepping stones dot radio's path of development. Some are milestones, such as KDKA's first broadcast. Others mark decisions made in corporate boardrooms, decisions that charted the medium's course. Still others represent developments from inside the laboratory.

Armstrong Applies the Principle

Frequency modulation (FM)—changing the frequency of a radio wave in order to modulate a signal—was not new to Edwin Armstrong. He had studied it, and did not believe the words of his predecessors that FM had no meaningful application to broadcasting. Armstrong agreed with David Sarnoff on the need for a device that would clear the static from radio transmission.[39] To Armstrong that challenge came to mean years of research at Columbia. Finally, in 1933 RCA engineers accepted an invitation to witness his latest efforts. Although the equipment worked, RCA was not enthusiastic. Still, it gave Armstrong permission to continue the experiments at the Empire State Building. There, he conducted successful tests ranging up to sixty-five miles. Armstrong was sure FM held the key to revolutionizing radio.

But the vision in RCA's eye was television. Tests had already been successful, and the company was undoubtedly thinking a few years ahead to the public-relations splash a television demonstration would make at the 1939 New York World's Fair. Armstrong grew increasingly suspicious of the intentions of Sarnoff and RCA. Deciding not to wait, he launched a lecture tour and demonstrated FM to dozens of audiences across the United States.[40] Selling his RCA stock and receiving encouragement from the Yankee and Colonial networks in New England, Armstrong built his own FM station in Alpine, New Jersey. There, after battling the FCC for a license, he continued his experiments and managed successful broadcasts of up to 300 miles while spending a personal fortune of between $700,000 and $800,000.[41] The World's Fair came and went, and Armstrong was left with his fledgling experiments. But even with the thrill of television, FM was beginning to catch on—so much so that on January 1, 1941, the FCC authorized commercial FM broadcasting. Although it might have seemed that Armstrong could look his old friend David Sarnoff in the eye with an "I told you so," that was not the case. In 1945, RCA won a victory when the FCC moved FM to a higher frequency in order to make room on the spectrum for television.[42]

In 1948, after seeing RCA get away without paying royalties on FM sound transmission for TV, Armstrong brought suit. The legal battle went on for five years, after which Armstrong finally agreed to a settlement. He died shortly thereafter, reportedly committing suicide. But in spite of setbacks, corporate lobbying, government tampering, and changed frequencies, FM has continued to develop and win audiences.

Factors Affecting FM Growth

The growth of FM broadcasting can be attributed to a number of factors. *First*, even though the development of FM was set back by World War II, the FCC gave its permission in 1941 for full-scale development of FM, which prospered during that brief prewar period. *Second*, the perfection of sound recording gave the public a new appreciation for quality reproduced music, which FM could provide better than *AM* (amplitude modulation). *Third*, FM was boosted by the development of stereo sound recording and the corresponding public demand for stereo FM. *Fourth*, the June 1, 1961, decision by the FCC permitting FM to broadcast stereo signals gave FM the ability to supply that demand. *Fifth*, crowding on the AM frequencies prompted new broadcasters to enter the industry on the FM spectrum. *Sixth*, FCC requirements that gradually eliminated the once common practice of simulcasting the same program on combination AM/FM stations under the same licensee forced licensees to develop the FM stations with diverse programming. *Seventh*, more and more radio receivers became capable of receiving FM signals.

FM's growth has been substantial. So, despite his frustrations, Edwin Armstrong opened a new era in radio, one that has had a profound effect not only on the industry but also on the radio programming we receive.[43]

THE TRANSISTOR

The story of the *transistor* begins in the Bell Laboratories in 1947, when Dr. William Shockley invited colleagues to observe an experiment he had conducted successfully by using crystals much like those in early radio receivers. What Dr. Shockley was experimenting with was the *transistor effect*. Using a small silicon crystal, scientists at Bell Labs discovered that the crystal could be made to react to electrical currents much the same as the vacuum tube did. Working with Walter H. Brattain and John Bardeen, Dr. Shockley perfected the transistor. Today, other scientists have perfected the transistor to the point where thousands of them can fit onto a tiny chip smaller than the end of one's finger.

Transistors function like a switch controlling an electrical current. The transistor in your portable radio consists of a wafer-thin crystal in three layers, with a wire attached to each. One wire detects the radio signal being sent through space. When it detects the signal the wire allows current to flow through the transistor in sequence with the incoming signal. By attaching a battery to the transistor, we can cause the radio signal to trigger a circuit and thereby release current from the battery. Because the current is released in exact sequence with the incoming signal, the transistor permits the signal to be amplified tens of thousands of times by the battery's current.

The small size of the transistor revolutionized radio. When the practical applications of the transistor were realized, radio receivers powered by nothing more than a tiny battery could be taken outside the home. Radio receivers were suddenly everywhere—on the beach, at the ball game, at picnics. There was a new gift-giving spree as transistor radios became the thing to own. We now take for granted the tiny pocket device that can put us in touch instantly with dozens of AM and FM radio stations. For William Shockley, Walter H. Brattain, and John Bardeen, their discovery won them the Nobel Prize in Physics.

REPRODUCING AN IMAGE

Even before Heinrich Hertz proved the existence of electromagnetic waves, scientists were working to find a way to reproduce images and to send them from a transmitter to a receiver.

Early Mechanical Reproduction

In 1843 Alexander Bain developed in theory a system for sending pictures by wire.[44] In 1862 Abbe Caselli developed a facsimile transmission system that could send examples of handwriting and simple pictures over telegraph wires.[45] A somewhat more modern system was demonstrated four years later by Frederick Collier Bakewell.[46] During the 1880s Paul Nipkow experimented with a mechanical television system consisting of a scanning disk. The disk proved that images could be transmitted electrically and mechanically by means of a series of wires between the transmitter and receiver. By punching holes in the disk, arranging the holes in a spiral, and revolving the disk, one could scan a picture placed behind the disk. If a series of pictures replaced each other in rapid succession, the illusion of a moving image could be transmitted over wires. The system worked even in 1884, yet it lacked many of the components necessary for making television a reality. First, the system was mechanical, not electronic. Compared with today's television, it was slow and cumbersome. Second, the wires limited the distance that the image could be transmitted, because stringing wires to many different locations was impractical. Third, the image was unclear, because coordinating the scanning disc with the changing pictures still had not been perfected. Experimentation on the scanning disk continued. Ernst F. Alexanderson, inventor of the Alexanderson alternator, worked on mechanical television, experimenting with both small-screen and large-screen systems.

Philo Farnsworth: The Basic Electronic System

Although he lacked the publicity that some of his more famous contemporaries enjoyed, an inquisitive schoolboy from Buckhorn, Utah, made some of the most important contributions to the science of television.[47] Philo Farnsworth (Figure 3–6) was born in 1906 into the Mormon family of Lewis Edwin and Serena Bastian Farnsworth. In 1918 the family was living in Rigby, Idaho, and Philo was becoming friends with his science teacher and school superintendent, Justin Tolman. Tolman provided the boy with science books in order to fuel a fire of intellect that Farnsworth had already exhibited in reading about the work of Einstein and other scientists. He was well acquainted with the experiments on electromagnetic energy and by 1922 had theoretically combined the components of the cathode-ray tube and the photoelectric cell into what he called a dissector tube. That year the family left Rigby, and after a short stint in the railroad yards at Glen's Ferry, Idaho, Farnsworth ended up in high school in Provo, Utah.[48] In Provo

FIGURE 3–6 Philo Farnsworth at work in his laboratory. A high school science teacher in Rigby, Idaho, provided Farnsworth with science books that spurred the future inventor to begin theorizing about television. Farnsworth eventually licensed some of his inventions to RCA, which later tried unsuccessfully to buy outright the Farnsworth system.

he had the run of the Brigham Young University laboratories and could continue his interest in science and in what would later become television.

After Provo, Farnsworth found brief employment at Feld Electric Company of Salt Lake City and then at the local Community Chest. At the Community Chest he encountered George Everson and Leslie Gorrell, who arranged $5,000 in funding for Farnsworth's research.[49] In partnership with one another, the three located a site for research near the California Institute of Technology. There Farnsworth developed a system consisting of an electro-light relay, a magnetic image dissector, a magnetic image builder, and a dissector-cell combination.[50] By 1926 the enterprise and the scientific developments had gained additional support, and preparations were made for patent applications. The system was found workable and moved to San Francisco in 1927. With trial-and-error modifications the first successful transmission of electronic television was achieved on September 7, 1927.[51] The following year demonstrations were made for the General Electric Company, and in 1930 the apparatus was seen by the Russian-born scientist V. K. Zworykin. On August 26, 1930, Farnsworth was awarded two patents: 1,773,980 for his television system and 1,773,981 for his receiving system.[52] Farnsworth continued his research and eventually moved east to the Chestnut Hill area of Philadelphia and an association with the Philco Company. The association later ended, and Farnsworth obtained independent support for his research.

Farnsworth and V. K. Zworykin

The visit of V. K. Zworykin (Figure 3–7) to Farnsworth's San Francisco laboratory was not the only link between the two men. Throughout the 1930s they were entangled in major patent litigation. Zworykin had also been developing an electronic television system and had associated with such industry notables as Westinghouse and later RCA, whose Electronics Research Laboratory he directed. Zworykin is best noted for his "iconoscope" television pickup tube, which he began developing in the early 1920s and continued to perfect into the late 1930s. It became an early standard for television production and remained in use until it was gradually replaced by the more advanced orthicon tube in the mid-1940s.[53] He applied for a patent for a television system on December 23, 1923, but it was not issued until fifteen years later.[54]

Two patent-interference suits developed during this time between Farnsworth and the Zworykin/RCA interests. Interference cases occur when there is a dispute over the priority between a patentee and a patent applicant or between two patent applicants.[55] In 1927 Farnsworth and four others brought a patent-interference case against Zworykin. After four and a half years of deliberation the priority of invention was awarded in 1932 to Zworykin. The same year, RCA in turn filed a patent-interference suit against Farnsworth. In the first suit Farnsworth and his colleagues had charged that Zworykin was "misdescriptive in his disclosure, his system was not operative, and the application was subject to change of new matter incorporated in the application after it was filed."[56] When RCA replied with its suit in May of 1932, it charged that patent 1,773,980 for Farnsworth's television system interfered with the Zworykin application filed in 1923. Testimony was heard from Farnsworth's old science teacher, Justin Tolman, who re-created the original drawings the young Farnsworth had made for him in 1918. Others from the Farnsworth organization testified; Zwory-

FIGURE 3–7 V. K. Zworykin developed the iconoscope television tube. Zworykin, a Russian-born scientist, directed RCA's research laboratories. Two patent-interference suits developed between Farnsworth and the Zworykin-RCA interests.

kin and members of the RCA staff testified on behalf of RCA. "The basis for the interference rested on a single claim in the Farnsworth patent. This was that the Farnsworth apparatus formed an electrical image, and means for scanning each elementary area of the electrical energy in accordance with the intensity of the elementary area of the electrical image being scanned."[57] Testimony ended in a final hearing in the United States Patent Office in April 1934. The examiner ruled in favor of Farnsworth. Zworykin and RCA appealed and lost. They did not pursue the matter in civil court.

Farnsworth Licenses RCA

Beyond the obvious issue of who owned the patent rights, RCA had a vested interest in the outcome of this case. Corporate policy at RCA was oriented toward purchasing and owning outright the emerging technology of television, not licensing it. Farnsworth, however, had no desire to sell his system, especially to RCA. For RCA to develop its television system, it had to enter into a licensing agreement with Farnsworth. For a fee of $1 million, RCA was licensed to use the devices Farnsworth had patented. Despite their patent dispute, both Farnsworth and Zworykin have been acknowledged for their contributions to the early development of television.

THE EXPERIMENTAL ERA

While the battles over patents took place, experiments in the application of television technology continued. The first United States television station to sign on was W2XBS in 1930, owned by NBC in New York. The following year an experimental RCA–NBC transmitter and antenna were in operation atop the Empire State Building. At RCA $1 million was earmarked for field tests. From these tests came the forerunner of big-screen television: RCA's electron "projection" gun made history by producing television pictures on an 8-by-10-foot screen. RCA–NBC mobile television arrived in 1937. The following year, scenes from the play *Susan and God* were telecast from NBC studios in New York, and RCA president David Sarnoff announced that television sets would go on sale at the World's Fair in 1939.

Franklin Delano Roosevelt opened that World's Fair and became the first American president ever seen on television by the general public. The fair-going public flocked to look inside the special television-receiver prototype displayed by RCA. An 8-by-10-inch screen reflected on the lid kept fairgoers asking questions about how it worked and how they could buy one. The same year, AT&T lines linked an NBC camera at a Madison Square bicycle race to a broadcast transmitter, proving that both wires and airwaves could complement each other in aiding television's growth.

By 1941 the FCC had come to realize both the potential and the demands of television and had authorized commercial licensing of television. But the glory was short-lived. War raged in Europe, and the United States needed skilled technicians to work in electronic plants and laboratories at home. There was little use for television. In fact, when the Japanese attacked Pearl Harbor, pushing the United States officially into World War II, it was radio, not television, that brought the sounds of bombs and gunfire into American living rooms. Television would have to wait for Vietnam in order to match that dubious distinction.

THE FREEZE, UHF, COLOR

Three events that occurred between 1948 and 1964 helped mold television's future. Although not directly related, they occurred somewhat simultaneously and represent an era best described as one of decision and indecision. After World War II had ended, the broadcast industry once again began to gear up for television. At the FCC, concern was beginning to mount over the signal interference that would occur if all the stations wanting to begin broadcasting were licensed to do so. Bombarded by requests, the commission instituted the famous television freeze of 1948, placing a hold on all new licenses. In 1952 the freeze was lifted. The FCC assigned twelve channels in the *very-high frequency* (VHF) area (channels 2 through 13) of the electromagnetic spectrum, and seventy channels in the *ultrahigh-frequency* (UHF) area of the spectrum (Channels 14 through 83).

In theory, UHF was on a par with VHF; in practice they were far apart. One big reason was the lack of receiving sets having UHF tuners. UHF simply could not compete in the marketplace. If people did not watch UHF, the UHF stations would find it difficult to attract advertising dollars. Finally, in 1964 the FCC began requiring manufacturers to install both VHF and UHF tuners on all TV sets. In 1976, the electronics firm of Sarkes Tarzian developed a device called a Uni-tuner that tuned both UHF and VHF channels with the same "click knob."

Although UHF still has a long way to go to reach its full potential, its future is beginning to brighten significantly. Many UHF stations are not network-affiliated, and expensive network advertising rates are sending many national advertisers to these independent stations. Moreover, a wider assortment of syndicated programming is permitting independent stations to capture a larger viewing audience once reserved for network affiliates.

At the same time the freeze was taking

place and the FCC was deciding how to allocate frequencies, two giants were battling over the future of color television. RCA and CBS competed over what type of color television system should become the national standard. CBS won the first round when the FCC approved a noncompatible color system for commercial broadcasting.[58] This meant that color signals could not be received on sets built for black-and-white reception. Meanwhile, RCA, which had been developing a compatible system, slapped CBS with a lawsuit. The appeals went all the way to the Supreme Court, which upheld the FCC's approval of the CBS system. Elated over the victory, CBS bought a company called Hytron Electronics and its subsidiary Air King. The new company manufactured receivers capable of picking up the CBS color telecasts. Unfortunately the joy was short-lived. Realizing the importance of compatible color, the FCC in 1952 reversed its decision, and CBS's venture into color television came to an abrupt halt.

TELEVISION TECHNOLOGY: IMPROVED CAMERA TUBES

The FCC eventually approved a 525-line resolution system for American television, meaning the picture would be scanned 525 times in rapid succession. It was a giant improvement over Zworykin's initial 60-line system and a considerable improvement over the 441-line system used in Europe before a 625-line system was adopted.

Some of the most important improvements in television technology occurred in the area of camera-tube sensitivity. We learned earlier about Zworykin's iconoscope tube and the work of Philo Farnsworth. Although the iconoscope increased picture clarity, its need for high-intensity lights made it uncomfortable at best to work with. Scene illumination of at least 1,000 foot-candles was necessary for even marginal quality.[59] In addition, the camera had problems in the way in which the image was scanned by the electron beam.

Gradually the image orthicon tube replaced the iconoscope. Developed by the U.S. military, it overcame some of the shortcomings of the iconoscope. It permitted the use of conventional camera lenses but still needed an illumination of 1,000 foot-candles. A better tube arrived in the form of the *image orthicon*, which reduced the needed light to 200 foot-candles and improved the electron-beam scanning process.[60] After the new tube's successful debut in a telecast from New York's Yankee Stadium in 1947, stations quickly put it into operation.

While the image orthicon was being developed, educational and industrial broadcasters were using a smaller, low-cost pickup tube called the *vidicon*. With some adaptation, it was used in 1948 in the network television comedy "I Love Lucy."[61] Endorsed by the show's director of photography, Karl Freund, it achieved enough acceptance to become an important part of broadcasting.

Color television presented its own set of problems and a new generation of pickup tubes. The image orthicon started the color television era, but in 1965 the *Plumbicon*—the registered trademark of a tube developed by Philips in the Netherlands—became the first to be used in live color cameras. The Plumbicon brought to television the ability to record color images with the sensitivity of the human eye. The development of camera-tube technology continued into the digital era where new transmitting and receiving systems promise a future of television pictures with photographic quality.[62]

RECORDING THE IMAGE

With the ability to capture images came the desire to record them for storage and playback. Engineers turned to both film and magnetic recording devices, which eventually evolved into videotape technology.

Adapting Film to Television

An intermediate film transmitter that "scanned" film was introduced at the Berlin Radio Exhibition in 1932. In 1933 an intermediate film receiver was demonstrated at the same exhibition. Kinescope recording, a quick-developing film-recording process, was used widely in the early 1930s. In fact, when a nationwide microwave link was completed in 1951, kinescope recording became popular for network transmissions until the conversion to videotape. Even after the conversion, the 16mm camera continued to be essential to the television news-production process and still remains a favorite of many television newsrooms. Super-8 film also become a favorite of some broadcasters. Less expensive than 16mm, super-8 used one-third more area on the film, and the development of improved camera design and electronic image enhancers made it adaptable to many broadcast uses.[63]

Magnetic Disk Recording

Recording television fascinated John L. Baird, and as early as 1927 he conducted successful experiments using a magnetic disk.[64] Although the quality was too unsatisfactory for future television-recording purposes, Baird's research efforts ushered in a new era in video recording for everything from full-length movies to electronic news gathering (ENG). Building on Baird's work, researchers spent the next twenty years trying to perfect a video-recording device by using such modes as a combination television camera and standard 16mm film, and even large-screen television using 35mm film, the unsuccessful brainchild of Lee de Forest.

Early color television recording concentrated on combining color clarity with picture clarity. Although most of the early attempts were marginal, in 1948 Eastman Kodak introduced a 16mm system developed in cooperation with NBC and the Allen B. DuMont studios. In February 1950 a Navy camera, Kodak film, and a CBS receiver were used in the first "completely successful" recording of color television.

Videotape

The birth of videotape came a year later. The Electronic Division of Bing Crosby Enterprises demonstrated a videotape recorder in 1951 and improved the quality a year later. In 1953 RCA demonstrated its version of a videotape recorder. The big videotape breakthrough and attendant publicity, however, came in April 1956, when Ampex engineers demonstrated their videotape recording to a CBS-TV affiliates' meeting. RCA demonstrated a color videotape in 1957, but Ampex was to carry the banner for some years to come. Ampex engineer Charles P. Ginsburg is credited with much of Ampex's videotape success, although a team of engineers worked on videotape development.[65] In 1964 the SONY Corporation of Japan introduced a system claiming improved recording-head design and simplified operation for black-and-white recording. Portable videotape units proved their worth in the mid-1960s as schools and businesses discovered the usefulness of the one-inch, reel-to-reel videotape, which could be easily stored and applied to instructional purposes. Next to arrive were the video cassettes. CBS introduced the first video-cassette system—EVR—in 1968. The following year, the SONY Corporation of America introduced the first color videotape-cassette recorder. Further refinements in videotape storage were developed by CBS under the direction of Dr. Peter C. Goldmark. The CBS Rapid Transmission and Storage (RTS) system, which became operable in 1976, permits up to thirty hours of programming to be stored on one video cassette. Different programs can be played back from the tape simultaneously over different transmission systems, such as different cable channels.

Changes in TV Receiver Design

Changes in television receiver design have been as dramatic as the rest of television's facelift. A comparison of the receiver displayed by RCA at the 1939 World's Fair with today's average home set illustrates the considerable difference in both size and design. The transistor's application to television permitted a vast reduction in size, and miniature computerlike processing devices called *microprocessors* constituted a further advance. Already, pocket televisions are rapidly becoming common. Scientists are experimenting with television screens about one-half inch thick and three-dimensional television receivers using holography are more than science fiction.

While some manufacturers are working to reduce the size of receivers, others are working to increase the size of the screen. *High definition television* (HDTV) is the new frontier of television technology. If developed fully, it will be commonplace for the average consumer to own a large-screen television with exceptional picture quality.

THE MODERN ERA: RADIO

The developments in radio and television, both in technology and programming, paved the way for a modern system of mass communication that circles the globe. Both radio and television have prospered and survived, but not without the need to adapt and change. One example is radio.[66] When you tune across the radio dial, the sounds you hear today are much different from the sounds you would have heard thirty to forty years ago. When radio evolved as a new medium, radio drama and network radio were the mainstays of programming. Television arrived and radio temporarily lost a large part of its audience. Fortunately it was able to adapt. FM radio developed because of the reasons discussed earlier. Many of the factors resulting in the growth of FM radio also contributed to radio's ability to survive television and prosper as a specialized medium.

The Specialized Medium

Radio is ideally suited to serve local communities with specialized programming. For example, in the 1970s, strong local radio news departments developed and competed favorably with television at much less cost. Reaching specialized audiences also meant new diversity in music formats. Stations also went beyond traditional formats—for example, regional formats such as beach music and foreign-language programming appealing to specific nationalities. Like the print medium of magazines, radio transformed itself into the specialized medium of the electronic age.

Confronting the Problem of AM Listenership

While FM radio continues to prosper, many AM stations are running into difficulty. The high quality of reproduction accompanied by stereo has made FM the listeners' choice for music. That leaves many AM stations faced with a bleak future. In the 1970s, FM stations passed AM stations in listenership. Loss of audience means loss of income. Some AM stations have been sold at a loss in the millions of dollars. Some of the more powerful stations in larger markets are succeeding with strong news, information, and sports programming, but in smaller markets the competition is just too great. It is especially severe for stand-alone AM stations that do not have an FM counterpart and therefore do not benefit from combined staffs and physical facilities. Teenagers who grew up with FM don't seem to know that AM even exists and therefore are not switching to the AM band as a source of informational programming. With cable television and news and information sources such as the Cable News Network (CNN) and CNN Headline News, other leisure time sources exist for news and information programming.

Strategies for Revitalizing AM

Solving the AM problem may involve a series of steps, including preferential treatment in government regulation. Other steps to increase AM's stature include conversion to AM stereo. While some AM stations do broadcast in stereo, having a larger number of AM stations broadcasting in stereo would permit trade associations such as the National Association of Broadcasters to promote AM stereo as an industry-wide campaign.

Community involvement and strong local news programming are also important. For example, the deregulation of radio in the 1980s resulted in a decline of radio news operations, especially on FM stations that didn't want their music formats broken up by talk. On the other hand, AM stations are in a position to engage in strong community involvement programming including local news coverage. Many FM stations still want to retain the strength of their music formats and shy away from such programming. The upscale audience with higher levels of education and therefore higher incomes are more prone to listen to news and information programming, and these high-income earners are the people advertisers want to reach.

Developing a pool of people who can successfully program AM stations will also help. Many of today's program directors are skilled and knowledgeable in music formats but unqualified to handle news and information programming. Not all AM stations, however, engage in news and information programming. Talk shows, game shows, and sports also fill up the AM dial. Other AM stations use syndication services that provide all of the station's programming, leaving staff members to concentrate on sales.

Radio's Future

Overall, radio appears to have an attractive future. While television is facing increased programming challenges from cable and other technologies such as video-cassette recorders (VCRs), radio will remain basically the same. It will continue to be the only broadcast medium able to reach consumers while they are driving. The increased use of syndication will permit stations to cut local programming costs.

The specialized nature of radio, especially FM, will permit the medium to find an audience, even in crowded markets. For some audiences, such as college students, radio continues to be the primary source of entertainment. The ability to reach specialized audiences accompanied by the cost efficiency of the medium means it will be able to thrive in poor economic times when other more expensive media tend to be too pricey for some advertisers.

THE MODERN ERA: TELEVISION

Television has gained wide acceptance as a mass medium of international importance and impact. A number of programming trends established its identity as a social force, which was both enjoyed and criticized.

Programming Trends

The image-orthicon television camera tube that helped televise "I Love Lucy" was capturing a program that became a national phenomenon and showed just how popular a medium like television could become. The 1950s in America saw Milton Berle's "Texaco Star Theater" so popular that restaurant owners who did not have a television receiver in their place of business faced bleak Tuesday nights when the program aired. Blue collar comedy arrived with such programs as "The Life of Riley," which starred William Bendix as a riveter in an airplane plant. Children's puppet shows such as "Kukla, Fran, and Ollie" and "Howdy Doody" were popular as were children's science programs such as "Mr. Wizard."

The 1950s saw the start of the acclaimed program "See It Now" and later "Person to Person" (Figure 3–8), both of which featured Edward R. Murrow. Murrow, whose World War II radio reports from Europe made him the conscience of broadcast journalism, brought a new level of credibility to

FIGURE 3–8 Edward R. Murrow, to many the father of broadcast journalism, had a distinguished career with CBS. His television program "See It Now" debuted in 1951 as an outgrowth of the "Hear It Now" radio program. Shown is his interview with motion picture actress Marilyn Monroe on the TV program "Person to Person," which aired between 1953 and 1959.

television news. That credibility was especially evident when Murrow, using "See It Now," challenged the tactics of Senator Joseph McCarthy, who was bolstering his Senate career by creating an anti-Communist hysteria, and accusing many Americans of subversive activities. Later in the spring, McCarthy took center stage in the famous Army-McCarthy hearings by claiming Communists were operating in the U.S. Army. But his exposure on "See It Now" had helped erode public support, and McCarthy's scare tactics and bullying eventually resulted in a Senate censure.

Politics occupied television in the 1960s as the medium turned its attention to protests over civil rights and the Vietnam War. The world also watched coverage of the assassination of President John F. Kennedy in 1963, and of the murders of the Rev. Martin Luther King, Jr., in 1968 and President Kennedy's brother, Senator Robert Kennedy, also in 1968. Television in space arrived with numerous coverage of flights of the Apollo missions, which eventually resulted in the landing of men on the moon in 1969.

The 1970s and 1980s saw a continuation of network entertainment programs, especially situation comedies. One of the most famous of the early 1970s was "The Mary Tyler Moore Show," which starred Mary Tyler Moore and Ed Asner and was set around the activities of a television newsroom. The mini-series also prospered. "Roots," (Figure 3–9) a mini-series that dealt with the history of a black family and that traced the family's origins from Africa

FIGURE 3–9 "Roots" did much to move ABC into a position of leadership in network programming in the early 1970s. The twelve-hour dramatic series was based on Alex Haley's novel about a black man who traces his ancestry back to Africa. A host of stars appeared in the program, including Madge Sinclair (on the left) and John Amos (on the right).

to slavery in Colonial America, was highly acclaimed and drew large audiences. Although entertainment programs dominated the medium, network sports coverage from the Olympics to "Monday Night Football" grew in popularity.

The 1990s promise to continue the era of sports programming. But the situation comedy also remains popular. Society and the role of women continue to change. In another situation comedy set in a television newsroom, "Murphy Brown," (Figure 3–10) unlike Mary Tyler Moore, has the status of anchorwoman and investigative reporter.

Satellites, Minicams, and ENG

Breakthroughs in satellite technology made it possible to beam signals from anywhere in the world back to local networks and stations. The minaturization of components

FIGURE 3–10 Candice Bergen, starring in "Murphy Brown," continues the programming tradition of the situation comedy. The program also helped CBS in the network ratings. The character of Murphy, played by Bergen, is a TV anchorwoman on a fictional investigative news program. To some extent, the program parallels the "Mary Tyler Moore Show," a popular sitcom in the 1970s, which also centered around a fictional television news program. "Mary," however, did not portray a woman with the professional status of an on-air investigative reporter.

meant smaller and lighter television cameras, which gave TV news more mobility and moved it into the era of *electronic news gathering* (ENG), a term used to describe the live relay of news video back to local stations for airing and editing. ENG changed much of news and public-affairs reporting; anchorpersons broke away to reporters doing live remote broadcasts in the middle of evening newscasts. To some degree, ENG technology gave television a mobility approaching that of radio.

The Challenge of Cable and VCRs

The way television programs are distributed also began to change. The change was especially felt in the 1980s when network viewership dropped off as people began viewing programs on cable television and VCRs. Cable television systems, which serve many communities with as many as three dozen channels, permit a wide range of viewing choices. Where local stations carrying network programs were once the primary suppliers of television to the home, viewers can now choose from distant stations outside their local area as well as programming on cable networks such as CNN, HBO, MTV, and many others. Hand-held channel selectors permit viewers to change channels rapidly and easily and to view many more programs for much shorter time periods than in the past.

VCRs permit flexibility in choosing when to watch a particular program. The ability to record programs means not having to be present when the program is aired. Remote-control devices permit viewers to fast-forward past commercials, a problem for advertisers who expect to reach a certain size audience. Competition for viewing time from video rental stores also means less time spent watching traditional television.

The Future of Television

How we receive television into our homes and the way we consume it may change. For example, later in this book we will learn about fiber-optic cables that use light waves to carry thousands of television signals. We have yet to determine what the results of having a very large menu of video alternatives will do to our viewing habits.

The types of programs we receive may also change. When a viewer can select among many different TV channels, broadcasters may need to offer more specialized programming to gain an audience. We have already seen this trend in cable television. C-SPAN is a cable television service that broadcasts the proceedings of Congress, and it also airs other public-affairs programs. MTV offers a channel devoted to music videos and news about the recording industry. Each is an example of specialized television.

SUMMARY

Charles David Herrold's station in San Jose, California, started in 1909, was one of the earliest of a string of pioneer stations. Such names as WHA at the University of Wisconsin, WWJ in Detroit, and KDKA in Pittsburgh were added to the list. The stations expanded in power and in audience and were joined by a host of other stations as radio matured.

One of the major developers was the Radio Corporation of America, formed in 1919 as part of a scheme to keep the Alexanderson alternator in the United States. RCA's direction was charted by former Marconi employee David Sarnoff.

The 1920s were marked by agreements and disagreements among the major radio powers. Although Westinghouse, GE,

AT&T, and some smaller concerns joined together to share inventive efforts, they competed in developing commercial broadcasting. AT&T's WEAF attempted "toll broadcasting," and when it tried to monopolize stations' use of the long lines, matters went to court. A corporate agreement resulted, and AT&T went back to the telephone business while the other companies forged ahead with broadcasting.

NBC, CBS, and Mutual emerged as the major networks. When NBC was required to dispose of half of its dual-network system, the American Broadcasting Company was born.

New technology has been important in radio's development. Through the work of Edwin Armstrong, radio gained a sizable "sound" advantage in the form of FM. The FCC's support of FM, requiring separate programming from AM, and the development of stereo FM broadcasting opened up new possibilities for this area of the spectrum. And just when it was needed, the invention of the transistor by Bell Lab scientists made radio a portable medium.

Early attempts to reproduce an image using a mechanical process were soon replaced by electronic reproduction. The work of Vladimir Zworykin and Philo Farnsworth resulted in improved picture quality. The early experimental era of television saw station W2XBS sign on the air in New York City, an event that was followed by a series of breakthroughs in television technology.

Television was introduced to the American public at the 1939 New York World's Fair. In 1941 the FCC approved commercial television. Then, during a television freeze that began in 1948, the commission spent five years deciding frequency assignments and standards for color.

Television cameras, meanwhile, improved from the iconoscope tube to the orthicon, the image orthicon, the vidicon, and the Plumbicon tube, which is based on the vidicon concept but provides color clarity equivalent to that which can be detected by the naked eye. Magnetic recording of television programs progressed from the early kinescope methods to videotape. Film advanced to super–8 technology.

The modern era of radio and television has resulted in the need for both mediums to adapt to change. Radio has become a very specialized medium. The network programs and radio drama of the past have been replaced by specialized formats. AM listenership has steadily declined and attempts are being made through regulatory relief and new programming strategies to revitalize AM stations. Overall, the future of the medium is bright because of its cost-efficiency.

Television moved beyond its technological hurdles to become a social force in society with programming ranging from situation comedies to news coverage of space, war, and politics. The development of satellites and the miniaturization of television camera components ushered in the era of electronic news gathering (ENG). Such new technology as cable and VCRs have changed the way we receive and consume television programs. The future of the medium will involve continued growth, although news distribution systems and programming may result in the medium continuing in the direction toward more and more specialization.

OPPORTUNITIES FOR FURTHER LEARNING

Barnouw, E. *Tube of Plenty: The Evolution of American Television*, 3rd ed. New York: Oxford University Press, 1990.

COMSTOCK, G. *The Evolution of American Television*. Newbury Park, CA: Sage, 1989.

FOLKERTS, J., and TEETER, D. L. *Voices of a Nation: A History of Media in the United States*. New York: Macmillan, 1988.

GRAHAM, M. B. W. *RCA and the Videodisc Player: The Business of Research*. New York: Cambridge University Press, 1987.

PAPAZIAN, E. *Medium Rare: The Evolution, Workings and Impact of Commercial Television*. New York: Media Dynamics, 1989.

PERSICO, J. *Edward R. Murrow: An American Original*. New York: McGraw-Hill, 1988.

STEPHENS, M. *A History of News: From the Drum to the Satellite*. New York: Viking, 1988.

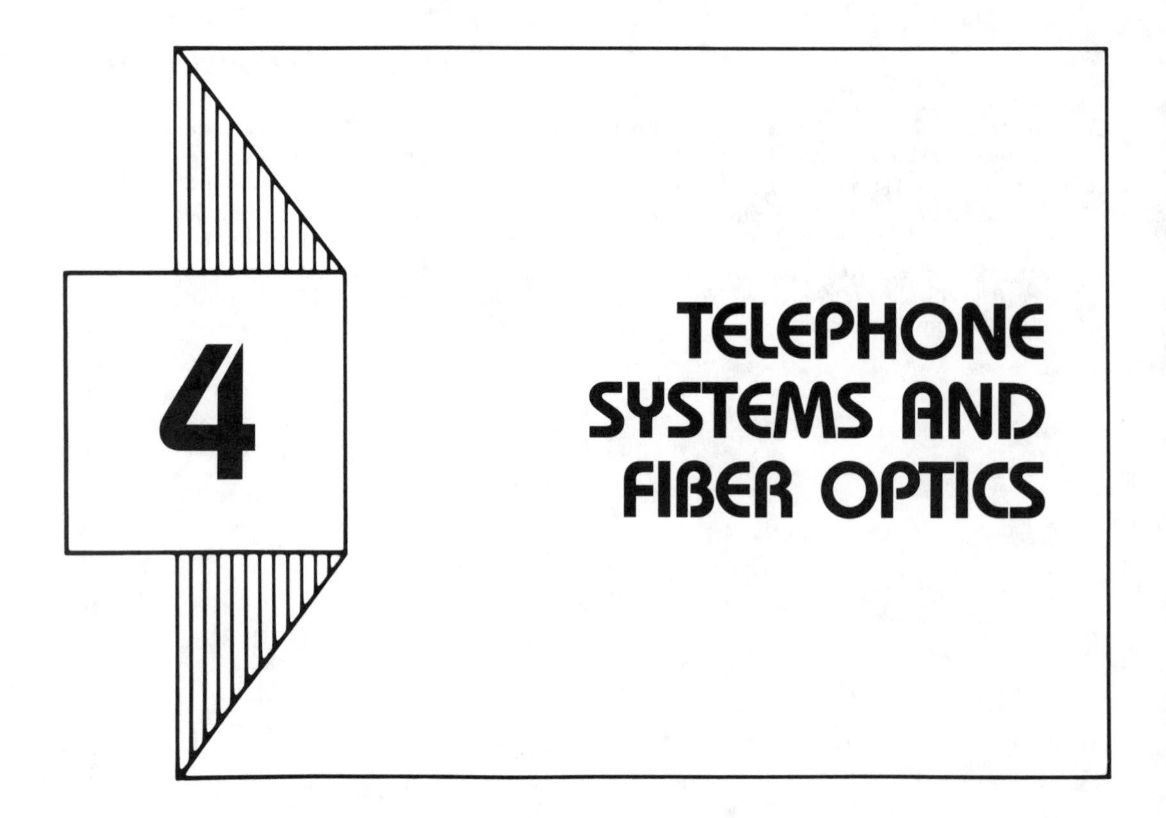

4 TELEPHONE SYSTEMS AND FIBER OPTICS

While the telegraph was making its impact on nineteenth-century communication, the telephone wasn't far behind.

ALEXANDER GRAHAM BELL AND THE IDEA BEHIND THE TELEPHONE

Alexander Graham Bell was a product of European culture and refinement. His father was a well-known speech professor whose specialty was teaching the deaf and whose major contribution to his field was a system of "visible speech" whereby he taught the deaf to talk. When Alexander began studying at the University of London, it was natural that he would follow his father's profession.

Tuberculosis struck the Bell family in the late 1860s, forcing them to move to Canada. There, Alexander Graham Bell himself had the opportunity to teach, and he became a respected practitioner of visible speech. Later he set up his own school for the deaf in Boston and was later appointed professor of vocal psychology at Boston University.

Bell had become captivated by the study of electricity while studying vocal resonance in London. He had read about experiments by von Helmholtz and the use of tuning forks to produce sounds. But Bell couldn't read German very well, and he concluded mistakenly that the German scientist was transmitting sounds from one tuning fork to another by using a wire.

That wasn't at all what von Helmholtz was doing, but the idea lit an experimenter's spark in Bell, and in Canada and America he began experiments on sound transfer. Bell envisioned a system whereby a transmitter would send different tones over a single wire to a "tuning fork" receiver.

Bell concentrated on designing a sending and receiving device based on a principle of vibrating metal reeds. The device would vibrate the way the human eardrum does in response to sound waves. Bell's idea was to invent a "harmonic telegraph" that would have direct and immediate application to the telegraph industry.

Bell's Association with Hubbard and Sanders

Bell's teaching gave him two important contacts that contributed both the financial and legal expertise he needed in order to continue his efforts. The first was Gardiner Greene Hubbard, a Boston lawyer who was president of the Clarke School for the Deaf. Hubbard provided Bell with money as well as legal advice on securing patents for Bell's inventions. He also gave him his daughter, Mabel, in marriage. In her youth, Mabel had been stricken deaf with scarlet fever, and Bell used visible speech to teach her to speak. Also providing Bell with financial backing was Thomas Sanders. Sanders had a deaf son, and Bell taught him to speak as he had Mabel. On February 27, 1875, the three men entered into an agreement to invent a harmonic telegraph. Bell would provide the inventive genius, Sanders the money, and Hubbard the legal advice and money.[1]

BASIC PATENTS OF THE TELEPHONE SYSTEM

One week later, Bell filed his first patent application in Washington, D.C. The patent, for "Improvement in Transmitters and Receivers for Electrical Telegraphs," was granted on April 6, 1875.

Bell set to work with Thomas A. Watson, an employee of a Boston electrical shop. The device developed by Bell and Watson was not the telephone, but it set the stage for the machine that would ultimately transmit speech over wires. What Bell and Watson did construct was an instrument that used the principle of variable resistance (Figure 4–1). It consisted of a membrane stretched over a small frame with a wire running from the center of the membrane perpendicularly into a small cup of acid water. When someone shouted at the membrane it would vibrate, and the wire, correspondingly, would move up and down in the water, thereby varying the resistance between the wire and the liquid. The following year's experiments culminated in another patent (Figure 4–2). Filed on February 14, 1876, and granted on March 7, 1876, it was titled "Improvement in Telegraphy."[2] Nothing in the original agreement of the three men mentioned the telephone. For this reason, as the device being perfected by Bell and Watson turned more and more toward the telephone, Hubbard offered to relinquish his rights to that invention; he saw the potential for substantial profit in the telegraph. Bell later wrote:

> My understanding always was that the speaking telephone was included in the inventions that belonged to the Messrs. Hubbard and Sanders from the autumn of 1874, but I found at a later period that they had not had this idea, which might account for the little encouragement I received to spend time on experiments relating to it. Even as late as 1876, when the telephone was an assured success, Mr. Hubbard generously offered to relinquish to me all right and title to that invention, as he was inclined to think it was outside our original understanding.[3]

FIGURE 4–1 Alexander Graham Bell's "liquid" telephone used variable resistance. A membrane was stretched over a small frame with a wire running to the center of a small cup of acid water. When words were shouted into the horn, the membrane vibrated and changed the resistance. The concept can also be seen in Bell's patent, which is illustrated in Figure 4–2.

Bell, however, consulted an attorney and agreed to include the telephone as part of the original agreement among the three men.

Experiments during the winter of 1875/76 resulted in a third patent, "Telephonic Telegraph Receivers," filed on April 8, 1876, and issued on June 6, 1876.[4] Although the second and third patents were also originally issued in Bell's name only, he later assigned them to Hubbard and Sanders as well on September 15, 1876. On January 15, 1877, a fourth patent was filed—"Improvement in Electrical Telegraphy."[5] These four patents—the foundation of the modern Bell System—all referred to the "telegraph" instead of the "telephone."[6]

HUBBARD AND THE BELL SYSTEM

The original agreement among Bell, Hubbard, and Sanders provided that if something of commercial value were to arise from the work of Bell, a company should be formed to develop the product. Consequently, in 1877 Hubbard was put in charge of what was officially called the Bell Telephone Company, Gardiner G. Hubbard, Trustee.[7] Whereas it was Hubbard's business sense that caused the company to grow, the scientific drive fell upon the shoulders of Watson.[8] Bell, meanwhile, had married Hubbard's daughter and traveled to England to help introduce the telephone there. Watson, under a contract dated September 1, 1876, received a beginning wage of three dollars per day.[9] Between 1877 and his resignation in 1881, he carried on the research and development of the company. Without scientific training he improved the device that Bell had left behind and molded it into a product suitable for a commercial enterprise. In 1877 Bell and Sanders assigned all of their rights to the telephone to Hubbard, who now became the person guiding the developing firm.

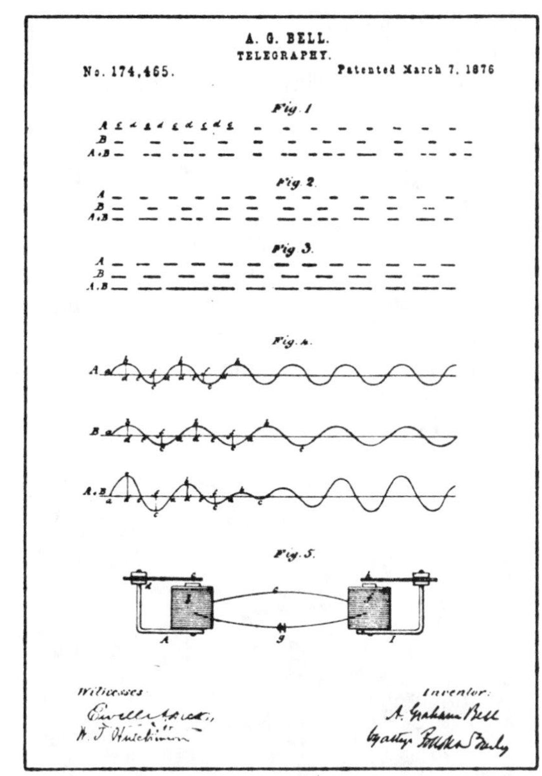

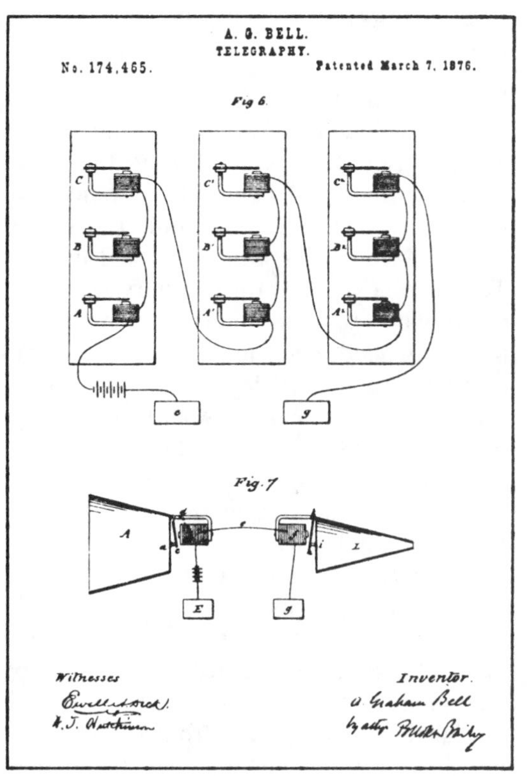

FIGURE 4–2 Alexander Graham Bell's patent of 1876. It was one of many patents credited to Bell, his associates, and the various companies that had their origin in Bell's work.

The Decision to Lease

Hubbard had to make a decision—to sell telephones or lease them. He decided on the latter, partly because of the example of one of his clients, the Gordon-McKay Shoe Machinery Company.[10] That company had leased its equipment to shoemakers and received a royalty for every pair of shoes sewn on the machines. This seemed an ideal arrangement to Hubbard, and was duly reflected in the Bell Company's declaration of trust: "The business of manufacturing telephones and licensing parties to use the same for a royalty, shall be carried on and managed by the Trustee, under the name of the Bell Telephone Company, under and in accordance with such general directions, rules and regulations as may be made for that purpose by the Board of Managers."[11] The declaration also provided for 5,000 shares of stock to be divided among Hubbard, his wife, his brother, Watson, Sanders, Bell, and Bell's wife.

Hubbard's decision to lease and not sell telephones put one constraint on the company—a serious lack of funds. It was clear that even though Hubbard and Sanders had committed considerable money to the venture, still more resources were needed.

Expansion in New England

Sanders, who had ties with the New England financial world, interested a group of Massachusetts and Rhode Island businesspeople in investing in the development of the telephone in New England. The result was the incorporation of the New England Telephone Company on February 12, 1878. Headquartered in Boston, the new investors controlled a considerable share of the company. The articles of incorporation stated that the company was formed "for the purpose of carrying on the business of manufacturing and renting telephones and constructing lines of telegraph therefore in the New England States."[12] Hubbard assigned to the new company the rights of the four original patents. In return, Hubbard, Watson, and Sanders received half the stock in the New England Telephone Company and an agreement that the company would buy all of its equipment from the original Bell Telephone Company. The New England Company would in turn lease the equipment to individuals wanting telephone service.

Moreover, the two companies agreed to cooperate in the event that it became advantageous to expand their telephone system beyond New England. This agreement alluded to what is today the "long lines" concept of telephone interconnection. Specifically, the cooperative provision stated that

> insomuch as said parties and their successors and assigns may have a common interest in the working of continuous and connecting lines extending outside of New England, the said parties agree that they will endeavor to cooperate in the establishing of connecting lines and in the joint working of the same, and the division of the expense and the profits thereof pro rata upon some equally fair and equitable basis.[13]

Five days before the agreement was signed, the first commercial telephone exchange opened in New Haven, Connecticut.[14]

THE BELL TELEPHONE COMPANY

Prosperity came quickly (Figure 4–3). The projections for the development of a telephone system beyond New England caused a new company to be formed on June 29, 1878, called simply the Bell Telephone Company. Hubbard, Sanders, Watson, and Bell (through his attorney) were all in-

FIGURE 4–3 The early prosperity of the telephone industry is graphically illustrated in this picture of telephone wires strung in New York City in the 1880s. By the late 1880s, the Bell Telephone Company had been formed, lines stretched into New England, and substantial progress had been made in linking together the telephone companies in the Midwest.

volved, as were principals from the New England Company. Hubbard received 3,000 shares of stock and in turn awarded his interests, including patent rights, to the new company. Theodore N. Vail, a former Western Union employee, became the company's general manager, and the control and guidance of the firm were invested in an Executive Committee of the Bell Telephone Company. The Bell Telephone Company, Gardiner G. Hubbard, Trustee, had come to an end.

WILLIAM H. FORBES AND THE NATIONAL BELL TELEPHONE COMPANY

On December 31, 1878, William H. Forbes was elected director of the Bell Telephone Company. Forbes immediately saw the advantage of consolidation. On March 20, 1879, the New England Telephone Company and the Bell Telephone Company assigned their rights under the first and third of the original Bell patents to the National Bell Telephone Company. The decision came none too soon. Thomas Edison had developed an improved carbon "transmitter," and patent litigation and challenges to the original Bell patents were filling the air.

The Era of Patent Challenge

One challenge came from Elisha Gray,[15] who had filed a *caveat* for a patent on a variable-resistance telephone just a few hours after Bell did. Although a caveat does not have the legal weight of a patent application, Gray went to court claiming credit for the concept of variable resistance and accusing Bell of looking first at the caveat and then altering his patent application. The court, however, was sympathetic to Bell.

Another challenge came from a country tinkerer named Daniel Drawbaugh. His lawyers managed to persuade a contingent of farm folk to testify that they had used a device invented by Drawbaugh to talk to each other before Bell received his patent. The U.S. Supreme Court ruled in favor of Bell by a one-vote margin. A victory in either case might have resulted in fleets of telephone-company trucks roving the country today with the name of Gray or Drawbaugh printed on their cabs instead of the familiar Bell symbol.

Bell also developed a machine whereby the voice could be transmitted over light waves. Called the *photophone*, it was patented in 1880 (Figure 4–4). Although it had no practical applications then, it was the forerunner of today's fiber-optics lightwave communication, which promises to revolutionize not only telephone communication but also data transmission and cable television and which we discuss later in this chapter.

The Battle with Western Union

Even with patents and public acceptance, the National Bell Telephone Company was not an instant commercial success. Western Union managed a nationwide system of telegraph lines and, with Gray's receiving device, the transmitter invented by Edison, and a handset created by Robert Brown, it had made considerable inroads into the telephone industry.[16] Missing the chance a few years earlier to buy the Bell patents, Western Union was now pouring millions into its own telephone system. A showdown was inevitable. To let Western Union continue would be disastrous for the National Bell Telephone Company. Western Union had the background, capital, expertise, and ambition to challenge its much smaller rival.

FIGURE 4–4 The photophone did not materialize as a commercial product the way Bell's telephone did. The device, which used light beams as a way of communicating between the sender and receiver, was based on the same idea that today is used in fiber-optic communication, which uses small strands of glass to carry thousands of "channels" of information over light waves.

What it did not have was the rights to the Bell equipment. After a long, involved court proceeding the two companies reached an agreement whereby Western Union would stay out of the telephone business if Bell would stay out of the telegraph business.

THE AMERICAN BELL TELEPHONE COMPANY

The implementation of the settlement required a special session of the Massachusetts Legislature and the passage of an act that gave the telephone company the following rights: "manufacturing, owning, selling, using and licensing others to use electrical speaking telephones and other apparatus and appliances pertaining to the transmission of intelligence by electricity, and for that licenses purpose constructing and maintaining by itself and its public and private lines and district exchanges."[17] The act also resulted in the formation of the American Bell Telephone Company on March 20, 1880. The new company formally purchased the stock of the National Bell Telephone Company. National Bell had served an important purpose beyond merely expanding telephone service: it had developed a system of telephone exchanges that permitted localized switching. The next hurdle in the development of the telephone system was to connect the various exchanges, and this became the hallmark of American Bell Telephone Company. The new company created the Long Lines system, which began with the construction of a telephone line between Boston and New York. Opened on March 27, 1884,[18] it was soon followed by lines linking Boston, New York, Philadelphia, and Washington and also New York and Albany. To finance long-lines development, the company issued $2 million in bonds.[19]

THE AMERICAN TELEPHONE & TELEGRAPH COMPANY

To develop the long-lines system, American Bell formed a subsidiary called the American Telephone and Telegraph Company (AT&T). Incorporated in New York State because of its favorable legal and financial climate, the company was entrusted with constructing lines throughout the North American continent, including Mexico and Canada, and "by cable and other appropriate means with the rest of the known world."[20] Theodore Vail became AT&T's first president and Edward J. Hall its first general manager.

Partly because of New York's importance and size and partly because of its business and legal climate, it became evident that AT&T would become the central organization of the telephone system. Although economic incentives by the Massachusetts Legislature attempted to favor American Bell, the enacted increases in capitalization were still not adequate. Thus, in 1900 the American Bell Telephone Company transferred all of its assets to AT&T through a somewhat involved trading procedure. AT&T now became the parent company, a coordinated federation that also included a number of associated companies.

THE BREAKUP OF AT&T

AT&T's structure remained essentially the same until 1974 when the U.S. government filed an antitrust suit against AT&T. The case traveled through the courts, eventually resulting in what became known as the Modified Final Judgment (MFJ) affirmed by the U.S. Supreme Court in 1983. Originally, AT&T had been regulated as a monopoly, for the primary goal of the company and the government was to provide a unified system of low-cost telephone service throughout the United States. As that goal was achieved, however, new forces gradually began to compete with it in long distance communication. The customer who once was satisfied just with telephone service began to see new uses for the telephone, specifically the linking of the telephone with some information systems such as personal computers.

THE SEVEN REGIONAL HOLDING COMPANIES

Under MFJ, which had been altered somewhat by the courts, AT&T divested itself of the local operations. The Bell companies were separated from AT&T, and long-distance communication was opened up to more competition. The twenty-two local operating companies regrouped into seven regional holding companies: Pacific Telesis Group, which covers California and the lower Western states; US West, which includes the Northwestern states as well as much of the upper Midwest and some of the lower Southwest; Southwestern Bell Corporation, which covers Texas and the lower Midwest; Ameritech, which also includes a group of telephone companies in the Midwest; BellSouth, which covers the Southeast; Bell Atlantic, which includes the Middle Atlantic States; and NYNEX, which includes New England.[21] The Bell companies could provide, but not manufacture, new equipment for use in the home.

A RESTRUCTURED AT&T IN THE MARKETPLACE

Equipment already installed at the time of the MFJ remained with AT&T, and AT&T could provide new equipment. Bell Laboratories and Western Electric—AT&T's research and manufacturing arms, respectively—remained a part of the parent company. AT&T shareholders retained stock in the parent company and were assigned a proportionate interest in the local companies. With the consent decree, the competitors of AT&T are given access to the local exchanges. AT&T in turn is free to offer consumers equipment that can be rented or purchased and used in connection with local or long-distance telephone systems. The new subsidiary through which AT&T sells electronic equipment to the consumers is AT&T Information Systems.[22]

INFORMATION SERVICES AND GATEWAY DEVELOPMENT

A 1988 ruling by federal Judge Harold Greene affirmed the provision of the MFJ, which prohibited the telephone companies from becoming information service providers. In other words, a regional holding company cannot, under Judge Greene's ruling, publish an electronic newspaper but can distribute an electronic newspaper published by another nontelephone company. Other services can be offered such as voice storage and retrieval and electronic mail.

Certain "gateway" services can also be marketed by the regional holding companies. For example, the telephone company can operate and market a service through which other individuals or companies can provide information. Specifically, a local travel agency, grocery store, or movie theatre can provide information that consumers can access through their telephone, which in turn is connected to a home terminal such as a personal computer or other video-monitor-type device.

EMERGING ISSUES AFFECTING TELEPHONE, BROADCASTING, AND CABLE

While at first glance there may not appear to be much similarity between a telephone company and broadcasting or cable TV, all three are involved in some ways in the transfer of information from one point to another. Moreover, no one can predict whether or not the courts will eventually lift the restrictions on regional telephone companies becoming information providers.

Technologies

New technology will also play a role. Consider the technology of fiber optics, which we will learn more about later in the text. For now, suffice it to say that a single fiber-optic cable can carry the information equivalent of hundreds of television programs simultaneously. As these cables replace the wired telephone cables of the past, the amount of information we as consumers could receive through our telephone company would increase dramatically.

INTEGRATED SERVICES DIGITAL NETWORK (ISDN)

On the horizon is the development of the Integrated Services Digital Network (ISDN), which involves the development of an international master plan for high-capacity standardized transmission and receiving systems, including telephones, which will carry audio, video, and data throughout the world using digital technology (Figure 4–5). Although some years away, and faced with political and technological obstacles, it would make us rethink the way information, including audio and video entertainment programs, will be "carried" from one place to another.

Everyone from policymakers to manufacturers are watching closely to see where consumer demand will take these emerging "media" of the future. Since people can only consume so much information in a single day, will our attention turn to these new information sources at the expense of other and more traditional media?

EMERGING CONSUMER APPLICATIONS

When Alexander Graham Bell's telephone company was just starting, thoughts of telephone conversations initiated from luxury aircraft, business meetings conducted by

FIGURE 4–5 The International Telecommunication Union exhibit of the technology of ISDN. ISDN, which stands for *I*ntegrated *S*ervices *D*igital *N*etwork, uses digital technology in the switching and distribution of telephone and data communications. International applications of ISDN will bring different countries under one standardized telecommunications system. ISDN will greatly increase the ease with which telephone, data, and video communications will be transmitted and received across international boundaries.

video telephone systems, and thousands of mobile telephones utilized in cities throughout the world were fantasies. Today, as the technology of the telephone has steadily become more sophisticated and as consumer demand for telecommunication services has increased, the telephone, itself changing in concept and use, has become a vital link in human and electronic communication (Figure 4–6).

Cellular Radiotelephone Systems

By approving space on the spectrum for *cellular radiotelephone systems*, the FCC has made it possible for thousands of additional mobile telephones to become operational.[23] In the past, limited frequencies and less sophisticated technology severely restricted the number of mobile telephones that could be licensed to any one area. Today, cellular radiotelephone systems have made it possible for a series of small transmitters with limited-coverage "cells" to make up the core of a mobile-telephone system. In earlier mobile-telephone systems, in comparison, a single transmitter served a much larger area.

With cellular radiotelephone service a person making a call from, say, an automobile automatically connects with the transmitter located in the same area in which the automobile is traveling. The two-way radio transmission that carries the telephone call does not interfere with conversations in other areas, also called "cells."

FIGURE 4–6 Making the telephone more accessible and consumer friendly has aided in its gradual transformation to a medium used for much more than just one-to-one conversations. An example of this transformation can be seen in two contrasting London phone booths belonging to British Telecom. The booth on the right alerts consumers to a multitude of credit cards that can be used to access the system.

As the automobile moves closer to another area or cell, a computer automatically switches the call to an available frequency in the next cell. The central transmit-receive switching center in each cell links with telephone land lines to complete the connection. Along with the actual number of telephones, the cost to the consumer is expected to decrease.

Plane-to-Ground Telephone Service

Imagine you are flying home for the holiday vacation and your flight is running late. You are airborne and will not have any time to call home when you change flights because the connection is too tight. Instead of having your friends spend two hours waiting for you at the airport, you call them from the airplane and inform them of the late arrival. The telephone call is automatically relayed to one of a series of ground stations. In your case it will be the ground station ahead of your flight path. Given the speed of the aircraft, your telephone conversation can last from fifteen to twenty minutes.[24]

Research suggests that approximately 20 percent of all people who travel through airports use the telephone. Consequently, the appeal of plane-to-ground calling is substantial.[25] For business travelers the service is particularly appealing, for while flying they are out of touch with their office. Although you will not find plane-to-ground telephone service on every flight, more and more airlines are including it on major routes and in large-capacity aircraft.

Teleconferencing

If a group of business executives wants to conduct a meeting, they have three choices: (1) to meet face to face, (2) to conduct a conference call, where everyone can talk to everyone else, or (3) to conduct a teleconference. The *teleconference* is the newest of the three formats and brings together the technologies of the telephone and television. AT&T's teleconferencing system utilizes full-color television facilities installed in major cities. Using strategically placed cameras and monitors, executives in different parts of the country can carry on discussions in full view of one another.

Modern teleconferencing facilities are equipped with graphic capabilities and devices such as electronic blackboards that permit participants to share sketches and other diagrams; computers capable of creating and transmitting data displays; distribution of documents to participants by means of facsimile transmission; random access and transmission of 35mm slides; and freeze-frame video (capturing a single frame of video and transmitting it to participants). Such services supplement standard teleconferencing and are called *audiographic* teleconferencing services.

The growth of teleconferencing will depend on such factors as (1) the cost of energy and, therefore, the cost of travel to remote meeting sites; (2) the cost of teleconferencing and, therefore, its availability to smaller companies; and (3) whether users will really find that the system can replace the effectiveness and naturalness of good interpersonal communication in a face-to-face situation.[26]

FIBER OPTICS

What the telephone companies are to communication, *fiber optics* are to the telephone companies. This relatively new technology is being applied to local and transoceanic communication systems. Fiber optics (Figure 4–7) promises to make major changes

FIGURE 4–7 Fiber-optic technology permits light waves to travel through strands of glass. Because of the frequencies of light waves, the amount of information a fiber-optic system is capable of handling is many times greater than conventional radio, microwave, or wired communication systems are capable of handling. The concept of electromagnetic energy is discussed in more detail in the next chapter.

in the way we communicate in the information society.

Early Theorizing and Experiments in Optical Communication

The theoretical work in fiber optics predated the actual use of "fibers." As we learned earlier in this chapter, Alexander Graham Bell's photophone used light waves to transmit and receive voice communication. But Bell's experiment was preceded by ten years through the work in 1870 of John Tyndall, who demonstrated a light phenomenon to the Royal Society in England.[27] Tyndall's experiment showed how light could pass through flowing water from one container to another.

British and United States patents were issued in 1930 for the use of fibers to scan and transmit television pictures, the same year researchers in Germany demonstrated light transmission through fibers. Bundles (Figure 4–8) of fibers were then used successfully in later experiments.

Realizing the Potential of Fiber Optics

Experiments continued until 1967 when two researchers, K.C. Kao and G.A. Bockman, at England's Standard Telecommunications Labs offered the suggestion that these small glass fibers could usher in a new era of communication. The result was a new awareness among telecommunication industries that not only a new era but also a new medium of communication was at hand. The telephone companies, some of which operated with large research budgets, saw an opportunity for new growth and development (Figure 4–9), primarily because fiber optics could greatly enhance the amount of information that could be transmitted and received over a single optical cable.

PIONEERING APPLICATIONS OF FIBER OPTICS

Early applications of fiber-optic technology occurred soon after telephone companies realized its potential. General Telephone

FIGURE 4–8 The physical characteristics of fiber-optic cables permit them to be bundled together much like wired cables. The difference, however, is in the fact that the same size fiber-optic cable can carry thousands of times more information and does not deteriorate over time in the same way that wired cables do.

and Electronics (GTE) installed a fiber-optic link in Long Beach, California, in 1977, and two years later in Tampa, Florida, a National Football League game between the Tampa Bay Buccaneers and the New York Giants resulted in a 5.6-mile link between Tampa Stadium and downtown Tampa.

Tests by GTE in Cerritos, California, as well as by Southern Bell and Northern Telecom in Heathrow, Florida, resulted in fiber-optic installations for the purpose of testing in-home services. A television network trial of fiber optics involving ABC, NBC, CBS, Fox, and PBS began in the early 1990s and gave the networks an opportunity to test fiber optics as an alternative to satellites.[28]

By the early 1990s, a bundle of transoceanic fiber-optic cable the thickness of a garden hose spanned both the Atlantic and Pacific oceans with a cable capacity of 40,000 simultaneous telephone conversations (Figure 4–10).

THE WORKINGS OF FIBER OPTICS

In the next chapter we will learn about the electromagnetic spectrum, but for now, understand that the spectrum is like a yardstick of physical energy and that your car radio shows a very small portion of that yardstick. You know that on your car radio you can hear, say, a dozen radio stations. Now imagine that much higher on the yardstick, where light waves instead of radio waves carry information, you could hear millions of radio stations. That comparison reflects the amount of space on the yardstick that is available to carry information when high-frequency light waves are used.

Very thin glass-like fibers can carry light, laser beams of light to be exact, and within these thin light-carrying fibers, which are small enough to pass through the eye of a needle, tremendous amounts of information can be transmitted and received (Figure 4–11).[29]

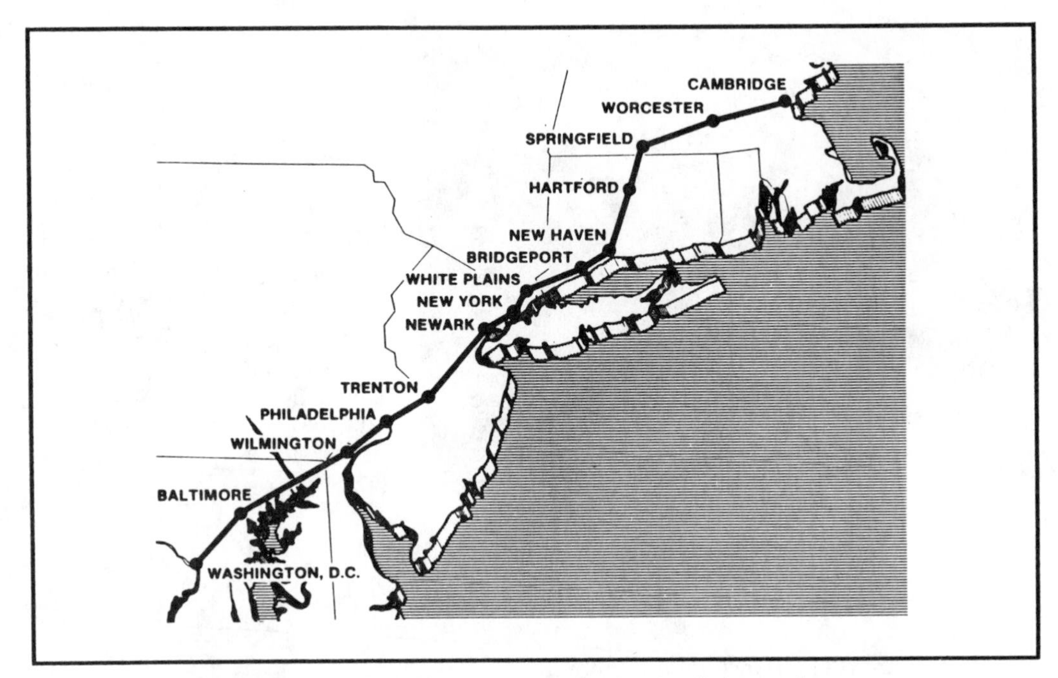

FIGURE 4–9 AT&T's fiber-optic communication link stretches from Washington, D.C., to Cambridge, Massachusetts. Thousands of telephone calls can be handled simultaneously through the link.

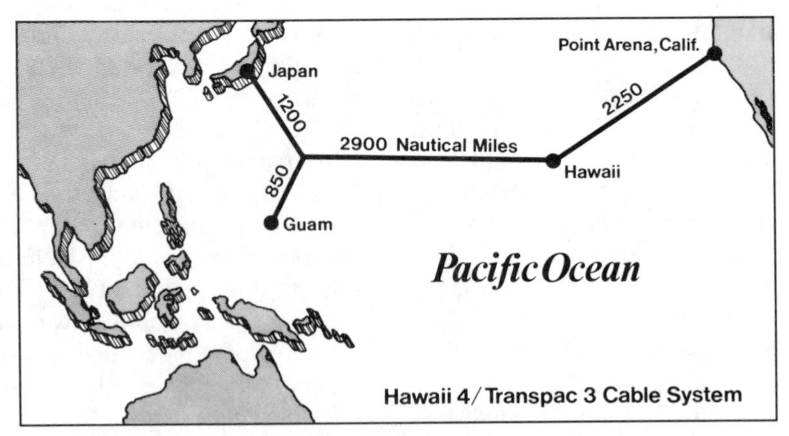

FIGURE 4–10 The Pacific fiber-optic system that links the western United States with the Far East. In addition to the routes shown here, a branch also connects Guam with the Philippines. The increased availability of fiber-optic links will permit users to choose fiber-optic technology in place of satellites for transcontinental and other long-distance communication.

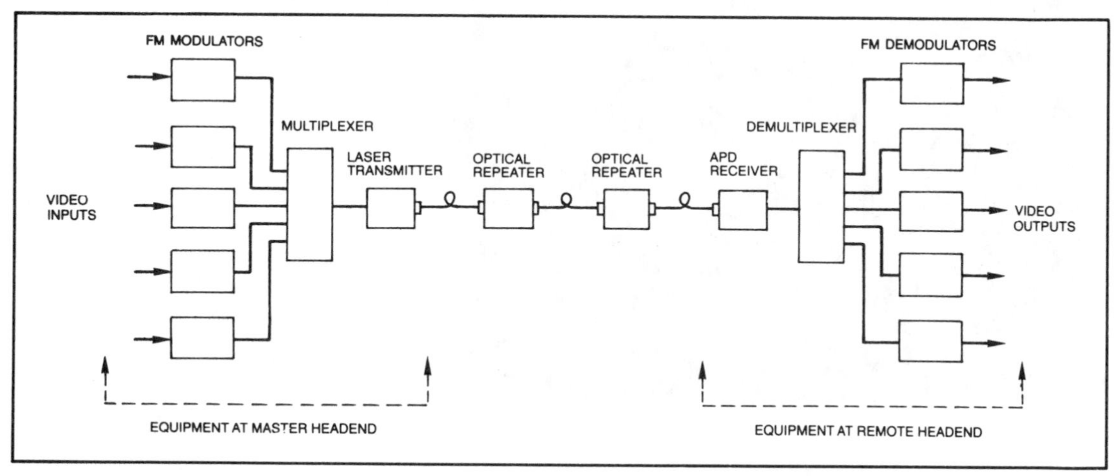

FIGURE 4–11 A fiber-optic communication link for video includes equipment at a master head end and a remote head end. Video inputs are entered into the system through FM modulators, are transmitted as laser beams by a laser transmitter, and eventually reach the remote head end where they are converted again to video outputs.

ADVANTAGES OF FIBER OPTICS

Fiber-optic technology offers some advantages over other types of wired and unwired systems.

Increased Information Capacity

While wired cable can carry television, voice, and data communication, the amount of information increases significantly when fiber-optic cables are employed (Figure 4–12). For the telephone company, this increased capacity means a single hair-thin fiber can carry many thousands of telephone calls simultaneously. Cable systems operate "pay-per-view" services offering cable subscribers the ability to "call up" different programming such as motion pictures. But the number of motion pictures available for call-up at any one time is limited because of the limitations of the wired transmission system.

Because fiber-optic cable can carry many hundreds of video channels simultaneously, the number of motion pictures that can be transmitted to different home receivers in a fully developed fiber optic system increases greatly, and the subscriber can access the equivalent of an electronic video store with hundreds of selections.

Increased Signal Quality

Many of us are familiar with the telephone company commercial showing a pin dropping as the announcer talks about the clarity of the telephone conversation. Such clarity is possible because cable is free from interference caused by over-the-air transmission. For example, your car radio may receive static which you hear in the speaker. The static is caused by the radio waves trav-

FIGURE 4–12 A comparison of wired cables and fiber-optic cables. The bundle of wired cables is necessary to approach the information capacity of a single fiber-optic cable.

eling through the atmosphere. With optic fibers, however, there is virtually no distortion because the atmosphere is not used and because *digital* technology reduces the interference.

Adding digital technology to the fiber-optic system permits even greater clarity. The consumer can use the telephone to access recorded music, which will be transmitted into the home with the same quality as the digital audio tape or compact discs.

Many other advantages exist beyond those noted here (Figure 4–13). Using fiber-optic systems to carry voice and pictures simultaneously may result in telephone callers of the future being as concerned with how they "look" over the phone as how they sound.

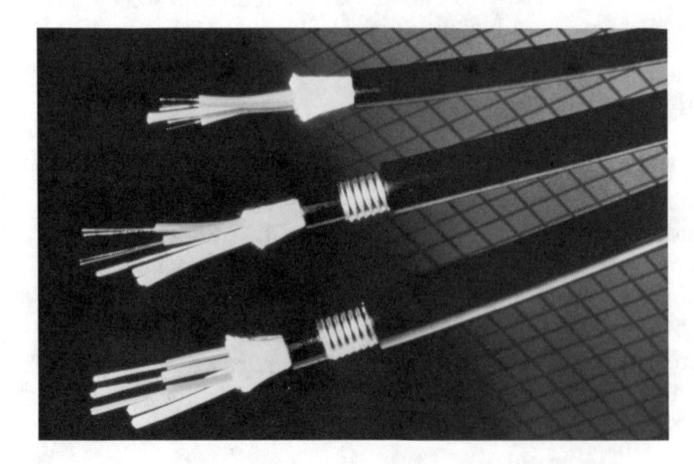

FIGURE 4–13 New materials used in the production of fiber-optic cables make it possible for the cables to have many of the characteristics of flexibility that wired cables have, but with increased information capacity. Fiber-optic cables can be bent and fed through other tubing, much like wires can. The flexible qualities of fiber-optic cable become especially important in installing the cables where other pipes and tubing already exist underground. For example, relatively small underground steam pipes once used for heating but no longer in service can be adapted to fiber-optic installations.

FIBER OPTICS AND THE FUTURE OF TELEPHONE COMPANIES

Important to reading this book is the knowledge that some of the entertainment and information services now offered through radio, television, and cable systems could someday be offered to the average consumer by the telephone company. Will these large telephone "utilities" change the way we receive and view television, how we listen to recorded music, or the manner in which we use our free time to communicate over the telephone? That question will become more and more important in the future as technology and our appetite for new entertainment and information services come together.

SUMMARY

The telephone traces back to Alexander Graham Bell, who with the legal and financial support of Gardiner Greene Hubbard and Thomas Sanders developed a "telegraph" that became a wired system for communicating over long distances by voice. The telephone has its scientific and commercial foundation in key patents issued to Bell between 1875 and 1877.

The association of Bell, Hubbard, and Sanders, accompanied by the scientific work of Thomas A. Watson, resulted in the first telephone company—the Bell Telephone Company, Gardiner G. Hubbard, Trustee. Using a strategy of leasing instead of selling telephone equipment, Hubbard sought to expand the company, and with a group of Massachusetts and Rhode Island investors he formed the New England Telephone Company. Shortly thereafter the Bell Telephone Company replaced the original telephone company.

When William H. Forbes was named director of the Bell Telephone Company he saw the advantages of consolidation, and he therefore joined with the New England Telephone Company in forming the National Bell Telephone Company. Through a series of patent challenges, some by Western Union, there resulted a settlement whereby Western Union would be concerned with telegraphic communication and National Bell with telephone communication.

To implement the conditions of the settlement, the Massachusetts Legislature enacted the formation of the American Bell Telephone Company. The new company was instrumental in developing a series of local telephone exchanges and beginning a long-lines division, which eventually became American Telephone and Telegraph (AT&T). In time, AT&T became the parent company of a coordinated federation that also included several associated companies.

In 1980 AT&T entered into a consent decree with the U.S. Department of Justice whereby AT&T was divested of its local companies but enabled to compete with other companies by providing electronic equipment to consumers under a new subsidiary called AT&T Information Systems.

A subsequent court ruling continued to restrict telephone companies from becoming information providers but did permit them to offer and market gateway and other services such as voice storage and retrieval and electronic mail.

Emerging issues confronting the telephone companies and their relationship to broadcasting and cable TV center around new technology such a fiber optics and the Integrated Services Digital Network (ISDN), whether the ban on providing information will be eventually lifted, and how consumers will accept and use these new sources of information.

Cellular telephones employ a series of "cells," much like small broadcast contours, then use a computer-switching device to move calls from one cell to another as the mobile telephone moves between cells. Plane-to-ground telephone systems use radio waves to link aircraft telephones into local switching centers, thus permitting calls to be made from the aircraft. Teleconferencing involves video-phone hookups among several telephones creating a video-phone meeting with two-way audio and video interaction possible.

Fiber optics promises some of the most radical changes in media-distribution systems. Glasslike fibers fed with laser light increase the amount of information that can be transmitted through either wired or unwired communication systems. Early pioneering applications included tests over short distances with residential test installations becoming commonplace in the 1990s. Advantages of the system include increased channel capacity and distortion-free transmission of voice, video, and data communication.

OPPORTUNITIES FOR FURTHER LEARNING

COLL, S. *The Deal of the Century: The Break Up of AT&T.* New York: Atheneum, 1986.

GONZALEZ-MANET, E. *The Hidden War of Information.* Norwood, NJ: Ablex, 1988.

HUDSON, H. *When Telephones Reach the Village: The Role of Telecommunications in Rural Development.* Norwood, NJ: Ablex, 1984.

JACOBSON, R. J. *An "Open" Approach to Information Policymaking: A Case Study of the Moore Universal Telephone Services Act.* Norwood, NJ: Ablex, 1988.

LUGGIERO, G. and others. *Inside AT&T: A Profile.* Alexandria, VA: Telecom Publishing Group, 1987.

MCLUHAN, M., and POWERS, B. R. *The Global Village: Transformations in World Life and Media in the 21st Century.* New York: Oxford University Press, 1989.

PEPPER, R. M. *Through the Looking Glass: Integrated Broadband Networks, Regulatory Policies, and Institutional Change.* OPP Working Paper No. 24. Washington, DC: Federal Communications Commission, 1988.

POOL, I. *Forecasting the Telephone: A Retrospective Technology Assessment.* Norwood, NJ: Ablex, 1982.

SLACK, J. 'D., and FEJES, F. *The Ideology of the Information Age.* Norwood, NJ: Ablex, 1987.

VIDEO PROGRAM DISTRIBUTION AND CABLE TELEVISION: CURRENT POLICY ISSUES AND RECOMMENDATIONS. Washington, DC: Department of Commerce, NTIA, 1988.

WEINHAUS, C. L., and OETTINGER, A. G. *Behind the Telephone Debates.* Norwood, NJ: Ablex, 1988.

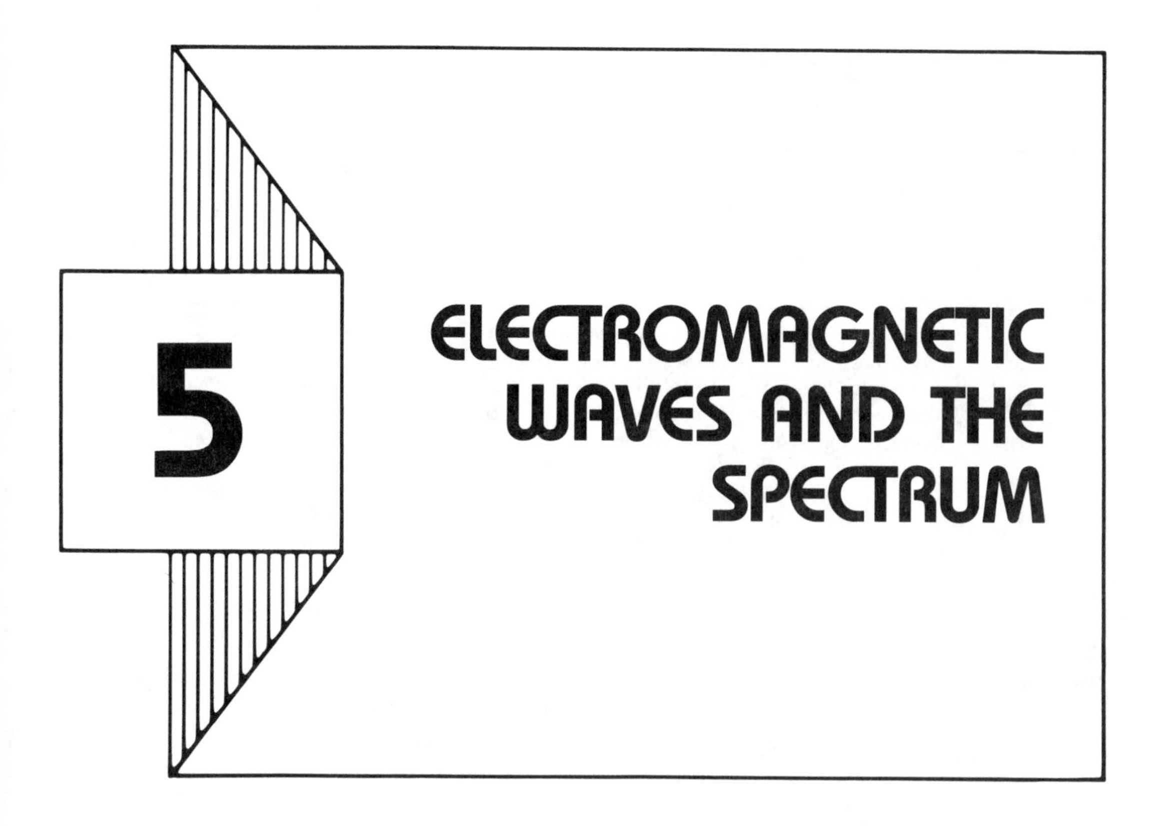

Heinrich Hertz probably never imagined that a century after his discovery of electromagnetic waves, those same waves would carry information around the world and even to planets millions of miles away. In this chapter we will learn how radio waves function and how radio and television signals and computer data travel between points. Knowing about the electromagnetic spectrum is critical to understanding the entire realm of telecommunication. Because the electromagnetic spectrum is a limited resource, the regulatory and economic policies governing any user of the spectrum are determined in part by this concept of a limited resource. In other words, there is only so much space on the spectrum and therefore only so many people can use it.

THE ELECTROMAGNETIC SPECTRUM

To understand how information such as data or broadcast signals are carried between transmitter and receiver, it is first necessary to understand the *electromagnetic spectrum* (Figure 5–1). Consider the spectrum as a measuring stick for electromagnetic energy. At the lower end of the measuring stick are radio waves. At the upper end of the spectrum we find visible light and

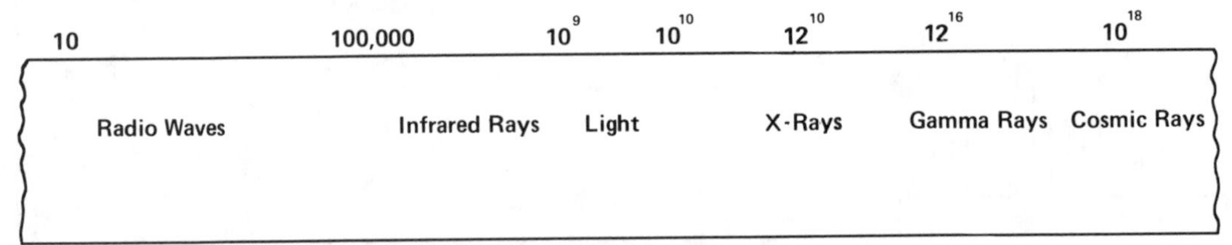

FIGURE 5–1 The electromagnetic spectrum. Notice that radio waves are at the lower end of the spectrum. Low-frequency waves, such as those employed in AM radio broadcasting around 580 to 1600 kilohertz (thousands of hertz per second), tend to bounce off the atmosphere, whereas higher-frequency waves, such as those used in FM radio broadcasting, around 88 to 108 megahertz (millions of hertz per second), adhere more to line-of-sight transmission. The numbers in the diagram are for illustration only. The "10" at the left end of the spectrum is in the kilohertz range. The "100,000" and above numbers are in the megahertz range.

X rays. For our purposes, we shall concentrate on radio waves.

DEFINING FREQUENCY

What differentiates radio waves from light waves or X rays? The answer is their *frequency*. Two radio stations in the same community operate on different frequencies so that they will not interfere with each other.

When current is applied to the transmitter of a radio station the antenna emits electromagnetic radiation. This radiation is actually a series of electromagnetic "waves." The next time you throw a rock into a pool of water, watch the series of waves that ripple one after the other in all directions from the point at which the rock entered the water. This is similar to what happens when electromagnetic energy travels through the atmosphere or the vacuum of outer space. The *number of waves* passing a certain point in a given interval of time is the *frequency*.

UNDERSTANDING WAVELENGTH

The *distance between two waves* is called the *wavelength* (Figure 5–2). If we take a stop-action picture of the ripples (waves) in our pond and then figure the distance between two ripples, that would be the wavelength.

Understanding the Term "Hertz"

The term *hertz*, in honor of Heinrich Hertz, is used when discussing frequencies and the electromagnetic spectrum.

When *one complete wave passes* a point we have observed *one hertz* (Hz), although the waves generated as electromagnetic energy travel much too fast to count and cannot be seen.

Understanding "Hertz per Second"

We know that all electromagnetic waves travel at the speed of light, that is, 186,000 miles per second. Since we measure the speed of light in miles per second or, using the metric system, in meters per second, the second becomes the commonly used time interval. The *number of waves passing a certain point in one second* is called *hertz per second*. One thousand hertz per second is called a *kilohertz* (kHz); one million hertz per second is called a *megahertz* (MHz).[1]

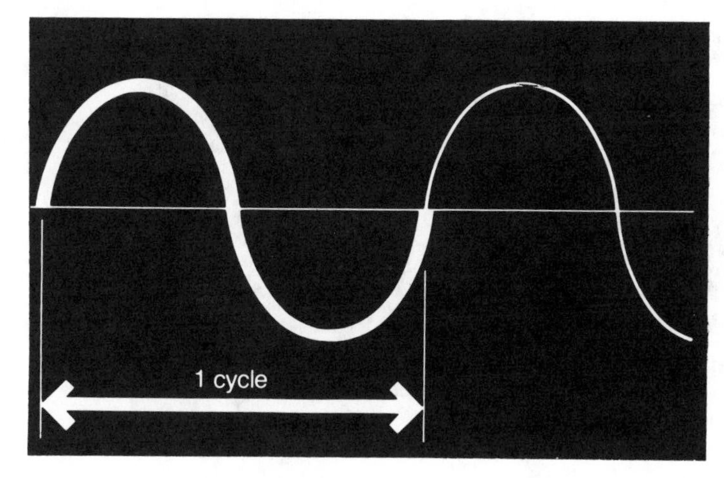

FIGURE 5–2 When one complete wave passes a given point, it is called a "cycle," or more commonly a "hertz," in honor of Heinrich Hertz. Remember, the term *kilo*hertz is used to represent 1,000 hertz (1,000 cycles per second). The term *mega*hertz is used to denote 1 million hertz (1,000,000 cycles per second). The term *giga*hertz represents 1 billion hertz (1,000,000,000 cycles per second).

COMPUTING FREQUENCY

Now that we understand hertz per second, we can easily compute frequency. If, for instance, 1,000 waves pass a given point in one second, the frequency, or location on the electromagnetic spectrum, is 1 kilohertz. Similarly, 540,000 hertz per second is represented as 540 kilohertz. On your AM radio, that particular frequency would be at the lower end of the dial at 540.

TUNING TO A WAVELENGTH OR FREQUENCY

In simplified terms, a radio receiver "counts" the waves or hertz per second to determine a frequency. Your radio does this when you tune from one station to another. Your receiver is picking up only those waves that are being transmitted on the same frequency to which you tune. Thus, different frequencies on your radio dial correspond to different positions on the electromagnetic spectrum.

AM BROADCASTING

Now that we understand the electromagnetic spectrum and how waves are radiated into space, we will learn how voice and music use those waves to reach radio listeners. We will begin with AM broadcasting.

AM stands for *amplitude modulation*. Amplitude is defined as *breadth of range*. Modulation means to adjust or adapt to a certain proportion. Now let us apply both of these words to radio waves.

Figure 5–3 illustrates an unmodulated radio wave. Notice how even it is. Now examine Figure 5–4, a radio wave that has been altered by adjustment of the amplitude, or breadth of range, of the wave. The wave characteristics of music and the human voice are transformed into the wave, which in turn "carries" them between the transmitter and receiver. Notice that the wavelength of each wave, the frequency, remains constant. The change takes place in amplitude, not in frequency. When the *wave is adjusted to carry changes in sound*, it is said to be *modulated*.

FM BROADCASTING

In FM broadcasting, instead of changing the amplitude of the wave, we change the wavelength or frequency. Figure 5–5 illustrates

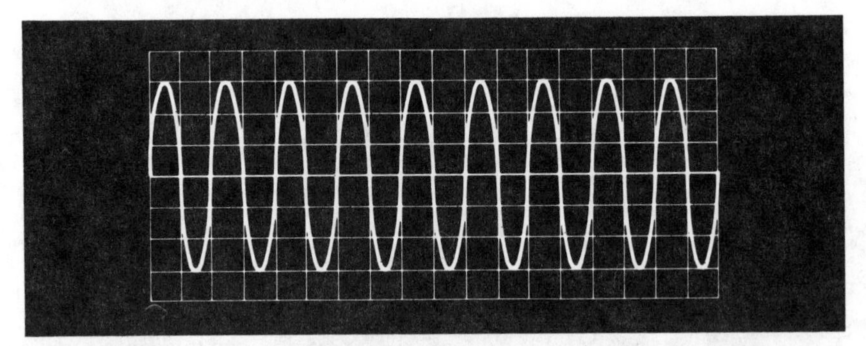

FIGURE 5–3 An unmodulated wave. Modulating the wave—the term "modulate" means to adjust—results in either adjusting the amplitude of the wave, or adjusting the frequency of the wave, which in turn changes the sound we hear in a radio receiver.

a *frequency-modulated* wave. Notice there is no change in the amplitude of the wave. Instead, the wavelength, or frequency, varies. Different sounds indicate different wavelengths (hertz per second). FM broadcasting to the general public operates between 88 and 108 megahertz.

FM STEREO

Within the 200-kHz space allocated to each FM station is ample room for the separation of broadcast signals, room that permits the same station to broadcast on two slightly different frequencies. There is also room for a tone that triggers specially equipped radios to receive this stereo signal. Radios equipped to receive stereo actually have two separate receiving systems, which, when triggered by the tone, separately receive the two frequencies being broadcast by the stereo station.

Many of us have seen a small light flip on in a stereo FM receiver when we tune it to a station broadcasting in stereo. This sig-

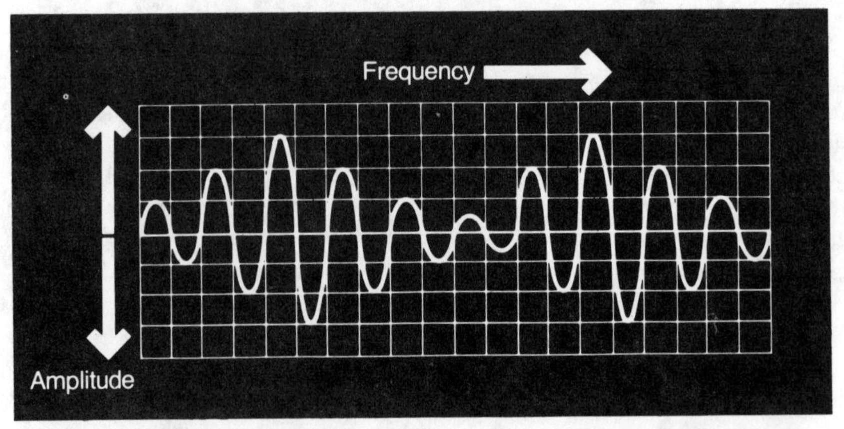

FIGURE 5–4 An amplitude-modulated wave. Notice that the frequency (width of the wave or, in other words, the distance between waves) remains the same, but the amplitude varies.

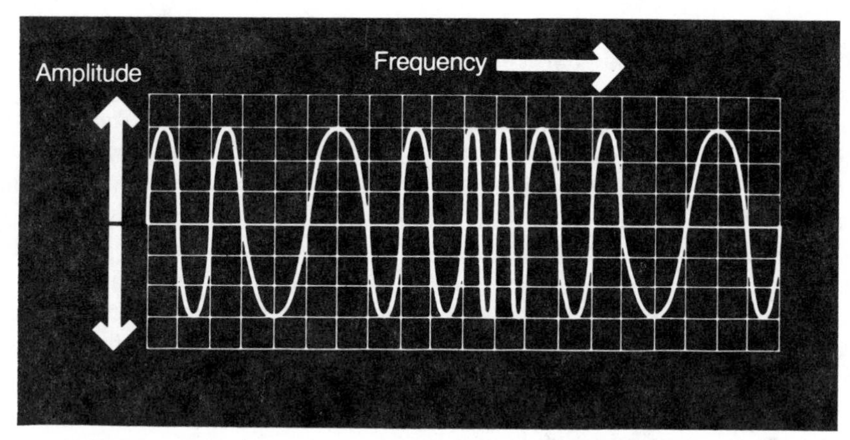

FIGURE 5–5 A frequency-modulated wave. Notice that the frequency (space between the waves) varies, but the amplitude remains constant.

nals us that our radio is tuned to a stereo station and that the station is broadcasting in stereo.

AM STEREO

AM stereo is now emerging from the experimental stage. For many years, AM broadcasting did not seriously consider stereo beyond laboratory ventures, primarily for two reasons. The first was the narrow channel width of AM stations—10 kHz compared with 200 kHz for FM. Second, as long as FM was not a serious competitive threat to AM, there was no widespread interest in the system. However, with FM gradually cutting into the AM audience, AM broadcasters began to search for something with which they could regain their competitive edge.

Active evaluation of all systems—undertaken by the National AM Stereo Committee (NASC) of the Electronics Industries Association—and the collection of field performance data on AM stereo was completed in 1977.[2] Since that time many stations have begun broadcasting in AM stereo as well as promoting it to their audiences.

TRANSMITTING TV SIGNALS

Television stations broadcast on frequencies located both above and below the standard FM radio frequencies of 88 to 108 MHz. It is important to understand that radio waves carry the television picture and sound. The width of the spectrum allocated for television transmission is established by the FCC at 6 MHz. Part of the frequency is used for transmission of the video portion of the signal, and part of it the audio portion.[3]

Processing the TV Picture

Earlier in this book we learned about Paul Nipkow's mechanical television, which consisted of a scanning disk in which holes were punched in the pattern of a spiral. When the disk turned, the holes would pass over a small opening through which could be seen a picture. In one complete revolution of the disk, the entire picture would be scanned and transmitted to another receiving unit, which would then reproduce the image. If the image were replaced with another in rapid succession, the illusion of motion would be created.

Further refinements changed the process, and with the work of Philo Farnsworth and the development of Vladimir Zworykin's iconoscope tube, electronics picked up where mechanics left off. Using electrons instead of a spiraling disk, one could scan the image with increasing clarity and speed. The result was the electron-scanning process as we know it today. For a simplified example, imagine a flag with a series of red and white stripes. The scanning process first scans the white stripes and then the red stripes. Now imagine this process taking place 525 times per second—the United States standard—while the picture (the flag) rapidly changes. The result is a series of rapidly scanned and broadcast pictures that appear on our television set as an illusion of motion.

Processing Color Television

Color television uses a similar process, except that the television camera separates the three primary colors of light: red, green, and blue.[4] All other colors are made up of a combination of these three. When the television camera scans an image, it separates the red, green, and blue hues. These are transmitted individually and then appear as tiny dots on our television screen. The dots are too small for us to see with our naked eye and thus tend to "run together," creating the color picture. This, plus the rapid scanning process, creates a picture in both color and motion.

HIGH-DEFINITION TELEVISION HDTV

High-definition television, called HDTV, is in the experimental stages. A larger screen is used in HDTV, made possible by more than 1,000 scans of the picture per second. The picture is exceptionally clear, of almost photographic quality. HDTV takes up more of the electromagnetic spectrum than conventional television signals. HDTV is in more widespread use in Japan and other countries but not yet fully developed in the United States. Nonbroadcast HDTV is gaining widespread use in the motion picture industry where motion pictures are shot and edited in HDTV then transferred to film for distribution to theatres. We will learn more about HDTV later in the text.

THE PATH OF ELECTROMAGNETIC WAVES

The path electromagnetic waves travel can be divided into ground waves, sky waves, and direct waves. Ground waves adhere to the contour of the earth. The area of the electromagnetic spectrum where AM broadcasting stations are located results in radio waves adhering to the earth's surface, thus the term "ground waves." Some waves in this area of the spectrum also travel into the atmosphere and are referred to as "sky waves." If you drive your car from the base of the transmitter of an AM radio station, the signal you'll receive when close to the tower is made up of ground waves. Farther away from the tower, you will receive a signal made up of sky waves that have traveled out into the earth's atmosphere and then back to earth. Higher on the electromagnetic spectrum where FM broadcasting and television signals are located, the waves travel in a straight line and are called "direct waves." The direct waves, free for the most part from interference caused by the waves traveling into the atmosphere and bouncing back to earth, result in the clear reception found in the portion of the electromagnetic spectrum assigned to FM radio stations. The development of digital technology and its application to broadcasting will reduce some of the interference problems that broadcast stations experience.

ANALOG VERSUS DIGITAL TRANSMISSION

Two types of transmission can exist in both wired and over-the-air communication systems: analog and digital. Think of *analog transmission* as an analogy, something like something else. For example, the voice of a radio announcer is modulated in a certain way, transmitted over the air, and received on a radio receiver. The transmission is continuous over a given frequency range, and unless the receiver or transmitter was turned off, the sound at the receiver is the same as the sound that was transmitted. A typical telephone conversation may consist of voice vibrations that are turned into analogous electrical vibrations. What a television camera sees is transformed into an analog signal, which creates in the television receiver a picture similar to what the television camera witnessed.

With *digital transmission*, however, the signal is not continuous; it is broken up into numbers. The signal consists of a series of on–off pulses transmitted in the same way that information flows in a computer circuit. The pulses are bits of information in a binary-number code. For all practical purposes digital transmission is noise-free.

In addition to standard AM, FM, and TV frequencies, much higher frequencies of the electromagnetic spectrum—in the thousands of megahertz range—are used. It is in this area that microwave transmission is found. We have learned that the higher the frequency, the farther the electromagnetic waves will travel in a direct line-of-sight path between transmitter and receiver. Thus, microwaves always travel by line-of-sight transmission.

MICROWAVES: THE CONCEPT

Microwaves also allow many more channels of communication to operate because of their shorter wavelength. Because the waves are shorter, many more will fit into the same space on the electromagnetic spectrum. When we realize that an AM radio station is allocated a width of only 10 kHz, it is easy to see how much more information can be transmitted at higher frequencies. Microwave technology has many applications in broadcasting and telecommunication.

MICROWAVE RELAY SYSTEMS

Microwave relay systems can be both ground-based and satellite-based.

Ground-Based Systems

A network television program may travel thousands of miles before it reaches your local television station. The path it follows may use a ground-based microwave-relay systems. Using high-frequency line-of-sight transmission, these systems can carry crystal-clear signals over long distances through a series of relay antennas approximately thirty miles apart. These dishlike antennas dot almost every landscape, from the roofs of skyscrapers to mountain peaks.

Keep in mind that the TV program you receive in your home does not arrive directly by microwave. The local television station receives the signal by microwave and then retransmits it to your home receiver at a frequency assigned to the television station.

Satellite-Based Systems

Because space is a vacuum, microwaves travel over long distances unimpeded by the earth's heavy atmosphere (Figure 5–6). This permits a transmitting station in London, for example, to transmit a television picture by microwave to a satellite thousands of miles in space, which relays it back to an earth-receiving station in the United States. Satellites are also used to bring television signals to many outlying regions in which even microwave links would be too costly or where the transmitting site is temporary.

APPLICATIONS OF MICROWAVE RELAY SYSTEMS

Because of the ability of microwaves to carry large amounts of information with clarity, many applications have been found for this technology.

FIGURE 5–6 Microwave technology provides long-distance communication without wires. Ground-based microwave systems are especially efficient in regions that are too remote to construct wired systems economically. Satellite-based microwave systems transmit signals from a ground-based antenna (shown on the left) to a satellite in space, which in turn relays the microwave signals to other ground-based antennas. The satellite on the right is positioned so that the "footprint," or signal contour, covers that area (shown is the United Kingdom) where ground-based satellite receiving antennas are located.

Cable TV

Although you may receive television programs through your local cable-TV system, the original signal, especially from such premium services as HBO, MTV, and many others, reached the local cable company through a satellite-based microwave link. A cable company may receive the signal from a television station, then retransmit it by microwave to a cable system in a community hundreds of miles from that station.

Telephone Systems

Microwave and other radio-wave technologies are applied extensively in telephone communication (Figure 5–7). Where once only wires carried conversations between two points, now microwave technology links central switching centers with relay stations. The increased use of telephone systems to transfer data over long distances, and other services such as facsimile and teleconferencing, add to the need for increased channel capacity which microwave technology helps to meet.

Electronic News Gathering

The application of microwave to electronic news gathering has given television the flexibility that only radio once enjoyed. A mobile van and portable camera can provide

FIGURE 5–7 Increased demand for telephone services, including voice, data, and teleconferencing, has resulted in the need for increased channel capacity. Some of the increased demand is met through the use of radio wave and microwave technology. Microwave relay systems are especially beneficial and cost-efficient over rough mountainous terrain. This telephone center and antenna system is located at Interlaken, Switzerland, in the Swiss Alps.

live programming from almost any remote location. Beaming the signal to a satellite, it is then relayed back to the satellite receiving dish at the television station where the signal is incorporated into local programming, such as news shows, and retransmitted to home television sets.

Educational and Industrial Television

Many schools are part of statewide television systems connected through microwave links. Such connections, both ground-based and satellite-based, permit many two-way interactive programs to exist. Teachers and students can interact through these television classrooms much the same way as if they were speaking to each other in person. Businesses are also using such systems to link offices and manufacturing plants and transmit data and video communication. A plant with two locations in the same city can use microwave for an intracorporate television link. For example, special sales-training seminars can be broadcast from the main corporate television studio to a special seminar room on the other side of town. Or, leasing the facilities of a national microwave system, a company can distribute a televised management-training program to plants throughout the country.

The existing, well-developed microwave systems will undoubtedly remain in service for years to come, although fiber-optic technology will also impact on the future of microwave systems. Because fiber optics can carry so much information, at the frequency of light and much higher than microwave technology permits, fiber optics will compete as an economical and efficient carrier of the information of the future.

SUMMARY

Radio waves can be thought of as a point on the electromagnetic spectrum, a yardstick of electromagnetic energy that includes such forces as microwaves, light waves, and X rays. Radio waves travel at the speed of light and at a frequency determined by the length of each wave. The waves vary in amplitude and frequency. Variations in amplitude are used to modulate radio waves in the AM broadcast band, and variations in frequency are used to modulate radio waves in the FM broadcast band.

Both AM and FM stations can separate their signals and broadcast in stereo.

Television signals are transmitted using radio waves. The 6 MHz space allocated to television stations carries both the sound and picture. The television picture is processed using a scanning system, and color television separates the primary colors of light, then transmits each separately. The "dots" of color on the television screen result in the color picture we see. High-definition television uses more scanning lines per second and a larger screen, thus producing an exceptionally clear picture of almost photographic quality.

Radio waves are of three types: ground, sky, and direct. Generally speaking, the higher the frequency the more direct the wave's line of travel.

Microwaves, which are found at very high frequencies on the electromagnetic spectrum, have many applications to broadcasting. They relay television programs between stations and networks. Microwaves are used by satellites to beam television signals around the globe, and by cable systems to import distant television signals for redistribution to cable subscribers. Microwave technology is also used in electronic news gathering and in educational and industrial broadcasting.

OPPORTUNITIES FOR FURTHER LEARNING

CHINESE INSTITUTE OF ELECTRONICS SOCIETY, eds. *Antennas and Em Theory*. New York: Pergamon; also Beijing: International Academic Publishers, 1990.

CHINESE SOCIETY FOR MEASUREMENT. *Electromagnetic Meteorology*. New York: Pergamon; also Beijing: International Academic Publishers, 1990.

FULLER, A. J. B. *Microwaves: An Introduction to Microwave Theory and Techniques*. New York: Pergamon, 1990.

MCMILLAN, A. S., and TUCKER, G. M., eds. *Proceedings on the International Conference on Millimeter Wave and Far-Infrared Technology*. New York: Pergamon; also Beijing: International Academic Publishers, 1990.

RWAT, B., and SIYONG, Z., eds. *Recent Advances in Microwave Technology*. New York: Pergamon; also Beijing: International Academic Publishers, 1989.

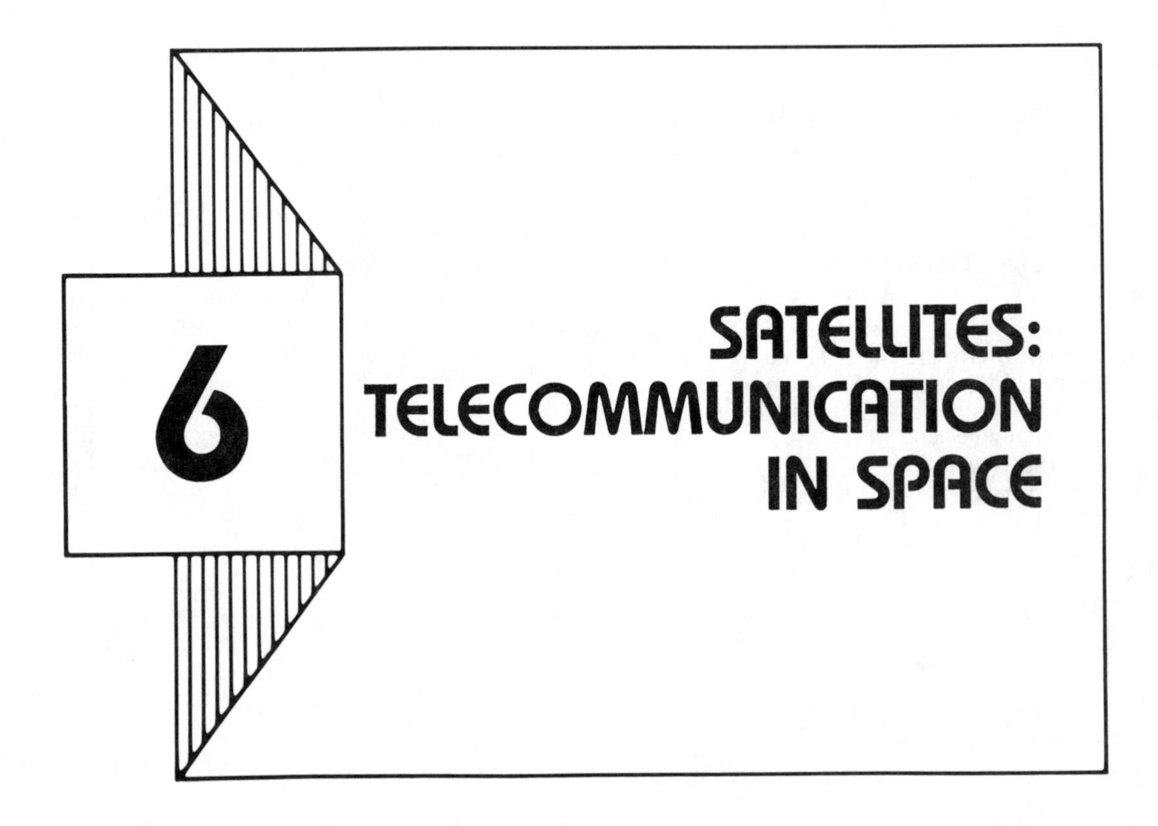

6 SATELLITES: TELECOMMUNICATION IN SPACE

Using microwave technology, satellites are the "relay" workhorses of broadcasting and telecommunication. They relay broadcasting and data transmission between ground-based stations, which can be as near as across town or as far away as the other side of the world. The satellite era began in the 1960s with the first communications satellite, Telstar.

SATELLITES: THE TELSTAR EXPERIMENTS

Although we cannot see them from earth, satellites hover above the planet and are the communication relay links that bring us everything from our telephone calls to radio and television programs.

The wedding between satellites and broadcasting took place on a warm New England evening in Andover, Maine. From this outpost the first television pictures were relayed by satellite across the Atlantic to Europe on July 10, 1962.

The hero of the evening was a 170-pound payload named *Telstar*, which had been launched into space in a cooperative effort by the *National Aeronautics and Space Administration* (NASA) and AT&T.[1] This beach-ball-size satellite received television signals from earth-based antennas and re-transmitted them to European receiving stations.

Powered by solar energy, on any given

pass over the United States or one of the European receiving stations, the satellite was in viewing range for only about forty-five minutes. The highest point in Telstar's orbit, its *apogee*, was 3,502 miles from earth. The closest point to earth during its orbit, the *perigee*, was 593 miles away.

The United States received signals from Europe via Telstar the following evening. From France came a seven-minute taped program with an appearance by the French minister of Postal Services and Telecommunications, and musical entertainment by French performers. The British signal followed shortly thereafter, consisting of a test pattern and a live broadcast.

The Politics of Telstar

It is not surprising that the breakthrough of transoceanic broadcasting brought with it a series of political issues, some based on age-old rivalries, others on contemporary concerns. The most intense rivalries were between England and France, long-time economic and political sparring partners. Telstar merely set the scene for their combat.

On the night of the first transoceanic broadcast, British pride was hurt when the British receiving station was not able to monitor clear audio and video signals from the United States. The French, meanwhile, using a station that was not supposed to be ready for tests, monitored signals with great clarity. The British did achieve a victory that night when they relayed a live television program to the United States in contrast with France's taped program.

In America, the press was quick to report that AT&T had paid NASA to launch Telstar. This agreement also called for the free availability of any inventions arising from the Telstar project to any company that wanted them. President John F. Kennedy formally called for a national corporation to oversee all satellite-communication developments in the United States. The Kennedy administration also emphasized the need for commercial broadcasters to participate in examining the potential of satellite communication.[2]

International Implications of Telstar

Telstar created much more than international television communication. The prospect of domestic television programming crossing national boundaries opened up a new arena for discussion and heated debate. The initial Telstar broadcast itself caused some concern. The program was produced by AT&T, which provoked CBS to break away from the initial network-pool coverage and not carry the remarks of AT&T's board chairman.

Jack Gould, television critic of the *New York Times*, said of the event, "The sight of Government dignitaries serving as a passive gallery for private corporation executives was not very good staging, particularly for consumption in foreign countries." Gould went on to predict, "The crucial decision that will determine the lasting value of international television—a willingness of countries and broadcasters to clear the necessary time on their own screens to see and hear other peoples of the world—cannot be made in laboratories in the sky but in offices on the ground."[3]

International programming gradually became commonplace. The funeral procession for President Kennedy in Washington, D.C., was seen in the halls of the Kremlin. By 1965 the ecumenical conference in Rome was being seen on both sides of the Atlantic.[4] European viewers saw and heard Washington dignitaries react to the unveiling of the Mona Lisa in the National Gallery of Art. Special programming from the 1964 Olympic Games in Tokyo traveled far beyond Japan.

STOPPED IN SPACE: THE SYNCHRONOUS-ORBIT SATELLITES

A little more than a year after the launching of Telstar another breakthrough in satellite technology occurred. At Lakehurst, New Jersey, in 1963, a technical crew waited for a satellite called Syncom II to "lock" into position for a transmission that would be heard halfway around the world on a ship stationed at Lagos, Nigeria. Out of a static-bearing receiver aboard the U.S. Navy's *Kingsport* were clearly heard the words "Kingsport, this is Lakehurst. Kingsport, this is Lakehurst. How do you hear me?" The words came from space, relayed back to earth from Syncom II, no ordinary satellite. This time the technical crew aboard the *Kingsport* did not have to adjust their receiving equipment just when the satellite passed within viewing range, because Syncom II was technically "stopped in space," the first successful *synchronous-orbit satellite*. A synchronous-orbit satellite is positioned at a point 22,300 miles in space where it rotates in the same path as the earth's rotation and at a speed that keeps it at the same point above the earth's surface, much like a 22,300 mile-high-antenna.[5]

COMSAT

By now the world was taking an active interest in satellite development. The morning after Telstar's broadcast, AT&T became the most active stock on the New York Stock Exchange. Less than a month later David Sarnoff, Chairman of the Board of RCA, proposed that a single company deal with international communication matters.[6] Western Union quickly supported Sarnoff's suggestion, saying it had proposed a similar concept before.[7] Having obtained the Kennedy administration's support, the concept was realized with the passage of the Communications Satellite Act of 1962 and the formation of the *Communications Satellite Corporation (COMSAT)*.

INTELSAT

COMSAT became the early planner of satellite systems on an international scale when it evolved as the manager of the International Satellite Consortium, a global cooperative effort to govern and develop world satellite systems (Figure 6–1). The consortium was established under two international agreements originally signed by fourteen countries and eventually ratified by fifty-four. In 1964 International Satellite Consortium became the *International Telecommunications Satellite Organization (INTELSAT)*, having a membership of more than eighty nations and presided over by a secretary-general. INTELSAT membership is open to any nation that was a party to INTELSAT's Interim Agreements, and to any other country that is a member of the International Telecommunication Union.

INTELSAT SATELLITES

On April 6, 1965, Early Bird became the first of a number of INTELSAT spacecraft to be launched. INTELSAT satellites have provided a worldwide system of communication, not only for broadcasting but also

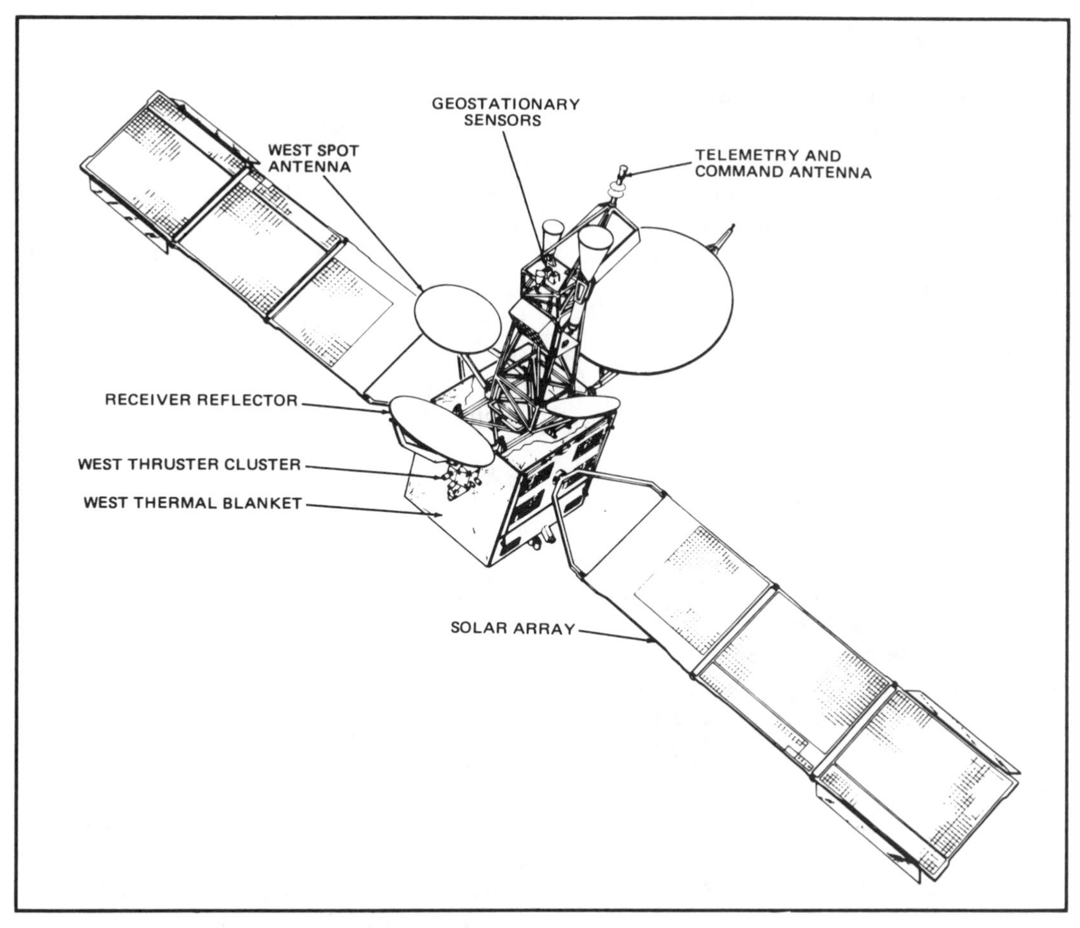

FIGURE 6–1 INTELSAT satellites provide international communication links. Positioned above the earth in synchronous orbit, making them stationary above a point on the earth's surface, the satellites can relay signals to any location. This diagram of an INTELSAT satellite shows some of the components of a synchronous orbit satellite. The solar arrays collect sunlight for solar power, and the geostationary sensors keep the satellite in a stationary position in relation to the earth's surface. The command antenna is used to receive signals that control the satellite's position and functions.

for data, telephone, two-way radio communication, weather monitoring, and other uses. Improvements in satellite technology were also introduced through the development of INTELSAT satellites. For example, improved beam separation increasing satellite capacity occurred with the INTELSAT IV–A satellites. The INTELSAT V series contained alternate frequency capability, thus doubling the capacity of the IV–A series. Even more sophisticated is the INTELSAT VII series scheduled for launch in about 1993, also with COMSAT's participation.

NASA SATELLITES

COMSAT and INTELSAT were not the only organizations trying to develop satellite communication. The National Aeronautics and Space Administration (NASA) has engaged in a series of satellite programs, including the use of the Space Shuttle as a launch vehicle.

Application Technology Satellite (ATS)

NASA developed the *Application Technology Satellite* (ATS) system that beamed educational television programming to schools from India to the United States, to Africa, to the Galapagos Islands.[8] The program generated a cooperative, although somewhat reluctant, effort among state governments. For example, issues affecting the local autonomy of schools usually formed a political thicket. Nevertheless, the Federation of Rocky Mountain States, which includes Arizona, Colorado, Idaho, Montana, New Mexico, Utah, and Wyoming, joined together to bring two-way educational television to the outlying communities in the region.

NASA's Space Shuttle

NASA's space-shuttle missions have shown the feasibility of using the shuttle as a satellite-launch vehicle. The first satellite launched from the shuttle was the Business Systems satellite in 1982. A Telesat Canada satellite was launched on the same mission.

In theory, the shuttle can be used not only for launching but also for repairing equipment in space. Plans for the shuttle involve a space-operations center where the shuttle could dock and then assemble and repair space hardware, including satellites (Figure 6–2).

PRIVATE INTERNATIONAL SATELLITE SYSTEMS

Private competition in the international satellite business began to occur in the late 1980s with the launch of the Pan American Satellite I (PAS I) developed by Alpha Lyracom. An advertisement for the PAS I touted ". . .broadcasters throughout the U.S., Europe, Latin America and the Caribbean can create broadcast networks a better way. With dedicated capacity on Alpha Lyracom's PAS I satellite."[9]

DOMESTIC SATELLITE SYSTEMS

A number of private domestic companies have been involved in the development of satellites in addition to AT&T and its continuing series of Telstar satellites (Figure 6–3). Some of the more familiar names in both past and present domestic satellite systems include RCA, GE, Western Union, Hughes Communications, and GTE.

Western Union entered the domestic satellite business in the early 1970s with the launching of three satellites in its Westar series.[10] When General Motors' subsidiary Hughes Communications found its Galaxy satellites at maximum capacity it bought out Western Union's satellites, and the Westar series became Galaxy satellites.[11] Hughes also announced the purchase of IBM Corporation's Satellite Transponder Leasing Corp. With additional satellite launches scheduled for the 1990s, Hughes Communications becomes one of the major players in the direct-broadcast satellite market.

General Electric entered the satellite business when it acquired RCA's Americom. RCA Americom became GE Americom. The early Satcom satellites made extensive use of solar power with large

FIGURE 6–2 Future planning for the space shuttle of the National Aeronautics and Space Administration (NASA) includes docking and repair of space stations, which will themselves serve as communication centers.

antenna arrays capturing the sun's light and turning it into power to operate the satellite. Joint ventures between GE Americom and Alascom, an Alaskan long-distance common carrier, are scheduled to result in launches in the early 1990s of the Aurora satellites, which will replace some of the Satcom satellites and provide service to U.S. broadcast and cable programmers.[12]

General Telephone and Electronics (GTE) began development of its domestic Spacenet satellite system in 1981 and eventually added the GStar series satellites, although some imbalanced loading of fuel caused the GStar III to miss its orbital slot, and using additional fuel to move it into position cut its life by five to eight years. Additional satellite capacity is scheduled through the launch of the GStar IV in the early 1990s.

SUPERSTATIONS

With the advent of satellites some radio and television stations have expanded their listening and viewing areas by beaming their

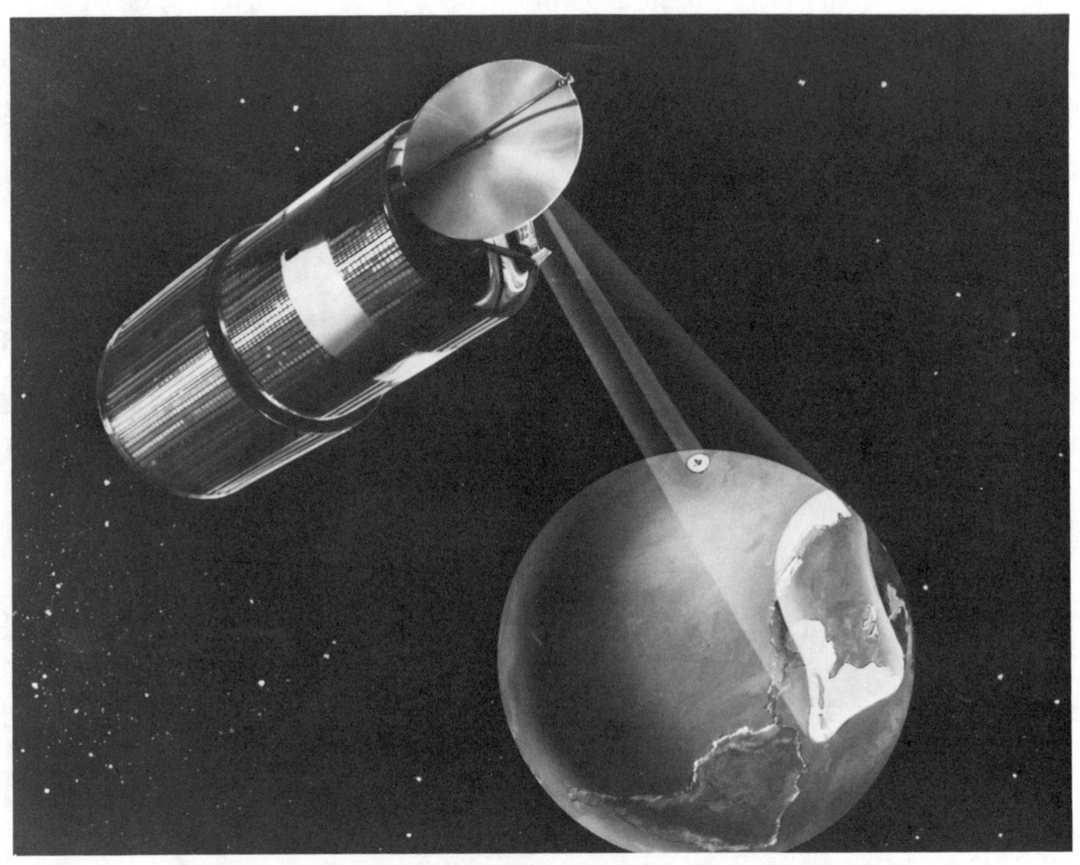

FIGURE 6–3 Domestic commercial satellite systems make up a large portion of the space-based communication networks. Even though commercial investments in satellite systems have slowed somewhat, partly because of uncertainties over the impact of fiber-optic networks, satellites will remain a major communications technology of the future. The potential for direct-broadcast satellites (DBS) to become a source of television entertainment for the home may create a new demand for satellite technology.

signals to a satellite and then having the signals relayed to cable systems, which in turn send the signal to subscribers. Undoubtedly the most famous superstation, and the first to make the concept workable, is WTBS in Atlanta. Owned by Turner Broadcasting, WTBS (formerly WTCG) first beamed its signal to cable systems on December 16, 1976. Programming has tended to stress family-oriented fare complemented by movies and sports, including Atlanta Braves baseball and Atlanta Hawks basketball games. Original features are also presented. Turner also operates the satellite-fed Cable News Network (CNN), the CNN Headline News, and Turner Network Television (TNT)

Another superstation is the independent

Chicago station WGN. Also oriented toward family programming, WGN carries original programming, its own news schedule and features, and syndicated programs. Radio superstations also exist.

DIRECT-BROADCAST SATELLITES (DBS)

Direct-broadcast satellite (DBS) systems employ the use of a high-powered satellite to beam television signals direct to small satellite receiving dishes. In Europe and Japan, DBS systems are more advanced, and investors have seen an economic and political climate that permits the development of DBS technologies. Approximately 2.2 million U.S. homes are equipped with satellite receiving dishes to receive DBS signals.[13]

Technology Issues

Three categories of DBS exist: (1) high-power satellites whose signals can be received on small receiving antennas that can easily be used in residential settings; (2) medium-power satellites capable of being received on larger antennas, typically one-meter in diameter, and (3) satellites used for commercial purposes where long-term interference-free communication is necessary and where the larger 10-foot to 12-foot antennas are in use.

The limited space on the electromagnetic spectrum and the limited number of available orbital slots have prevented full-scale development of DBS.

Economic Issues

At the very heart of DBS has been the fact that it is very expensive. Getting a satellite and programming system operable costs in the hundreds of millions of dollars and not many companies have had that kind of capital. Moreover, because the acceptance of DBS by the consumer is an unknown, not many companies want to take the risk.

Joint ventures, however, could prove fruitful. A venture with Hughes Communications, General Electric, Cablevision, and media magnate Rupert Murdoch's News Corporation joined together and announced a DBS system with more than a hundred channels to be on the air by the mid-1990s.[14]

Where the revenue will come from when systems such as Hughes' become operational is uncertain. Advertisers want guaranteed results. If quality programming is not available and no one watches, then advertisers will stay away. Even if consumers are charged for programming, the programs must be worth the price. In addition, alternative media are also available such as videotape and videodisk rentals, and cable TV.

To have the consumer purchase a dish antenna may prove a stumbling block if the antenna is too expensive.

Cable Television as a DBS Distributor

While the above issues make the immediate DBS picture unclear, the possibility exists that the cable television industry may itself get involved in DBS. For example, because of the ability to buy large blocks of programming and with an installation and maintenance structure already in place, the local cable company may offer subscribers in unwired neighborhoods the opportunity to "subscribe" to satellite programming. The cable company could rent and install the receiving antenna much in the same way that cable decoders are marketed. Moreover, large cable companies have business relationships with program suppliers, which might prove valuable in providing programming that appeals to a large segment of the population.

International DBS

DBS systems are emerging in a number of European nations and Japan. More than 2 million homes receive two DBS channels in Japan. Additional DBS systems are scheduled to be operational by the early 1990s. France and Germany in a joint government venture brought DBS to Europe in the late 1980s.

Whether these systems remain economically viable is uncertain. Although the potential for huge European audiences does exist, only a small percentage of the potential audience is reached by DBS. Equipment manufacturing problems as well as localized programming tastes have complicated DBS development in Europe.[15]

GLOBAL ISSUES IN SATELLITE COMMUNICATION

The development of satellites has resulted in a number of emerging issues centering around the impact of satellite technology on society. What will happen if world communication systems are developed to the point at which rooftop dish antennas are as common as the television set in the living room or the radio in a car? Already television programs from one country are impacting on the cultural and entertainment tastes of other nations (Figure 6–4).

An evening television news report is beamed across national boundaries, and world leaders respond almost immediately to events on the other side of the globe. News of an economic crisis in one part of the world influences financial markets thousands of miles away. The evening news reports a world leader's visit to another country, resulting in a new interest in foreign fashions. With instant global communications we learn more and more about different cultures while at the same time cultures fuse together and are less defined. What impact do these events have on the future of world cultures? Such questions are at the heart of the future of satellite communication.

SUMMARY

Satellites provide equally important functions. Telstar, launched in 1962, was the first satellite to broadcast international TV signals. Since then, many satellites have been used in broadcasting. Synchronous-orbit satellites provide twenty-four-hour communication among almost any points on earth, making continuous, live television coverage of international events a reality.

In the United States much of the administration of international satellite communication is undertaken by the Communications Satellite Corporation (COMSAT). COMSAT manages the International Telecommunications Satellite Organization (INTELSAT), a cooperative effort of member countries using satellite communication and the network of INTELSAT satellites. Early Bird, launched in 1965, was the first INTELSAT satellite. It was followed by a series of more advanced INTELSAT satellites, the latest being the INTELSAT V series. Membership is open to any nation that was a party to INTELSAT's Interim Agreements, and to any other country that is a member of the International Telecommunication Union (ITU).

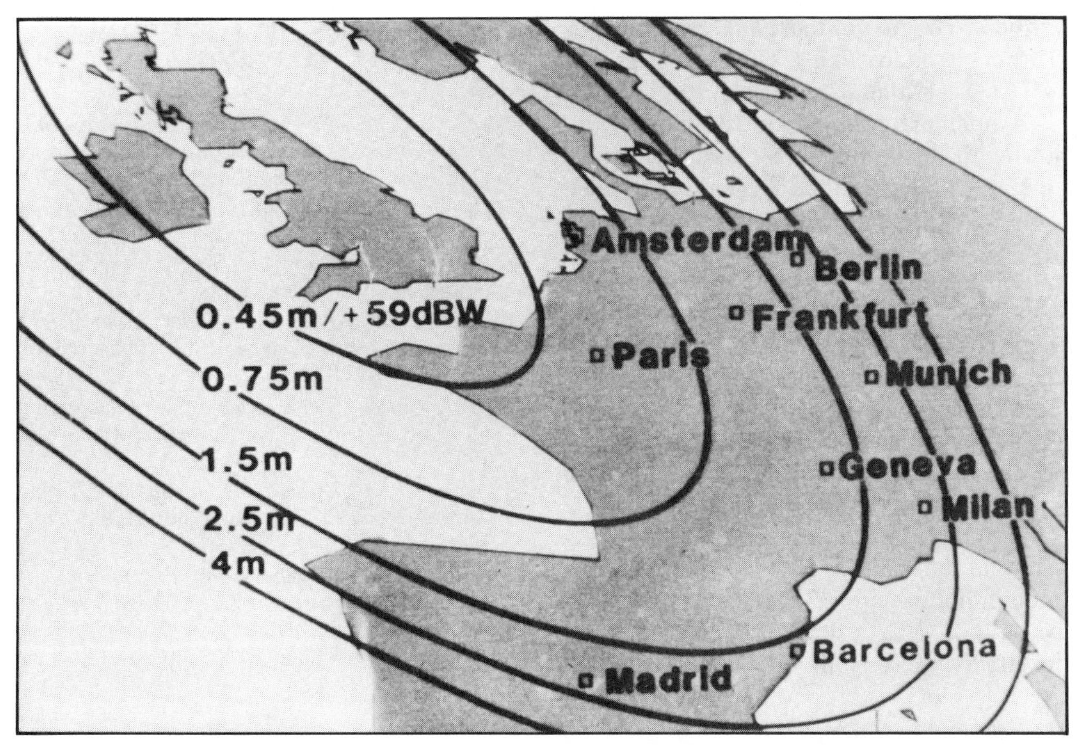

FIGURE 6–4 Ground-based communication systems, such as domestic radio and TV stations, operate mostly within national boundaries. Signals from satellites, such as those carrying television programming, transcend national boundaries and open up many social and political questions about the impact of satellite communication on other cultures.

Some of the first domestic satellites used for educational purposes were the Application Technology Satellites (ATS) developed by NASA. School districts in outlying regions of the Rocky Mountains and in Appalachia were among the first to be served by the ATS system. Later, after being repositioned, the satellites served such areas as India and conducted demonstrations under the auspices of the United States Agency for International Development (USAID).

Rockets have traditionally been used to launch satellites, but in the future NASA's space shuttle will provide an alternative launch vehicle. Moreover, the shuttle can be used to service space stations, where it could dock and repair communication hardware.

Domestic satellite systems are receiving attention as the amount of data communication and the preference for satellite over ground-based networks increase. Western Union's Westar system consists of five satellites, and more are planned.

Radio and television stations wanting to extend their coverage areas are beaming their signals to satellites and then back to earth stations owned by cable companies, which in turn distribute the signals to subscribers. The first of these *superstations* was Turner Broadcasting's WTBS (originally WTCG) in Atlanta.

A number of American companies, following the lead of nations that already have operable systems, have applied for permission to operate direct-broadcast satellites (DBS). Services such as pay TV, high-definition television (HDTV), and master-antenna systems for apartments, motels, and similar dwellings would be offered through direct-broadcast satellites. Satellite business networks provide video, voice, and data-communication services for companies.

The growth of satellite communication opens up new political and economic issues. The transfer of information across international boundaries is one concern, although not as serious as once thought, because of the slow acceptance and insufficient accessibility of DBS systems.

OPPORTUNITIES FOR FURTHER LEARNING

Direct Broadcasting Satellite Communications: Proceedings of a Symposium. Washington, DC: National Academy of Sciences, 1980.

HOWELL, W. J., Jr., *World Broadcasting in the Age of the Satellite: Comparative Systems, Policies, and Issues in Mass Telecommunication*. Norwood, NJ: Ablex, 1986.

LUTHER, S. F., *The United States and the Direct Broadcast Satellite*. New York: Oxford University Press, 1988.

MATTE, N. M., *Aerospace Law: Telecommunications Satellites*. Vancouver: Butterworth, 1982.

TAYLOR, J. P., *Direct-to-Home Satellite Broadcasting*. New York: Television/Radio Age, 1980.

WINNER, L. *The Whale and the Reactor: A Search for Limits in an Age of High Technology*. Chicago: University of Chicago Press, 1986.

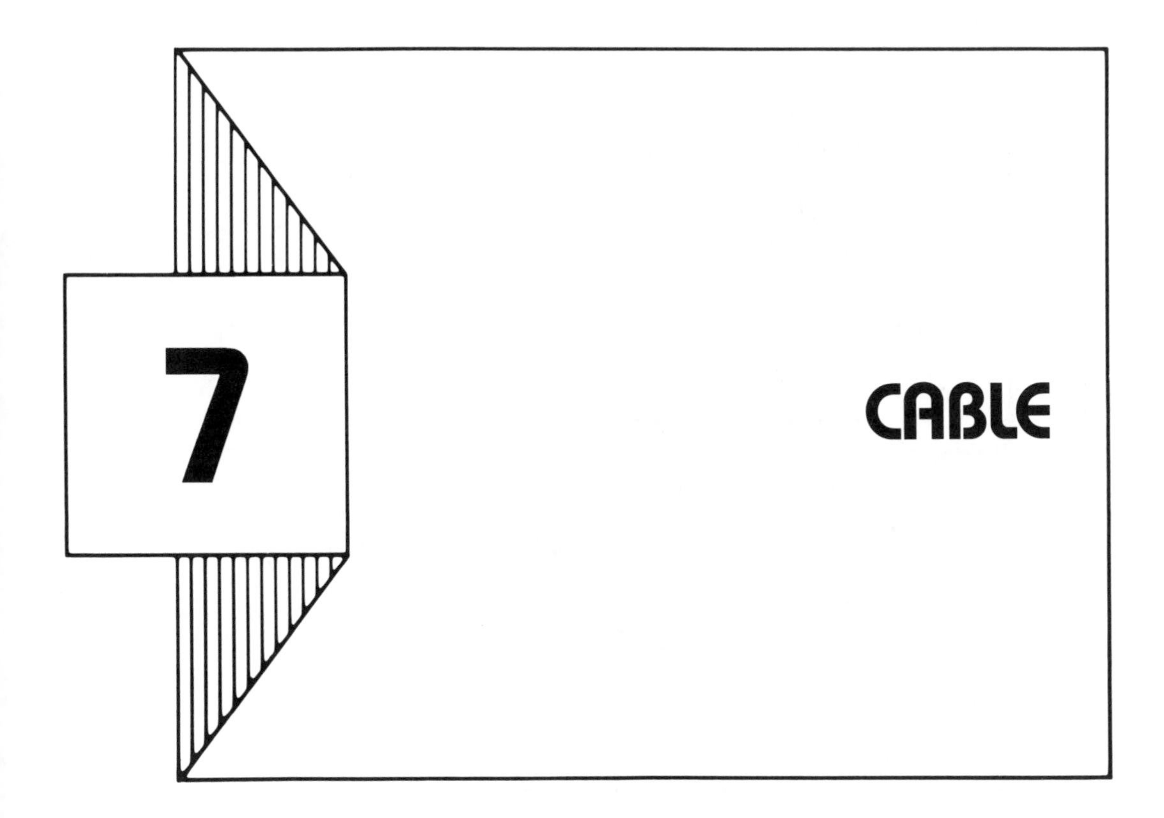

It is the late 1940s, and television is in its infancy. Large, bulky home receiving sets strain to tune in to the few television stations broadcasting the magic of pictures over the airwaves. If you live in a remote area, it takes a large, well-directed rooftop antenna to receive a local television station.

Still, the excitement of this medium does not dampen your spirits. Instead, you purchase numerous antennas. Newspapers are filled with ads claiming a certain model antenna will give you clear reception. Stores even sell "rabbit ears," two telescopic rods about three feet long connected to a base that sits atop your television set.

THE CABLE CONCEPT

Finally, someone realized there must be a better way. That way is community antenna television, commonly called *CATV* or *cable*. The concept is simple: Erect a tower with an antenna and locate the tower on a hill or mountain top (Figure 7–1). Then run cables from the tower to individual homes. The result is clear reception without the housetop clutter of antennas. Companies were created that owned the antennas and charged a fee to people who wanted to hook up.

Cable was especially attractive to people

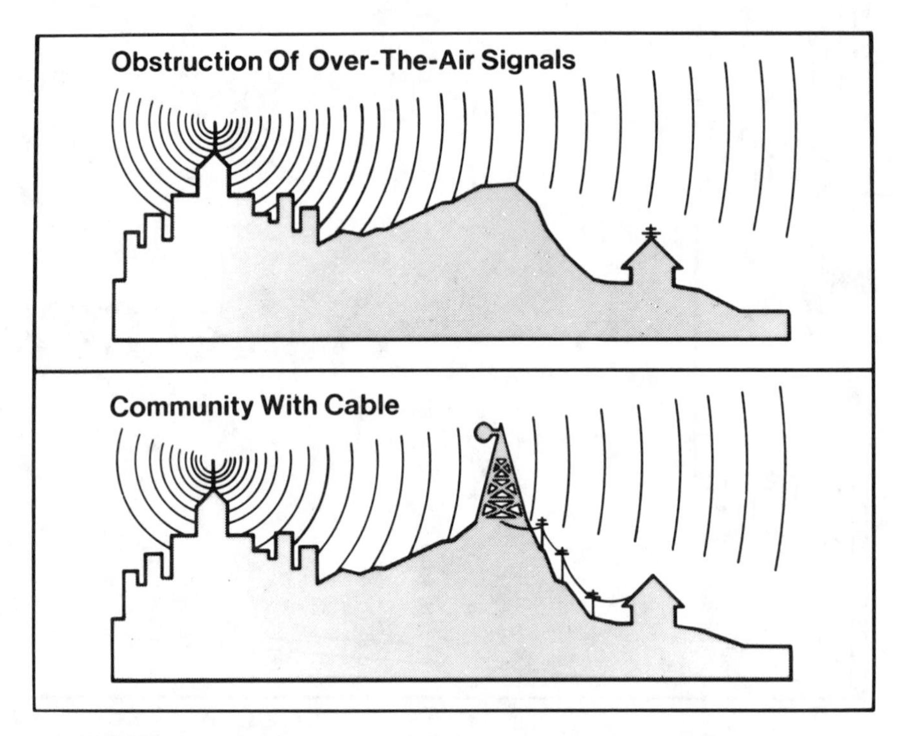

FIGURE 7–1 Cable systems originally developed as a means to improve TV reception of distant signals. Developed first in Oregon and Pennsylvania, cable is now found in virtually every major urban area, and its original function has been supplemented with two-way communication services ranging from public-opinion polling to electronic banking.

in hilly or mountainous country. Because both television video and audio signals are broadcast at a relatively high frequency, they travel in an almost straight line from the transmitter. When there is a mountain between the television station's transmitting antenna and a home antenna, it blocks these signals.

Today, cable is advantageous to people other than those living in mountainous regions. Over-the-air reception necessitates one-channel spacing between stations so that interference can be avoided; cable does not. Also, structures such as high-rise apartments, tightly spaced row houses, and clustered office buildings can obstruct even local television signals. Consequently, cable has become popular for urban reception as well as for distant signals.

CABLE'S BEGINNING

The beginnings of cable television can be divided into two distinct eras: (1) the 1940s when the original concept developed; and (2) the 1970s, which saw the birth of premium channels.

Starts in Oregon and Pennsylvania

There was a bit of friendly rivalry in the origin of cable television. Two individuals claim that famous first, one in Oregon and one in Pennsylvania. L. E. Parsons is credited with a working cable system in Astoria, Oregon, in 1948 as part of a test of reception for a new station that is now Seattle's KING-TV.[1] John Walson had a cable system operating in Mahoney City, Pennsylvania, that same year. Walson worked for Pennsylvania Power and Light in addition to selling television sets at his appliance store. After tiring of taking customers to the top of a nearby mountain to demonstrate television reception from Philadelphia, Walson erected a pole with an antenna and then strung "twin lead wire" down the mountain to his warehouse. From there he ran the line over Pennsylvania Power and Light poles to his store.

The Premium Channel Concept

The second era occurred in 1972 when the Federal Communications Commission modified its rules governing the ability of cable systems to import and redistribute distant television signals.

At about the same time, Charles Dolan and Gerald Levin of New York's Manhattan Cable discovered subscribers would pay for cable if they could receive programming not available over-the-air. The idea evolved into the Sterling Movie Network, Inc., which became Home Box Office (HBO), one of America's most popular premium channel services. Started in Wilkes-Barre, Pennsylvania, in 1972, HBO became a national distributor of programming using satellite technology to distribute programming direct to cable operators, who in turn sell it to subscribers.

THE SIZE OF THE INDUSTRY

Ever since those beginnings in 1948, cable has grown considerably. Although it is still a long way from being connected to all of the television sets in use, it has developed to the point where there are approximately 45 million subscribers in the United States. Pennsylvania has the most systems, and California the most subscribers. If you live in the mountain resort community of Kimberland Meadows, Idaho, the cable system is owned by the homeowners' association, has a small antenna system, distributes twelve channels, and has a handful of subscribers. If you live in a major metropolitan area, your cable system may be owned by a company that owns many cable systems, called multiple system operators (MSOs), and have more than a million subscribers. Cable has become an extremely powerful force in the industry.[2] Some systems have as many as 100 channels. Many operators originate their own programming, and some sell advertising or lease channels to other services, which pay a fee to the cable company.

COMPONENTS OF THE CABLE SYSTEM

To understand how a cable system operates, let us examine its components. The center of any cable system is the *head end*, a combination of human beings and technology (Figure 7–2). The human side includes the personnel who actually operate the system. The technical components include the *receiving antenna*, which receives the signals from a distant television station. The receiving antenna system is usually a tall tower on which a number of smaller antennas are positioned for receiving the distant signals. The tower can be located on a hill outside of town or a mountaintop far from a residential area. Installing the tower and antennas entails major construction and everything from lumber-cutting crews to giant helicopters.

The head end may also consist of tele-

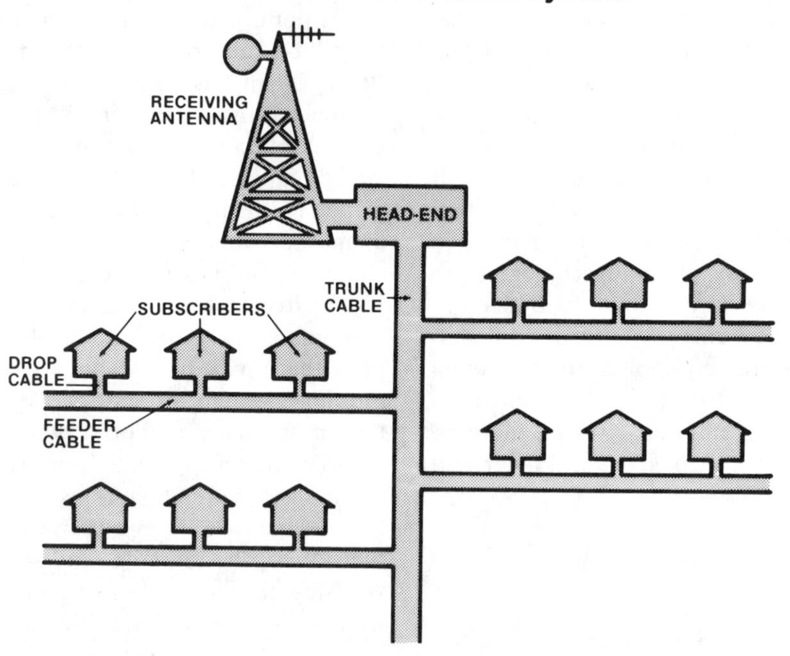

FIGURE 7–2 A basic cable-TV system consists of a head end where signals are received and processed, then fed through a trunk cable to feeder cables, which, in turn, are connected to drop cables and eventually home terminals.

vision production facilities such as cameras, lights, and other studio hardware, depending on the size of the cable system and how much local programming originates in the studio. The facilities can range from a small black-and-white camera to full-scale color production equipment. With all of this in mind, we will define the head end as the human and hardware combination responsible for originating, controlling, and processing signals over the cable system.

Another important cable-system component is the *distribution system*, which disperses the programming. The main part of the distribution system is the cable itself. The *coaxial cable* (Figure 7–3) used in most cable systems consists of an inner metal conductor shielded by a plastic foam. The foam is covered with another metal conductor, which in turn is covered by plastic sheathing. This protected cable may either be strung on utility poles or buried underground. The primary cable, the main transmission line, is called the *trunk cable*. It usually follows the main traffic arteries of a city, branching off into a series of smaller feeder cables, or *subtrunks*. The feeder cables usually travel into side streets or apartment complexes.

The actual connection to the home is made with a *drop cable*. This coaxial cable goes directly into the house, where it connects with a *home terminal*. The home terminal, in turn, connects directly to the back of the television set. In most cable systems, the home terminal is simply a splicing con-

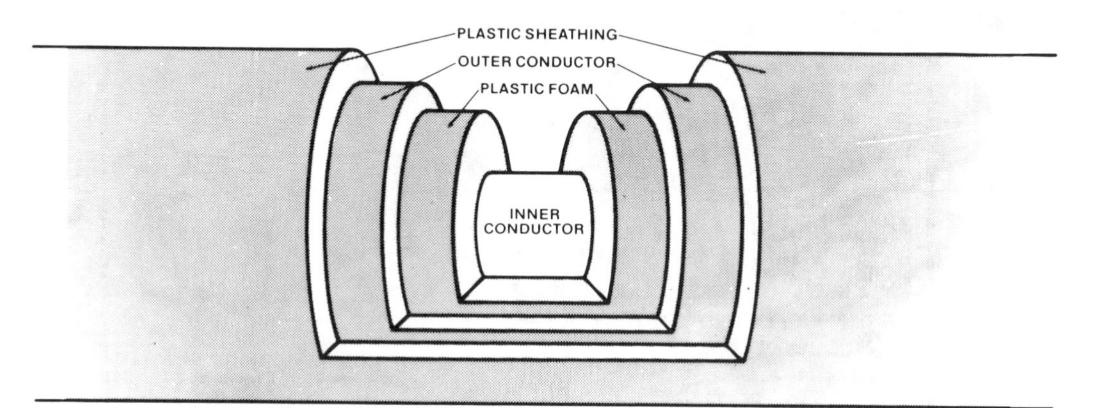

FIGURE 7–3 Although fiber-optic cable is in use in some areas, the basic hardware component of a cable system remains coaxial cable. An inner conductor is shielded by a series of outer metal and plastic layers.

nector that adapts the drop cable to a two-wire connector that fits onto the two screws on the back of every television set. In *two-way* cable, which we shall learn about next, the home terminal is more complex and may include a small keyboard. Some cable systems install these more sophisticated home terminals even if two-way cable is not yet operative. When it does become operative, the system and the subscriber will be ready.

TWO-WAY CABLE TELEVISION

Two-way cable systems, sometimes called *two-way interactive television*, permit the subscriber to feed back information to the head end. They can bring a wide variety of services into the home, and they are quickly becoming popular (Figure 7–4).

Programming reaches the two-way subscriber just as it would on one-way cable systems. The two-way subscriber, however, can communicate by means of a feedback loop. Feedback loops are generally of three types (Figure 7–5). One is a single cable used for both transmitting information to and receiving it from the subscriber. Another uses two separate cables. Incoming signals reach the subscriber through one cable, and signals from the subscriber return to the head end on the second cable. A third kind, the round-robin cable loop, is an adaption of the single cable but has separate drop cables.

PAY-CABLE CONNECTION AND FEE ARRANGEMENTS

Different arrangements exist for providing *pay-cable* service to subscribers. The three that follow, or combinations of them, are the most common.

Simple Fee

The subscriber pays a monthly fee to receive a special channel not offered as part of the basic service (which usually consists of up

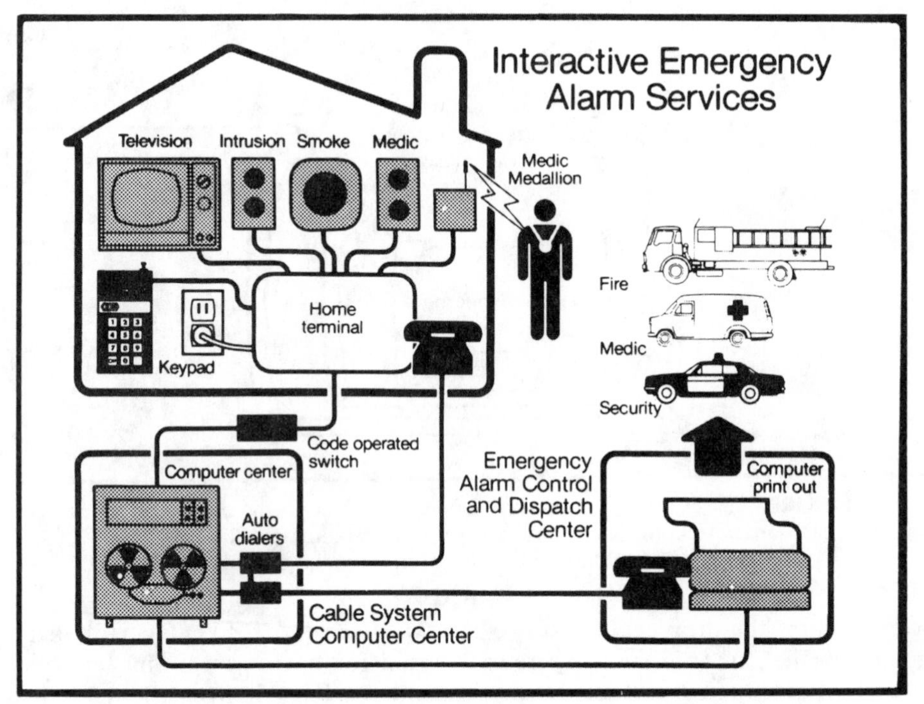

FIGURE 7–4 Interactive emergency alarm services are just one of the ways in which two-way cable provides the subscriber with more than just TV programming.

to twelve channels). In some communities the basic service is free and all of the income is generated from pay services. In other cases, as we have seen, subscribers pay a fee for the basic service and then an additional fee for another channel.

Tiering

An adaptation of the simple-fee service is called *tiering*. Here the cable operator lumps into tiers different channels or services and the subscriber pays an additional fee to receive the channels offered, by a particular tier. For example, a system may charge a monthly fee for the basic service and an additional monthly fee for the first tier of service beyond basic service. This first tier might include some of the satellite superstations we discussed in Chapter 6. It might also include an all-sports channel and a channel that shows first-run movies. Multiple tiers can exist where the programming is available. A second tier might include a channel showing adult movies and a local educational channel. A special rate may be available for those who want to buy everything the system offers.

Pay-per-View

A third pay-cable arrangement is *pay-per-view*.[3] Pay-per-view charges the subscriber on a per-program basis. The system operates in most markets where two-way cable capacity exists, and the subscriber can automatically choose to watch or not watch a special program, such as a movie or sporting

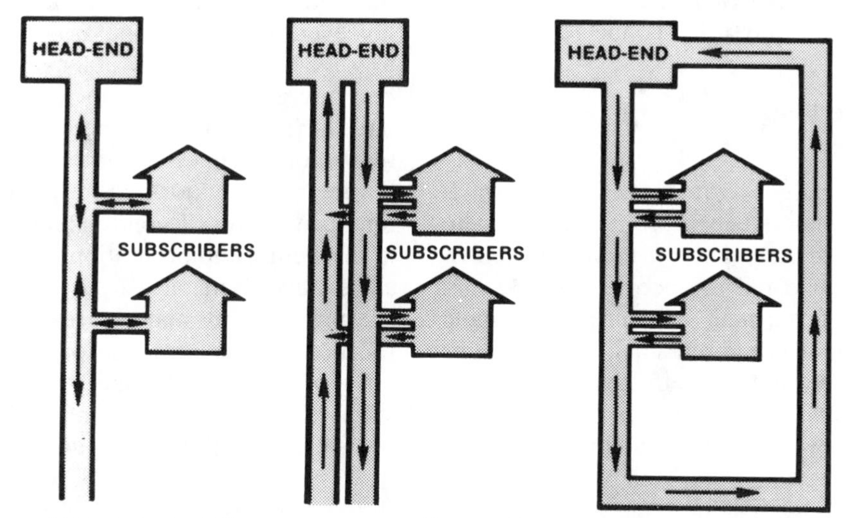

FIGURE 7–5 Different two-way cable installations can provide different system capabilities.

event. The subscriber enters a choice through the home terminal attached to the television set. The choice is automatically registered in the computer at the head end, which then channels the program to the subscriber, entering the additional charge on the subscriber's bill. Pay-per-view also operates in noninteractive one-way systems, but subscribers must call in their choices in advance and then are billed for their selections.

Cable Radio

The principle of cable radio is the same as television: distant station programming is cabled into a local community. A deviation from this concept is to program local stations as part of a promotional package for the station.[4] For example, a local station may install a television camera in its studio and air its programming live over the cable system where the viewers can watch the announcer and disc jockeys work. Such cooperative arrangements can benefit both the station and the cable company with an increase in the audience *share*.

CABLE INTERCONNECTS

Cable systems serve individual communities, but many systems are connecting with one another in order to capture some of the advertising dollars that in the past have been spent exclusively on radio or television.

Types of Interconnects

Interconnects are of two kinds: "hard" and "soft." Hard interconnects are cable systems physically connected by wires or microwave (Figure 7–6). Soft interconnects are associations of cable operators who work together to attract advertisers and air commercial programming but whose systems are not physically connected.

The Advertising Advantages

Interconnects developed as a way of attracting to cable systems those advertisers who were purchasing time on radio and television but who felt the accounting and mechanics of reaching audiences by buying time on many different small cable systems were simply not cost-effective. Moreover, independent cable operators working alone did not always have the ability or promotional and marketing expertise to attract major advertisers to their systems.

Although cable advertising has still not reached its potential, by working together, however, many cable operators are working to solve both of these problems.[5] First, with one buy an advertiser can reach the audience of the combined systems, an audience total more in line with the large audience numbers that major-market television systems can deliver. Second, talents are shared in marketing and promotion. Furthermore, since advertisers are accustomed to purchasing an entire market area, such as that reached by television stations, fragmented buys of advertising time across different suburbs served by different cable systems was simply not practical. But with intercon-

FIGURE 7–6 Microwave interconnections permit cable systems to link together and provide programming to larger audiences from a single head end installation.

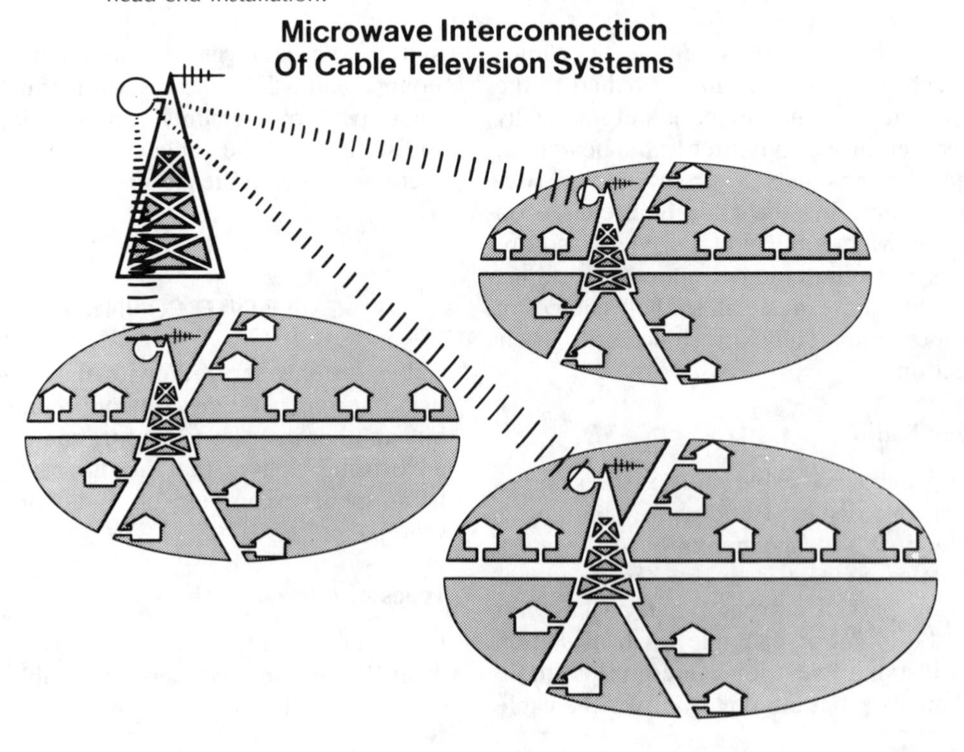

nects, advertisers can buy time on all of the systems serving a geographic area and thus receive blanket market coverage.

THE CABLE FRANCHISE

Whereas the head end is the hardware core of the cable system, the cable *franchise* is the political and economic core. Unlike radio and television stations, which are awarded a license to operate by the Federal Communications Commission, an arm of the federal government, cable systems are awarded a franchise by local government.

Franchises differ from community to community, but most are for periods of fifteen years and specify certain requirements the cable operator must meet when the system is constructed and operating. The franchise is a valuable commodity, and in major markets it is hotly contested. Construction costs for a system are great, but the promise for revenue is even greater. Franchise fights sometimes result in lawsuits ranging from charges of antitrust violations to allegations of bribes of city officials.

The Operator's Promises

In any franchise the cable operator must enter into a contract with the grantor of the franchise. Theoretically, if the contract is broken the cable operator loses the franchise and another operator takes over the system. Under such circumstances it would seem that what a cable operator promises will be delivered and that what the city or municipality expects to get will in fact be provided. In most cases, however, it is not that simple. In a fiercely contested franchise area, cable operators, in an effort to capture the franchise, may simply promise more than they can deliver.

Unfulfilled promises are not always intentional. Nevertheless, they happen. For example, an operator may not be able to meet a construction schedule. Running cables under streets and obtaining rights-of-way from utility companies are just two of the obstacles that confront an operator. Even after the system is built the projected number of subscribers may not be reached; this forces a cash-flow problem that prevents other services from being offered.

Because a sizable investment is made in the early stages of the cable system, simply throwing out one operator so another can take over cannot be accomplished without a major legal fight. Even then there is no assurance that another operator could provide any better service. Moreover, the company that owns the system may have a sizable group of local investors who carry local political clout.

The Municipality

A second element of the cable franchise is the municipal pride that comes into play in franchise deliberations. Cable franchises tend to create emotional issues. First, politicians view the awarding of the franchise as more than a simple decision. Some view the franchise as a channel of communication for reaching voters. Second, because cable is a medium of communication that will compete with existing media, such as radio and television, it makes news. Third, local officials are sometimes less concerned with what their own community will receive than with what a neighboring community has already received. The desire to be "one better" shows up in cable franchising just as it does elsewhere in society. Unfortunately, in trying to obtain the best system, some communities lose sight of what is a workable system.

The Uncertainty of Emerging Technologies

Added to the complexity of any cable franchise is the technological wonderland that is embodied in the communications industry. Today such emerging technologies as fiber optics, high-definition television, direct broadcast satellites, teletext, videotex, and interactive two-way cable promise much but are untried. Communities deciding on a cable franchise must sort out what they want the system to do and what technology will deliver the service, and then try to answer some questions. Will the technology work? Will it become outmoded? Will the public accept and pay for it?

The Role of Consultants

In an effort to sort through the barrage of technical information received when a request-for-proposal is issued, many communities hire a consultant. Consultants can provide a valuable service, since most politicians know very little about cable. Also, a consultant tends to isolate the city officials from what is still a political process.

Some cable operators hire local public relations consultants to help their company keep a favorable image in the community. In tough franchise fights, play sometimes gets dirty, and more than one cable operator has used unfavorable publicity against an opponent in order to sway public opinion and local officials. As professionals who can create a positive image as well as defend against smear tactics, public-relations consultants find their services in demand.

Political Pressures

The political climate can affect a franchise in different ways. For example, a rent-a-citizen lobbying strategy may be employed. Rent-a-citizen is a process in which a cable operator wanting to capture a local franchise provides a financial incentive to a well-known citizen who can make contacts and lobby for the operator. The "citizen" is given shares of stock in the company with the option of selling back the stock at a profit if the company is awarded the franchise.

Unless the citizen is a politician holding office and voting on the franchise, there is nothing illegal about the practice. Nonetheless, it generates controversy. One opinion is that such tactics insult city officials. Another is that without such political clout a cable operator does not stand a chance of being awarded a franchise. In either case, the operator has little to lose. The given-away stock is practically worthless if the franchise goes to another company. On the other hand, if the operator wins the franchise, the shares can be bought back.

Most of us at some point in our lives will live in a community that already receives cable or is considering the installation of a cable system. Knowing how to analyze the forces at work in the awarding of a cable franchise is a necessary step in becoming a responsible member of a community.

CABLE SERVICES

Cable systems today offer more than they did in the late 1940s, when the medium was in its infancy.

Superstations

Without cable, *superstations*, which are beamed to satellites and back to cable systems for distribution to subscribers, would not exist, much less have the national audience that some of them enjoy. Superstations, such as Atlanta's WTBS and Chicago's WGN-TV, provide the attractive, inexpensive programming that many cable systems need in order to round out a basic service and complement the local stations that are available. Superstations, additional channels, and improved reception are often the programming choices that prompted a subscriber to order cable in the first place. Once successful contact is made between the subscriber and the cable company, there exists the opportunity for selling the subscriber additional tiers on the system.

Basic Services

In addition to stations in one's own viewing area, a number of basic services, received as part of the subscriber's basic monthly fees, are offered to many cable subscribers. For example, The Weather Channel offers 24-hour weather information including local forecasts. Special regional forecasts and extended local coverage of storm conditions are also provided. The Discovery Channel specializes in programming with a heavy emphasis on science, especially nature and animal fare. A&E, the Arts and Entertainment channel, programs fine-arts programming and feature-length motion pictures. The Nashville Network concentrates on country music including special performances and interviews with country-western entertainers. Turner Network Television (TNT) programs motion pictures but also supplements these with numerous original productions. Regional channels may also be available such as a regional sports network that carries games from a specific part of the country.

News and Public Affairs Programming

Expanded news and public-affairs programs are offered as basic service by a number of cable operators. C-SPAN, the Cable-Satellite Public Affairs Network, provides commercial-free public service programming that provides live coverage of Congress and other events, which show the raw workings of the political process. The Cable News Network (CNN) is a news and public affairs channel, and the Headline News Service airs 24-hour news on the half hour with stock market reports scrolled over the video during the main part of the day.

Premium Channels

Cable expands the entertainment and other programming choices provided by radio and television stations.[6] For example, Home Box Office (HBO), the first pay-cable service, offers movies, features, and its own productions. Showtime offers similar programming. The Movie Channel programs full-length feature films with sprinklings of star profiles and capsule features. The Disney Channel programs primarily family and children's programming featuring Disney movie and television features.

Instructional Television Through Cable

Cable companies often contract with a local university to fill one of its cable channels with instructional programming. Under such an agreement the university can offer a complete curriculum that can be taken in the living room instead of in the classroom. The continuing-education and outreach functions of colleges thus gain a whole new perspective through cable. Students can enroll in one college while taking courses by cable television from another.

Cable is also becoming a means of communication between individuals and institutions that in the past have been distant and unable to communicate. The public that looks inside a school board meeting, watches teacher-training programs, and hears school officials discuss issues learns more about and participates more in the institution. This social and political responsiveness in turn permits the school officials to be more accountable to those forces affecting policy.[7]

Programming Local Arts

Commercial broadcasting has not been able to program the arts successfully, for two reasons. The first is the lack of viewers for such specialized programming. The second, caused by the first, is an inadequate profit foundation upon which to produce and program these arts. Although public broadcasting has inched toward this type of programming, it still must produce programs that appeal to a mass audience. Cable provides an alternative. Now local fine-arts programming can be produced and funded on a local level.

Fine-arts programming provides another benefit for cable. If a city has a good local symphony orchestra, for instance, it is not unusual for a group of local sponsors of the arts to contribute to its development with considerable zeal. This can spill over into the cable system. The local symphony can play to its audience over cable even during prime-time hours. On a commercial broadcasting station the only profitable time to air such programming might be in the wee hours of the morning.

Shopping Services

The growth of catalogue sales has spurred merchants to look into other direct-marketing techniques, including the use of cable to deliver electronic catalogues of merchandise. Shopping channels display goods and services that subscribers can purchase either by keying in their account information on two-way cable systems or by telephoning direct orders for merchandise seen on one-way systems. Washing machines, encyclopaedias, and cookies are just some of the products that have been advertised. In another system, a cooperative arrangement exists with a department store, which supplies the merchandise; the cable company, which distributes the channel; and a credit card, which handles the billing.

Although cable has been available since the late 1940s, the great potential for cable programming is just now beginning to be felt. Technology that permits more channels, the increasing number of subscribers, who form a financial base for cable's development, and an industry that is beginning to mature—all have improved cable as a medium. At the same time, many ambitious cable ventures backed by big names in the entertainment industry have not succeeded. CBS's fine-arts cable programming was stopped when CBS management determined after a short experimental period that it would not be profitable. This and similar examples show that even though technology can offer multiple channels, only so many channels can find an audience large enough to make them profitable.

LOCAL POLITICAL PROGRAMMING

In addition to the services and cooperative arrangements just discussed, cable offers unique services to local politics.

Broadcasting Public Meetings

Cable gives the public access to even the smallest governmental body. Television cameras and microphones can be placed in the audience of a school-board meeting, a city-council session, or a zoning-commission meeting. Those with an active interest attend the meetings, and those with a borderline interest watch from home.

Candidate Access

Commercial broadcasting is limited in the amount of advertising time that can be given to candidates for political office. It is just not economically feasible to turn all of a station's available advertising over to the politicians. The reason for this is that the FCC has decreed that political advertising be sold at the lowest rate the station charges to an advertiser. Some stations simply appropriate an amount of free time to candidates and dispense with selling political advertising altogether. In addition, election laws now restrict campaign financing, and thus the budgets that used to produce lucrative television campaigns.

Cable television provides a number of alternatives to these dilemmas. First, in small communities without a television station, cable TV permits candidates to reach the voters through a visual medium. Second, most local-access rates permit even the poorest candidates to campaign on the local cable channel. Third, when commercial access is not available, cable access may be. Fourth, cable permits the candidates to reach highly specialized audiences not normally reached by commercial television.

CABLE'S LOCAL-ACCESS CONCEPT

Most cable systems allow any member of the public access to a cable channel. Some larger cable systems also provide equipment at a nominal cost for people to produce local programming. Local-access programming is not the glittering lights of Hollywood. Nor is it the elaborate production studios of a major network. It more than likely materializes as one program did, on an October day in a small community of 8,000 people and 1,500 local cable subscribers. It is evening, and on a drive in the country a local resident spots a poster tacked to a utility pole: "Halloween Parade—Everyone Welcome—6:30 P.M., October 31, The Fire House Parking Lot." The perfect opportunity for local-access programming.

This is what local-access cable is all about. It is the grass-roots side of television, one that is not possible to incorporate in standard broadcasting stations.

LOCAL ACCESS AND SPECIALIZED AUDIENCES

One advantage of cable is its ability to reach specialized audiences. For example, cable-company hardware makes it possible for a series of "mini-hubs," which are local origination points along the cable route. A cable program can thus be limited to a few city blocks. This highly localized access gives programming opportunities to small neighborhoods having similar ethnic or religious backgrounds or other common ties.

Another example of local-access programming to specialized audiences is programming for the elderly. Cable can be connected to housing complexes for the elderly, such as nursing homes. It is popular with the elderly because of the added leisure time that retirement permits. Local-access cable programming permits the elderly to communicate with one another, to alleviate loneliness, and to feel more a part of the community. Special programs about social security benefits, Medicare information, transportation, and shopping bargains for senior citizens are all possible through local-access cable programming.

CABLE VERSUS THE BROADCASTER

Given cable's ability to carry broadcast messages beyond the coverage area of the over-the-air station, it would seem that broadcasters would support cable's efforts more. But the two have seldom coexisted harmoniously, and at times their opposition has been bitter.

The Broadcaster's Arguments

Place yourself in the position of a commercial broadcaster in a medium-sized community.

First, the importation of broadcast signals slashes your audience. You used to be able to offer a substantial audience to advertisers for a healthy profit. Second, as a broadcaster you are providing that service free of charge. On the other hand, a large interconnection of cable systems can successfully negotiate exclusive programming with a college football team, for example, and charge viewers to see the games. You, in turn, because of the cable systems' exclusive contract, would not be permitted to carry the game. Third, economics usually dictate that cable, and especially pay TV, be installed only in densely populated areas, where most of the potential subscribers are. Yet you, while competing with cable, are also serving the rural public, regardless of the population density.

Your fourth argument—a more general one—is that cable has developed as a parasite industry of broadcasting and is now trying to compete with it. Fifth, you argue that cable has been favored by the FCC with a general relaxation of rules, which permits it to compete more favorably with you. You compare this to fighting with one hand tied behind your back while your opponent's hands are free. Sixth, you claim that since cable companies have the ability to interconnect their systems, the theory of local accountability and service has been destroyed. Finally, while you operate in a limited spectrum space, cable can carry large numbers of channels, many of them programmed by the local cable systems themselves.

Cable's Rebuttal

Now put yourself in the cable operator's position.

First, you contend that over-the-air broadcasters are severely restricted in serving their viewing audiences, since even in the largest markets only a few stations can operate within a limited-spectrum space. You feel that cable serves its audiences far better with its variety of channels. Second, you point out that precisely because of their limited-spectrum space, broadcasters have made giant profits. You state that those profits are sometimes at the expense of viewers, who long for more innovative, though perhaps more costly, programming. Third, when broadcasters criticize cable's exclusive contracts with program distributors, you remind them of their exclusive contract advantage with the major networks. Fourth, although commercial broadcasters answer to only one master, the FCC, you sometimes face regulatory control by three levels of government—local, state, and national.

All of these arguments, in varying detail and intensity, are used throughout the broadcasting and cable industries. They have been presented in cloakrooms to members of Congress, at special legislative hearings, and in public-relations literature. Meanwhile, consumers are living in both worlds, unaware that the future is bound to bring some dramatic changes in how they receive their daily television fare.

Broadcast–Cable Cooperative Arrangements

Despite the atmosphere of competition that exists between cable systems and over-the-air broadcasting, some cooperative ventures have developed. For example, CNN Headline News in some markets contains local headline summaries of news featuring the anchorperson from a local broadcast station. Headline News gets the benefit of a localized insert in its newscast and the anchor gets publicity for the local station. In some markets the cable system itself will enter into an agreement with a local broadcast station and air special news feeds and features from the station in addition to carrying the station's regular programming. In addition to publicity, programming expertise from the broadcast station benefits the cable system, which, in most cases, still sees itself as an information carrier instead of an information provider.[8]

REGULATORY ISSUES

Despite the respectable growth and impact of cable, it is in its infancy as a social force. Standard broadcasting, both in size and influence, makes cable small in comparison. Still, cable contends with significant regulatory issues. As we have seen, cable and commercial broadcasting are subject to different levels of regulatory control. A cable system can find itself regulated by three systems—local, state, and federal. Standard broadcasters answer only to the FCC. Moreover, with cable, some regulations conflict with each other, creating a maze of court cases ranging from rate structures to local-access programming.

Another problem cable faces is its relationship with laws indirectly affecting its operation. Consider the case of local access. A local community group decides it wants to use the local-access channel to broadcast the school-board meeting live and in its entirety. The state's open-meetings law permits public access to all public meetings. But the school board says no. The school board's attorney contends that cable television cameras are not persons and can therefore be barred. The cable company reminds the school board that it permits the local television station to film and videotape portions of its meetings. In fact, when major issues are being discussed, the board even allows the station to broadcast live reports. But the school board replies that cable is not considered a bona fide news-gathering organization and does not come under the protection afforded a free press.

This is just one of many gray areas that cable faces. Many laws, such as open-meeting statutes and shield laws for reporters, have yet to define their applicability to such situations. Until they do, cable has an uphill climb for its legal identity.

THE ECONOMICS OF CABLE CONSTRUCTION AND OPERATION

The future of cable and how we interact with and use it is tied directly to its economic aspects. Thus it is important to understand these factors. You may find yourself voting in a local referendum on whether to raise the rates charged by your local cable system. Or your community may determine whether the local cable company should install its cable underground or, more economically, attach it to telephone poles. It may even decide what supplementary services, such as electronic funds transfer or bank-from-home, should be added to the local television fare. To make intelligent decisions, you will need to understand the economic forces affecting cable.

The Capital-Intensive Factor

Cable is a *capital-intensive business*. By capital-intensive we mean that maximum costs occur immediately. A standard radio or television station can go on the air with a minimum amount of equipment—some of it of marginal quality—and a skeleton staff.

Cable does not enjoy this luxury. Cable systems are designed for permanency, and the system must be taken to the total potential audience before it can even begin operation. Therefore, hiring skilled technicians, installing miles of cable, constructing elaborate antenna systems, and purchasing head-end equipment must all be done before the first subscriber is hooked on.

Construction Costs

Starting a cable system requires construction of the head end, production facilities, distribution system, and the purchase of subscriber home terminals. Costs for underground construction of the distribution system can easily cost hundreds of thousands of dollars per mile in crowded metropolitan centers. Even above-ground pole-attachment systems are expensive. The location and type of antenna can also raise the cost. An antenna that must be constructed on top of a mountain is going to cost much more than a tower built in a level field outside of town. All of this determines how much the subscriber must be charged, how long it will be before the cable system makes a profit, and where financing can be obtained.

Operating Costs

Construction costs are followed by the costs of operating the cable system. The main cost is system maintenance. Although cable operators usually install the best possible equipment for long life and maintenance-free service, nothing is infallible. Breaks due to storms and equipment repair at the head end are just part of the regular maintenance schedule.

Second, a subscriber cannot simply turn on a television set and tune to the cable channel without first having the set connected to the cable. That requires a service call, and service calls are responsible for much of the cable company's personnel time.

A third expense is vehicle operation. Unlike a radio or TV station, whose entire operation may be under one roof, the cable company can literally be spread all over town. In larger markets, servicing this territory can require a fleet of trucks, many requiring aerial ladders. Future developments in technology, however, will permit more and more switching and connecting functions to be done at the cable's head end.

Utility-pole and underground-duct rentals are a fourth large operating expense. If pole attachments are used, the cable company must rent them from the telephone company. When a major cable company rents hundreds of thousands of poles, the cost is considerable.

Fifth, local municipalities may charge franchise fees—money the cable company pays the local government for the privilege of operating.

Sixth, cable companies are charged copyright fees by program distributors.

Seventh, although construction of the antenna and other head end facilities is usually figured into the construction costs, the expense of bringing in distant signals may require separate lease agreements with telephone companies or private microwave carriers.

Eighth, local origination costs can also be substantial. Here a local cable company can incur some of the same studio expenses that a small television station does. Yet despite these construction and operating costs, cable remains a growing and profitable industry.[9]

INCOME FOR THE CABLE SYSTEM

Income from the cable system can be classified into four broad categories: (1) subscriber's monthly rental fees for basic service; (2) pay-cable fees for special programming, much of it exclusive, which usually consist of either a set charge above the monthly rental fee or a per-program assessment; (3) revenue from such special services as at-home banking; and (4) revenue from advertising.

For the cable company to make a profit, it must receive income in the form of subscriber fees. The number of subscribers and the amount of the fee are the key components. Additional fees can be charged for such services as pay-TV programs. As the services increase, the subscriber fee increases. At some point subscribers resist.

Another source of income is advertising. Cable companies sell advertising in the same way that standard broadcasting stations do. Moreover, the interconnection among cable systems through satellite and microwave links makes the concept of a cable network a reality. In such a network several cable companies carry the same programming and derive income from sponsors who buy space for advertising that will be seen throughout the network. As cable networks develop, a larger percentage of cable's income will be from advertising.

APPROACHING THE PROFIT MARGIN

A cable system has no set formula for success. But the enterprising operator does follow some basic guidelines. Among them is the delicate balance between the amount of money that can be charged to a subscriber and the number of subscribers needed to make the system profitable.

For example, if a cable system has 1,000 subscribers, each of whom pays a $30 monthly subscription fee, the total income would be $30,000 per month (if there is no income from other revenue sources). Now assume that the cable operator decides to increase the subscription rate to $35 per month. Theoretically, this raise would net the company $35,000 per month. But what if the rate increase drove away half of all subscribers? The income to the cable operator would then drop to $17,500 per month ($35 × 500), a significant loss to the cable operator.

THE PROBLEM OF DISCONNECTS

In wiring the home of a subscriber, a cable company incurs an expense. Vehicle and technician time, the time to contact the subscriber and sell the service, the time to set up the subscriber's account—all cost money. Thus, it is a serious matter when even one subscriber cancels service.[10] In many systems, the lack of an automatic connect and disconnect system means sending the technician and the truck back to the subscriber's home, disconnecting the service, and instituting another set of bookkeeping chores to take the subscriber off the account records.

Overselling Services

Disconnects cut into the profits and raise the operating costs of the system. Part of the problem stems from the cable company trying to sell too much. A customer may be enticed, cajoled, even pressured into buying as many services as possible. Some cable companies hire crews of professional, straight-commission door-to-door salespersons to canvass neighborhoods and sign up subscribers. High-pressure tactics can close a sale but entail the risk that the subscriber will call and cancel upon realizing that he or she does not want or cannot afford the service.

Sales and Marketing Strategies

To solve the disconnect problem, cable companies are becoming more sophisticated in their sales and marketing techniques. By not trying to sell as many services, or by holding customers to fewer services so they do not feel they have "bought the store," companies can reduce the number of eventual disconnects. Another method is to package services in such a way that buying a lesser number of services does not represent much savings. A more expensive basic fee with less expensive pay services is one way of accomplishing this. But pushing the basic service fee too high can result in a loss of premium-channel business, and if public opinion turns against the system because of poor service, mass disconnects of even the basic service can occur.

The cable company can also arrange to have retail establishments sell cable service at the same place that videodisks, videotape recorders, home computers, and other consumer electronics products are sold. The atmosphere is different with a retail salesperson than with a door-to-door salesperson. The buyer has most often decided to talk with the salesperson, and the salesperson is usually someone who understands electronic hardware and software.

Solutions to Disconnects

Part of the solution will come from the cable-programming services themselves. To get a hold on the market many cable-programming services have been trying to be all things to all people. Although they are touted as having an exclusive programming identity, many similarities exist among them. Even the pay channels offering first-run movies have come under criticism for their duplicate programming of only the choice movies that everyone wants to watch. As more and more cable services find their identity in the marketplace and less and less duplication exists, subscribers may develop channel loyalty and be less likely to disconnect.

REACHING THE NONSUBSCRIBER

Some potential subscribers choose not to have their home hooked up to cable. Estimates vary but industry sources claim that between 40 and 50 percent of the households that could be wired for cable are left unconnected because the subscriber simply chooses not to buy the service. A negative attitude toward watching television is one reason for the high proportion of unconnected households. Another is that families with small children sometimes worry about being able to control the viewing habits of youngsters who may be exposed to adult programming. In other cases the basic rates for cable have become so high that people cannot afford to be a subscriber or simply feel cable service is not worth the price. Clearly some of the problems associated with reaching the *untouchables* can be solved by better advertising and marketing techniques. But deep-seated resistance toward television is more difficult to overcome.

CUSTOMER SERVICE

Regardless of how successful a cable operator is in marketing, the business will suffer if customer service is not satisfactory. Horror stories about customer service abound in the cable industry. It is cable management's nemesis. Poor customer service has many origins. It can start at the level of the person who answers the telephone. Many cable operators hire untrained employees to handle customer inquiries. Such persons cannot articulate clearly, use improper grammar when answering the phones, are discourteous, lack adequate information, and in general are a disaster to the business. Other companies install computerized answering services that are worse than direct personal contact since the customer usually has to sit through a series of telephone touch-tone choices and then becomes irritated when they do want to talk to someone but can't bypass the computer. The next call the customer makes may be one of complaint to a city council member or the city manager's office. Until cable companies manage to hire and train interpersonally competent people skilled in customer relations, both the business and image of the cable industry will continue to suffer.

SUMMARY

Cable began in the 1940s in Oregon and Pennsylvania as a means of bringing distant television signals to outlying communities. Antennas were installed on mountaintops and a cable led to the community. Individual households paid a fee to have their television connected to the cable. Today, the system still operates on the same principle.

A cable system contains a head end, the combination of people and hardware responsible for originating, controlling, and processing signals over the cable system. A trunk cable leads from the head end to main feeder streets in a community and subtrunks feed to smaller traffic arteries. A drop cable runs from the subtrunk to the household, where it is connected to the television set through a home terminal. Cable systems are of two types: one-way and two-way.

Many cable systems are joining together in cable interconnects. These permit broad coverage of a market and are especially appealing to advertisers, who by making a single buy can reach a larger number of viewers, much the same way they would with television or radio time.

At the heart of the cable system is the cable franchise. Landing a franchise is an effort in political strategy, technical knowledge, and promises. In some communities the cable operator promises more than can be delivered in an effort to win the franchise. Surrounding any franchise is uncertainty as to how new technologies will affect the system's operation and profits. Consultants, local lobbying groups, and investors are all part of the franchising process. Cable performs the same function of improved reception and increased channels as it did in its infancy. Since the 1940s, however, many new services have been added. These include entertainment, news, sports, and weather channels. Instructional television and local arts programming are also available on some cable systems.

Local political programming and cable access are two additional services. Each finds a highly specialized audience. Local-access programming is also becoming more common as cable systems grow and the number of subscribers increase, permitting an audience for local, highly specialized programming.

Fundamental issues still divide broadcasters and cable operators. Both economic and political, the split may be healing as more and more broadcasting stations and cable companies are owned by the same communication conglomerates. Moreover, both cable managers and broadcasters are realizing that they face similar problems, which can best be addressed through a cooperative instead of an adversary relationship.

Among the regulatory and economic issues facing cable are the different levels of control exercised by government at the local, state, and federal level. Moreover, cable is a capital-intensive industry and therefore sustains formidable costs before it can deliver service to a community.

Managing a cable system means dealing with such problems as disconnects, nonsubscribers who may have negative feelings toward television or cable, and poor customer service.

OPPORTUNITIES FOR FURTHER LEARNING

GARAY, R. *Cable Television: A Reference Guide to Information*. Westport, CT: Greenwood Press, 1988.

HEETER, C., and GREENBERG, B. S. *Cableviewing*. Norwood, NJ: Ablex, 1988.

HOLLINS, T. *Beyond Broadcasting: Into the Cable Age*. London: BFI Publishing, 1984.

KAATZ, R. B. *Cable Advertiser's Handbook*. Lincolnwood, IL: National Textbook, 1985.

PARSONS, P. *Cable Television and the First Amendment*. Lexington, MA: Lexington Books, 1987.

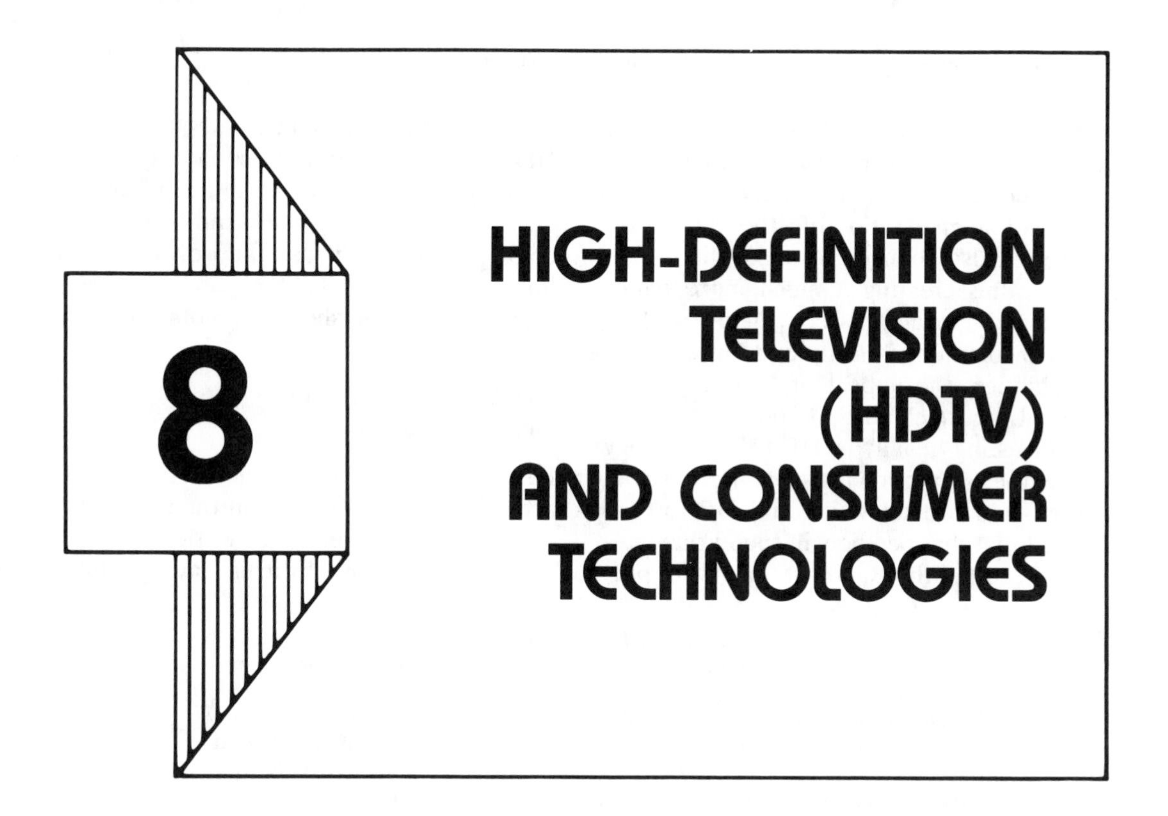

Other technologies are calling attention to the full potential of telecommunication. Investment capital is finding its way into these technologies, and terms such as *high-definition television (HDTV)*, and *cellular radio telephone* are becoming as familiar as the words "radio and television." We will begin this chapter by discussing high-definition television, which promises to affect the very way video is produced and presented. We will then divide other consumer technologies into distribution systems, telephone systems, radio services, television receiving and transmitting technology, and consumer electronics.

HIGH-DEFINITION TELEVISION (HDTV): THE CONCEPT

Imagine a television screen 8-feet-by-8-feet with picture clarity equivalent to that produced by the best motion-picture studio and without the grainy effect seen on many large-screen televisions. High-definition television (HDTV) provides that level of picture quality.

The Technology and Consumer Appeal of HDTV

The standard United States TV transmission system of 525 lines cannot effectively be enlarged without distortion. HDTV, however, uses a system of more than 1,000 lines and achieves picture-like quality. The width-to-height ratio of the television screen is also changed to make the screen wider, thus creating a much more panoramic and appealing effect (Figure 8–1).

With the added clarity and large-screen capabilities, the appeal of HDTV is substantial, so much so that some companies are investing heavily in HDTV technology under the assumption that the difference between standard television and HDTV will be so great that people will spend the extra money, in the thousands of dollars, to purchase the equipment necessary to receive HDTV. The difference, some say, is so great that it can be likened to the difference between watching black-and-white and color television. Thus, the revolution in home viewing habits that will occur with HDTV is nothing short of revolutionary.

HDTV Regulatory Issues and Standards

Part of the difficulty in the development and acceptance of HDTV lies in the regulatory framework of different countries. For example, while some antitrust laws prohibit

FIGURE 8–1 High-Definition Television (HDTV) uses a height-to-width screen ratio that is different from standard television. Along with significantly greater picture quality, the screen dimension creates a more panoramic effect, which is ideal for motion pictures that were originally filmed for viewing in theaters but are now seen on TV. When televising sporting events, HDTV permits a larger portion of the playing field to be seen. Shown is the Panasonic 20-inch HDTV color monitor.

cross-company cooperation, the development of HDTV technology demands more financial resources and brain power than can usually be found in any one company's research and development arm.

In addition, trying to achieve a world standard in HDTV technology is further complicated by the different broadcasting systems worldwide and the different companies involved in HDTV research. Gradually, however, these obstacles are expected to be eliminated, a world standard will be agreed upon, and different companies and countries will be able to compete in the manufacturing and sales of HDTV technologies.

The television technology currently in use for standard television transmission must be integrated into policy and technical considerations in HDTV's development (Figure 8–2). It is too costly and unrealistic to dump one technology for another. For example, a local television station not broadcasting in HDTV cannot change its transmission system to HDTV when the frequency is not available and no one in the

FIGURE 8–2 The clarity of HDTV makes it possible for a video picture to equal the quality of 35mm film, the standard used in most motion picture theaters. The future may see satellites or fiber-optic technology used to send motion pictures from central transmission points directly to theaters, thus bypassing the need to ship large reels of film. The clarity of HDTV makes it possible for large-screen television in the home to equal the quality of theater projection systems.

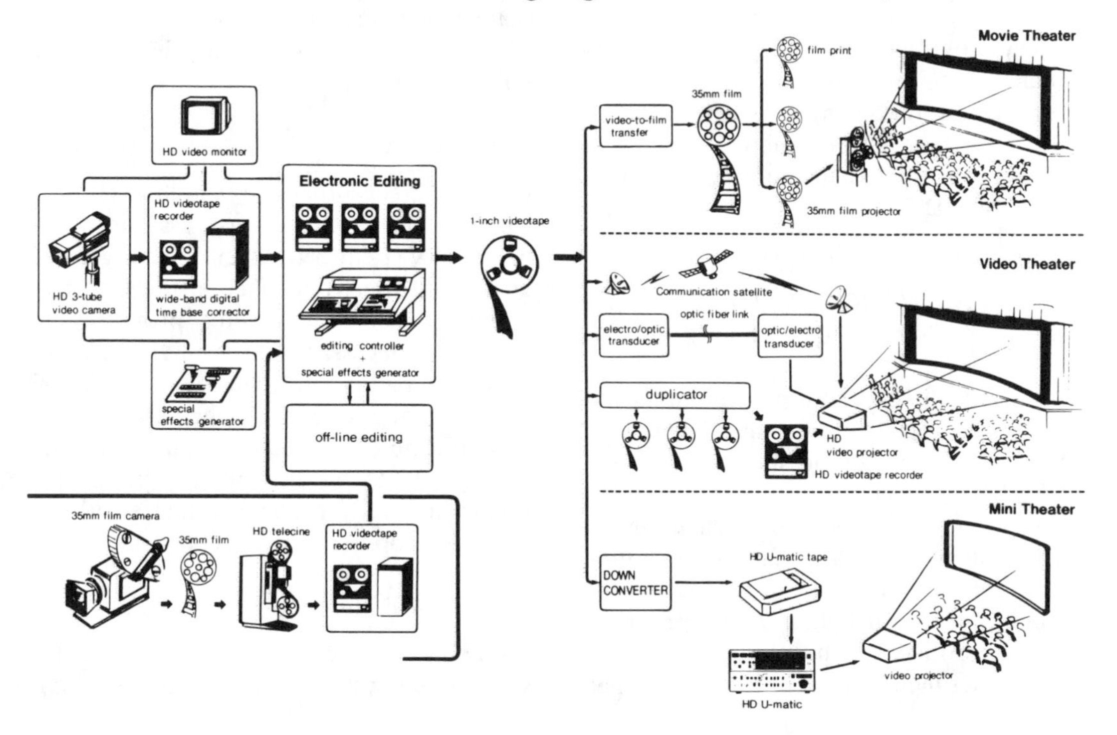

viewing audience has HDTV receivers. The key will be to develop ways whereby the station and the consumer both benefit, where the station can transmit the HDTV signal and the consumer can receive it without eliminating television reception for those who do not own HDTV receivers.

In addition, federal regulatory bodies can be slow to approve a new HDTV standard when it is not compatible with existing technologies. Gradually, regulatory agencies will come to grips with the technical standards necessary for HDTV to develop, at first in more technologically developed countries. Such issues as the ability of a country to maintain a system of free television; the protection of the balance of trade if one country should begin to monopolize the HDTV technical marketplace; and how to allocate, even recapture, the spectrum space needed for HDTV transmission—all are important. Equally important is how to protect local service to communities if HDTV programming sources are mostly national, even international. One solution may mean moving some services at one location on the electromagnetic spectrum to another position so as to free spectrum space for HDTV systems.[1]

HDTV PROGRAMMING AND PRODUCTION

The availability of more programming will also be necessary, although live coverage of major events, including sporting events—which have particular appeal for HDTV because of the dimensions of such things as basketball courts and football fields fitting the HDTV screen ratios—will carry the technology forward until more programming is produced. The earliest programs were experimental productions developed for use at trade shows, where HDTV was exhibited, and for satellite experiments. Gradually more and more productions took place. By the early 1990s a number of television programs and also feature-length films were being produced in HDTV.

For broadcast stations, HDTV would mean equipping studios for HDTV production, an expensive process. If a technology is not developed that is compatible with existing TV transmission systems, a possibility because of the technical considerations already discussed, then television stations would need to invest in two transmitters, one for HDTV and one for non-HDTV transmission.

DISTRIBUTION OF HDTV PROGRAMMING

In addition to the issues already discussed, there is the matter of how HDTV will eventually reach the marketplace since other distribution systems are available to carry HDTV programming.

The Potential of DBS

One is direct broadcast satellite (DBS), which Japan is already using as an alternative TV delivery system. The greater amount of spectrum space in the high frequencies used by satellites provides the bandwidth necessary to transmit HDTV signals. On the ground, however, a special *dish*-receiving antenna is necessary. While such technology is commonplace, getting enough HDTV viewers to justify committing the money for launch vehicles and satellite development remains an issue.

Cable Distribution

Cable systems are also interested in HDTV since wired systems may be able to eliminate some of the concerns over spectrum space. At the same time, however, the cable operators are watching the telephone companies who have the edge in fiber-optic applications and potentially could, in theory, deliver HDTV directly to the consumer. Video technologies such as laser disk offer still another distribution possibility that would bypass the other technologies completely.

ALTERNATIVE VIEWING APPLICATIONS FOR HDTV

In addition to television, alternative viewing possibilities exist for HDTV (Figure 8–3). Imagine you want to take a date to the movie but you want a more intimate atmosphere. Instead of going to a movie theatre, you decide to solicit an HDTV parlor. Here you watch an HDTV-produced and -delivered motion picture or television program in a living-room atmosphere with a comfortable couch instead of the theatre seat.

If your town does not have an HDTV parlor then you go to the motion picture theatre, but instead of the picture being reproduced by a 35mm projector, it appears with even greater clarity through HDTV. The motion picture theatre operator may have even received the feature film via a satellite dish rather than in a large can containing the reels of 35mm film.

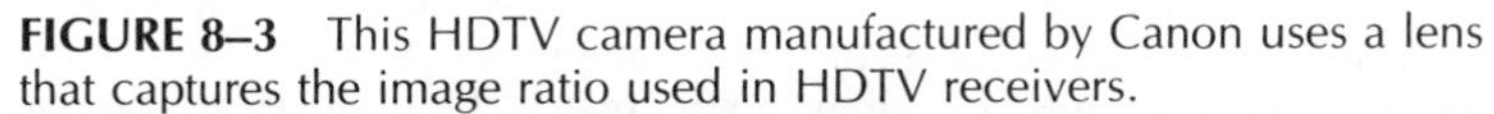

FIGURE 8–3 This HDTV camera manufactured by Canon uses a lens that captures the image ratio used in HDTV receivers.

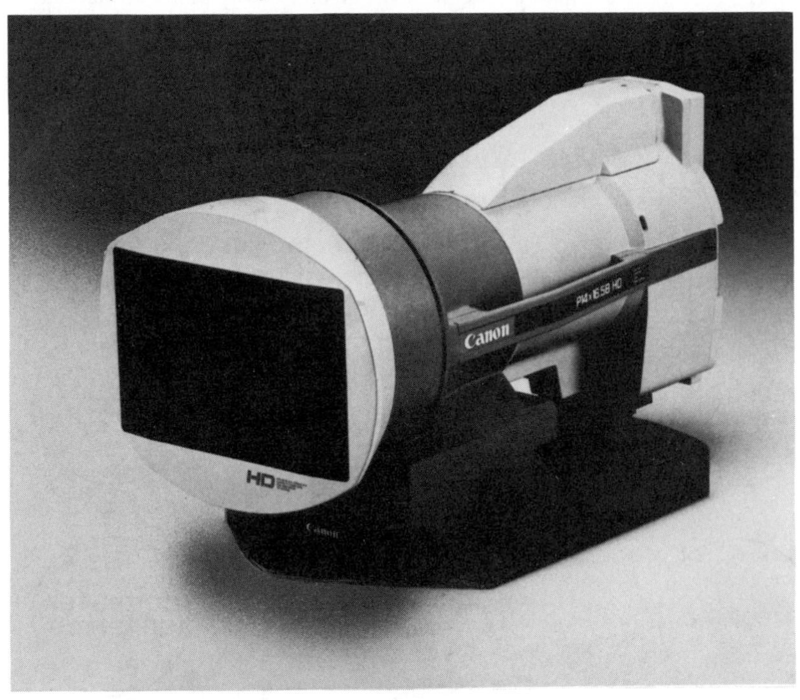

CLOSED-CIRCUIT AND INDUSTRIAL APPLICATIONS OF HDTV

If you are a medical student studying anatomy you will probably watch numerous operations televised to a viewing room. The clarity, however, is restricted by the technical limitations of standard television technology, a significant consideration where small, almost microscopic parts of the anatomy are involved. With HDTV, however, the smallest detail can be magnified and the resulting image will have almost picture-like quality. Similar applications of HDTV technology are possible in other educational settings (Figure 8–4). The larger screen ratio and the increased clarity and intensity of the picture offer major advancements in instructional technology.

HDTV APPLICATIONS IN FACSIMILE AND PRINTING TECHNOLOGIES

The technology already exists to make a hard copy of what appears on a television screen or computer video monitor. The clarity of that hard copy, however, is limited by the quality of the monitor. HDTV provides superior clarity and therefore can enhance the production of hard-copy prints made from video-based graphics.

Advancements in color copiers, color facsimile transmission systems, and printing

FIGURE 8–4 Closed-circuit applications of HDTV open up new opportunities for instructional television and other applications. For example, an HDTV closed-circuit TV picture originating from a hospital operating room can show the anatomy with much greater clarity than can standard television, thus increasing the instructional value of the transmission for medical students.

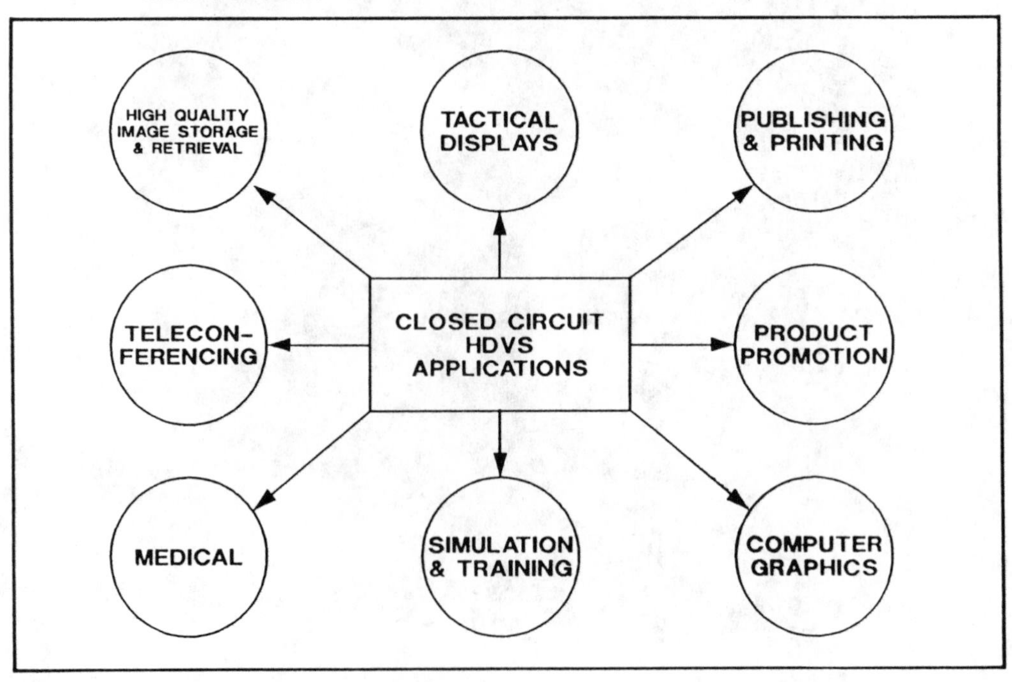

technologies that will incorporate HDTV processing could produce a front-page photograph on the regional edition of a local newspaper with the quality of a photograph developed at the local camera shop.

THE FUTURE OF HDTV

The future of HDTV will rest primarily on consumer demands (Figure 8–5). Will consumers be willing to pay the thousands of dollars more that will be required to obtain HDTV receiving equipment? Those who have seen HDTV say yes. This is not just a new and novel technology. It represents to many who have seen HDTV nothing less than a major advancement in video art, equivalent to the introduction of color television, and will increase our appetite for the quality that HDTV offers.

MULTIPOINT DISTRIBUTION SYSTEMS

Microwaves are very short waves high on the electromagnetic spectrum. *Multipoint distribution systems (MDS)* use microwaves

FIGURE 8–5 Consumer demand for HDTV will increase as the technology becomes available and the public becomes aware of the comparative picture quality offered by HDTV. This diagram shows a comparison of HDTV with 35mm film and other projection systems. Notice that HDTV quality, with scanning capacity of at least 1,100 lines per second, equals the quality of 35mm film and greatly surpasses the quality of 16mm and 8mm film. Current non-HDTV television systems scan images at between 500 and 600 lines per second, depending on the type of technology in use.

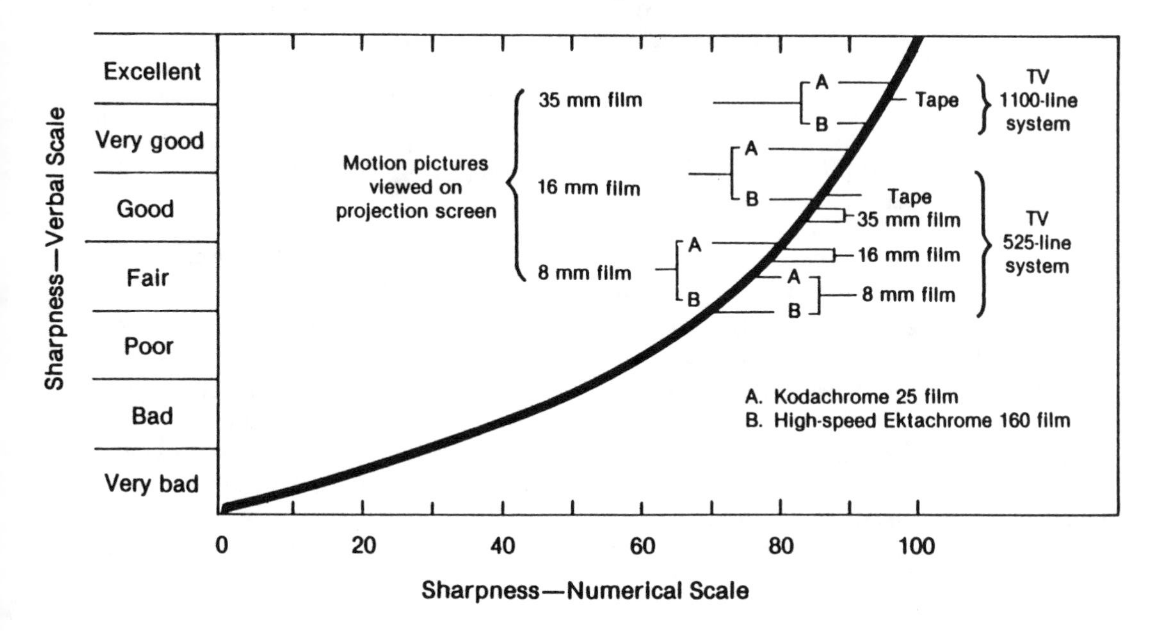

to transmit television signals from a master omnidirectional microwave antenna to smaller microwave antennas, which are usually located on apartment houses and hotels, although some private homes are also attached to MDS. The programming is similar to pay-cable programming, and the subscriber pays a fee to hook up to the system. The advantage of the MDS is that it can operate without the installation of cables.

Multipoint distribution systems have considerable potential where cable systems are not operable, especially in high-density metropolitan areas. The cost of equipment has steadily declined, and most systems operate as common carriers, leasing channels to programmers who then buy programming from such sources as HBO and sell it to multi-unit dwellings. Construction of an MDS is not cheap, but the subscriber antenna need not be purchased until the subscriber requests service. With cable, the upfront investment of wiring a community is substantial and the cable-franchise requirements can constitute a serious obstacle. A disadvantage of the MDS is that microwaves travel in line-of-sight paths. This can be a problem where numerous high-rise dwellings exist.[2]

SATELLITE MASTER-ANTENNA TELEVISION

Closely related to the MDS is *satellite master-antenna television (SMATV)*. Whereas MDS antennas receive signals from a central ground-based microwave transmitter, SMATV antennas receive signals from satellites. The SMATV system operates entirely on private property. For example, a hotel might install a SMATV receiving antenna and then install cable from the antenna to each hotel room in much the same way an MDS system would be installed. Like the MDS, SMATV is particularly attractive in high-density population areas where cable service has not penetrated or is less than satisfactory.[3]

Problems arise when SMATV operators try to link buildings having different owners. Some form of connection is necessary, the most economical being cable. However, many municipalities believe that the stringing of cable brings the SMATV system under control of the local cable ordinance. Franchise fees, requirements over which signals can be carried, and other restrictions immediately begin to apply. To get around such restrictions, SMATV operators sometimes resort to microwave hookups between buildings after obtaining from the FCC a cable-television relay-service (CARS) microwave license. The CARS license, while to some extent enabling the operator to avoid local control, brings the SMATV system under the FCC's cable-programming rules.

An SMATV system does not appear to need large numbers of channels in order to appeal to subscribers. One study found that a

> well chosen four to five channel service can satisfy about 90 percent of the demand for nonbroadcast programming and gain a significant market share at the expense of cable.[4]

SUBSCRIPTION TELEVISION

Subscription television (STV) is a system whereby the signal from a television station is scrambled and the subscriber pays a monthly fee for a decoder that unscrambles the signal. Decoder rental fees provide revenues for the system, which can then offer special programming.

Four elements are important in a successful STV operation: (1) equipment manufacturers, (2) STV stations, (3) STV franchises, and (4) program suppliers.[5] Equipment manufacturers must produce reliable, tamper-proof equipment to decode the STV signal. When decoders are not secure, unauthorized reception takes place, stripping the STV operator of revenue. Decoder prices vary, as does decoder reliability. Unsecured or less expensive decoders mean lost revenue and high maintenance costs. Expensive decoders mean high start-up costs. STV stations consist primarily of regular television stations that devote a portion of their programming day to STV. The STV franchisee runs the STV operation, leasing time from an STV station. Marketing, installation of decoders, warehousing and inventory control, billing and collections, and promotion are some of the functions a franchisee must undertake when operating an STV system.

Subscription television was given legitimacy in 1968, when the FCC established it as a broadcasting service. Its biggest competition has been pay cable, which had a distribution system in place before STV fully developed and had a head start in developing relationships with program suppliers.

An STV system is much more costly than either an SMATV or MDS operation. Consequently, the investment capital has not been put forward to support STV the way it has cable television and other services. FCC restrictions on the number of STV stations and the amount of their programming have also had an effect on STV's development.

LOW-POWER TV

Low-power television (LPTV) stations are much like translator stations, which rebroadcast the signals of television stations to outlying areas (Figure 8–6). The difference between them is that the LPTV station can originate programming.

The procedure for obtaining an LPTV license is less complex than that for a full-service station, specifically in the areas of proposed programming, community needs, and ascertainment surveys. A full-service station, however, is more protected from interference. Should a full-service station decide to apply for the frequency used by a low-power station, it will be given priority even if an LPTV station is in operation.

Low-power television has its greatest potential in areas not fully served by existing full-service stations. LPTV stations are restricted in their coverage area by transmitter power ranging from 100 to 1,000 watts.

RADIO SERVICES

Because of the dominant role television plays in our society, the application of radio technology sometimes takes a back seat. Still, research and development into radio communication continues. The medium plays an important role in mobile communication, especially by police and fire services. New uses of frequencies in the *gigahertz (GHz)* range are becoming possible.

Satellite-Aided Land-Mobile Radio

Two-way radio is a vital communication link for government and business. Consider the oceangoing vessel we learned about at the beginning of this book. Ship-to-shore communication guards its very survival. The police officer on patrol, the firefighter on an emergency call, the dispatcher at the taxi

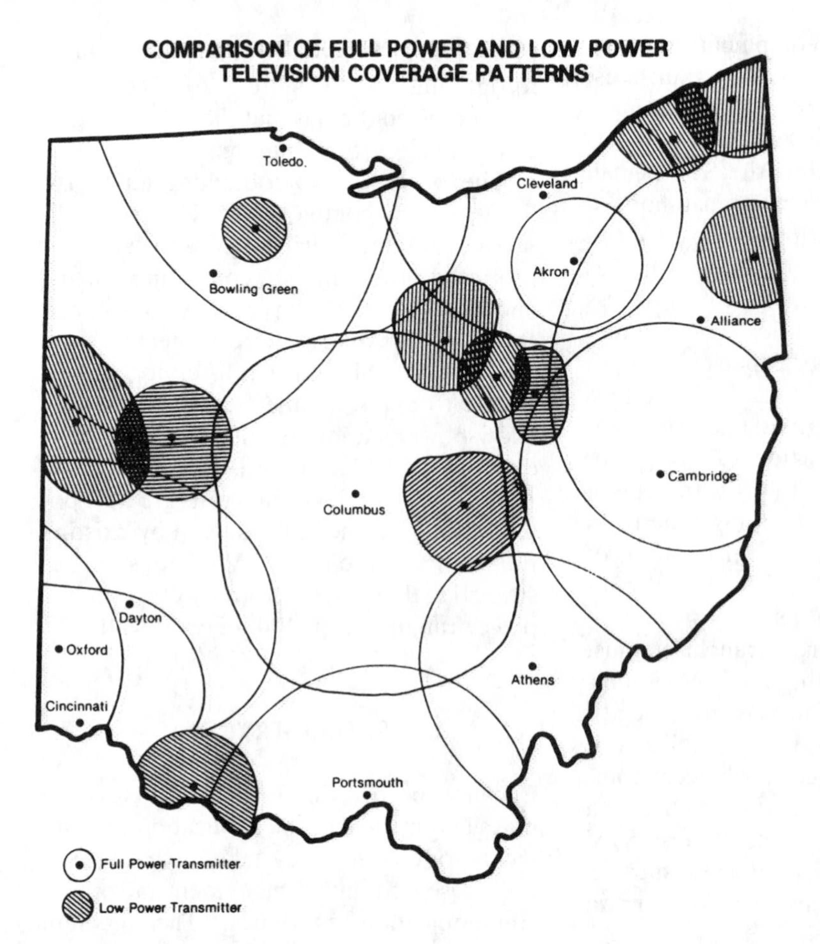

FIGURE 8–6 Low-power television (LPTV) consists of transmitters broadcasting with low power to relatively small geographic areas. LPTV permits a larger number of stations to exist in the same geographic region and to serve smaller communities with individual TV stations.

company—all rely on two-way radio communication. For much of government and business, this radio communication takes place at high frequencies where line-of-sight transmission occurs. As we have learned, skyscrapers, hilltops, and other obstacles often block the path of these high-frequency radio waves. Even in flat terrain, moreover, line-of-sight communication is limited by the curvature of the earth and the height of the transmitter tower.

To help alleviate some of these interference problems, scientists and businesspeople are using satellites as an aid to land-mobile radio. Because of the satellite's position in space it can successfully relay radio communication and escape the interference that occurs at ground level. Pioneering tests by NASA in cooperation with such industries as shipping and trucking have illustrated the utility of satellites in aiding land-mobile radio. In addition, the truck dispatcher or shipping executive can instantly locate a vehicle or vessel by using the satellite to pinpoint the source of its radio transmission.[6]

Private-Frequency Business-News Service

Imagine you are a business executive who manages a large accounting firm. Your clients include a wide variety of companies. You need to know what is going on inside those companies, not only to understand their business but to develop and maintain satisfactory relationships with their executives. How do you receive news of these businesses? You can read daily newspapers and subscribe to monthly magazines. If you are a busy executive, however, time is at a premium, and reading all of these sources of news takes up too much of it.

An alternative to the massive amount of printed material is to subscribe to a private radio news service tailored to the business community. To do so you rent a special radio receiver equipped with an audiocassette recorder and a small keyboard with just enough keys to call in the individual codes of the companies you want to monitor.

Your receiver is automatically tuned to a single frequency that receives one signal, that of a special business-news service sent by FM radio. The signal arrives on the subcarrier of a local FM station by satellite from the network studios of the business-news service. Before each story is broadcast by the network it is assigned one or more codes, which are transmitted along with the story. The codes are inaudible, but they trigger your receiver to tape-record those stories whose code matches one of the codes you have keyed into your receiver. Thus, if a story about AT&T is broadcast by the news service and you have keyed in AT&T's code on your receiver, the receiver will automatically tape-record that story. If you wish, you can also monitor it live.

The advantage of the system is that you do not have to continually monitor the receiver for the latest news about your clients. If you are out of the office or even away on vacation, the tape-recorded stories will be waiting for you when you return.[7] Instead of reading hundreds of pages of news articles you can be developing new business.

Increased Utility of the Electromagnetic Spectrum

Although in theory the electromagnetic spectrum has space for many more radio services than are now available, the usable space on the spectrum is limited. Extremely high frequencies in the gigahertz (GHz) range exist, but the technology to use them successfully has not been developed. The new "frequency frontier" is in the range of 30 to 300 GHz. If fully developed, such frequencies would provide more usable space on the spectrum than all of the current radio frequencies combined.

Millimeter waves, a generic term referring to waves between a centimeter and a millimeter long, have uses in space communication, where satellites can converse with each other in the 60-GHz range because there is no atmosphere to absorb the waves.[8] The signals can then be converted to more usable frequencies and sent back to earth. NASA is also developing millimeter-wave technology that will open up the 20- to 30-GHz area of the spectrum for transmissions between earth stations and satellites.[9]

New Uses for Subcarriers

The frequencies assigned to FM radio stations can carry the signal of the standard FM-radio programming we hear in our home and also leave room for an additional signal, which can be used for supplementary purposes. This subcarrier signal is the same signal that broadcasts the private-frequency business-news service.

FM stations are using frequency subcarriers for other purposes as well. In some paging services, for example, a small radio receiver is activated when a special code is transmitted. Each individual receiver has its own code. Messages are transmitted on the subcarrier frequency preceded by the code. Although a verbal message can be transmitted through the system, a simple tone suffices to alert the wearer of the receiver to call home or call the office. Other stations use the subcarrier for broadcasting commercial-free music to subscribers who pay a monthly fee to rent a special receiver capable of tuning in the subcarrier signal.

Future uses of FM subcarriers will be determined by consumer demand. For some marginally profitable FM stations such supplementary uses are particularly attractive as additional sources of income.

Preprogrammed Car Radios

Manufacturers, looking for ways to apply computer technology to basic consumer products, have developed preprogrammed car radios. The frequencies in a car radio are preprogrammed based on the radio format found on those frequencies in different cities. For example, the person who lives in Chicago and wants to listen to classical music can push a preset button on the car radio and the car radio will automatically scan the frequencies that play classical music.[10]

Digital Audio

The recording industry strives to bring realistic reproductions of sound to the consumer. What started as the cylinder recording of the 1880s produced by Thomas Edison's early phonograph has evolved through such advances as the 45-rpm record, high fidelity, stereo, and quadraphonic sound. The most advanced form of electronic reproduction of sound, however, belongs to *digital* technology.

Earlier in this book we learned about digital technology, the use of binary codes as a means of reproducing a signal. In digital recording this same computer technology is used to reproduce sound with a clarity unequaled in past recording technologies. The technology works like this. The sound to be reproduced—for example, a performance by an orchestra—is sampled as much as 50,000 times per second. Every detail of the sound, from the highest to the lowest notes, is captured and stored as a binary code. The sound that is captured at the performance of the orchestra can then be replayed with virtually perfect clarity, since each sampling is assigned a numerical value that is repeated exactly the way it was recorded. Moreover, since a recording usually goes through a series of editing and mixing steps before it is released to the public, any distortion that is detected during these steps is also eliminated.

What effect will all this have on radio? Radio's primary source of programming has traditionally been recorded music. But already radio finds competition in such devices as car stereos, portable stereos, and cassette tape decks. Part of the popularity of these new devices is their superior audio quality: certain distortions are inevitable when recordings must be transmitted over radio. Some industry executives predict that as the consumer demand grows for the high-quality sound that digital reproduces, radio will need to follow with digital transmitters and receivers.[11]

OTHER CONSUMER ELECTRONICS

The technologies that have played a part in developing more traditional forms of telecommunication have also played a role in the burgeoning field of consumer electronics. From sophisticated computer-based media to simple novelty gadgets, the mass media are becoming more and more specialized and more and more personal. Cameras that once used silver-based technologies now capture still photographs on electronic disks. The mechanical pinball machine has been replaced with a sophisticated electronic flight simulator in a video-game parlor. Some of the more well-known consumer electronics devices include home video recorders, videodisc players, electronic still cameras, and video games.

Home Video Recorders

At the CBS affiliates meeting in Chicago in 1956, Ampex unveiled what stockholders were told was a practical way to record and reproduce TV pictures on magnetic tape. The new system brought wide acclaim and launched the era of the videotape. In 1957 RCA introduced its version, which could reproduce images in color as well as black and white. Then in June 1962 Machtronics introduced a portable videotape recorder, and many manufacturing companies, such as SONY, Memorex, Arvin Industries, and Panasonic, began manufacturing videotape components and systems.

The next revolution came in April 1969 with the introduction of the video cassette by SONY. The video cassette has entered every facet of television, from libraries and instructional resource centers to the newsroom. A perspective on how we use home video-cassette recorders was offered by researcher Mark Levy, who found they were used to complement and not replace the use of broadcast television.[12] Levy also found that we use the video recorder to rearrange our viewing schedule to avoid viewing conflicts.[13]

The great number of people owning home video recorders resulted in ABC announcing plans for a network of stations that would broadcast scrambled signals late at night to homes equipped with special decoders. The decoders would permit the television set to present a clear picture that would in turn be recorded on the video recorder for viewing at the subscriber's convenience. The late-night programming would be sent from network origination points by both wired and satellite distribution systems to ABC affiliate stations, which would broadcast the signals at times when commercial sponsorship was very low or the station had been signing off the air.

Videodiscs

Videodiscs are much like long-play records except they produce a television picture as well as sound. Using a laser-beam-based playback system, the discs are free from the distortion that can occur with videotape systems. The potential of the videodisc lies with its ability to bypass television stations and distribute television programming directly to homes. Pushed to their full potential, videodiscs can be duplicated in mass quantities and shipped inexpensively anywhere in the world for playback on home players. They are particularly attractive when integrated with computer systems as educational tools.

Electronic Still Cameras

Still-camera photography has merged with telecommunication to create electronic still cameras the size of 35mm cameras and able to produce instant electronic pictures on a video screen. The SONY Mavica system employs a single-lens reflex camera and records the image on a small spinning magnetic disc. The disc is then transferred to a player that reproduces the picture on a screen.

Since the quality of the Mavica approaches print standards, inherent advantages exist for its use in such fields as print journalism. For example, photographs can be developed instantly and sent electronically over telephone systems to an editor in a distant newsroom. The electronic photograph can be reproduced on a printing plate in fifteen minutes and be ready for the press run.

VIDEO GAMES

At first in arcades, and now in home interactive television systems, video games have become the consumer game of the information society.

Pong: The Beginning

In 1972 a computer designer who wanted to improve the mechanical version of pinball developed an electronic game called Pong. Instead of the metal ball and the spring-loaded plunger it employed circuitry, a video screen, and a set of hand controls. The game became popular and it wasn't long before the company that produced it was bought by Warner Communications. Through its Atari subsidiary, Warner began to plow money into the development of a series of electronic games, which in 1980 resulted in one-third of the total earnings of the Warner conglomerate. By the mid-1980s the video-game industry had surpassed the recording/motion-picture industry in revenues.

Taito's Space Invaders

Although Pong is considered the forerunner of video games, other companies in other countries have also contributed to their growth. The Japanese, long important in the electronics market, entered the international video-game market in 1978 when Taito marketed Space Invaders. In Japan the video-game explosion created a kind of financial culture shock, and the Bank of Japan had to triple production of the hundred-yen piece used to feed the thirsty Space Invaders machines. In the first year of production Taito placed 100,000 Space Invaders in operation and the Japanese spent the equivalent of $600 million playing them. In the United States a company named Bally, through its Midway Division, began licensing Space Invaders in 1978, and by 1980 it had 60,000 units in operation.

The Game Arcades

Game arcades began popping up in shopping malls and converted gas stations—anywhere operators could find space. Atari obtained the rights to offer Space Invaders through home video devices that could connect to television sets. The popularity of the arcade version whetted consumer appetites, and Space Invaders became the most successful of Atari's early home video-game offerings. Other early video-game classics were Galaxian, Asteroids, Donkey Kong, Missile Command, Pac Man, and Battlezone.

NINTENDO AND THE RESURGENCE OF HOME VIDEO GAMES

The burgeoning of video games leveled off in the early 1980s. Still, they remain a popular and profitable part of the consumer-electronics industry. One of the biggest challenges in maintaining this profit is developing new games to replace those that have lost their novelty. That development began to affect the home video-game market in the late 1980s and continues into the early 1990s. The kingpin in this resurgence was Nintendo's Super Mario Bros. video game played on Nintendo's "famicom" computer, which transforms a television into a home video game.

Nintendo's success is credited to strict software quality control, and regularly replacing old games in retail outlets with new offerings. Nintendo's advertising is also credited with renewing interest in video-game technology.[14]

PERSONAL COMPUTERS

When Lee de Forest perfected the vacuum tube and the Bell Labs team of Shockley, Brattain, and Bardeen developed the transistor, they made contributions that far transcended the technology of the superheterodyne receiver and the transistor radio. The vacuum tube and the transistor would later transform the way machines "think."

The tiny transistor that won the Bell Labs team the Nobel Prize was the size of a thumbnail. Today, through the technology of integrated circuits and microprocessors, a single "chip" tiny enough to slide through the eye of a needle holds the power of millions of transistors. Those same chips have revolutionized the electronics industry, from radios to radar, but nowhere has the impact been greater than with computers. What once were cumbersome machines weighing tons and filling entire rooms today weigh but a few pounds and exceed the computing power of their mammoth ancestors.

Peripheral Technology and Networks

Adding peripheral technology such as the videodisc has expanded personal computers far beyond mathematical processing. In the future you may buy this text not as a printed book but as a videodisc, the pages of which you will read on the monitor of your personal computer. As you read material on the development of early television, you will stop, key in the correct information on your keyboard, and then sit back and watch examples of early television programming—not still pictures but the actual programs the book is discussing.

The word *network*, which has traditionally meant a radio or television network, also refers to a computer network. The evening edition of a national computer-based electronic newspaper is available through a computer network accessed via a personal computer just as the evening edition of the television news is available via a television network. These computer networks offer tremendous technical and social implications for society. Computer networks such as CompuServe, and Dow Jones are becoming as familiar as NBC, ABC, or CBS.

Friendly Technology

Few people will spend money on something they fear. Many people still fear computers. College and high school courses in computer literacy, new advances in software (Figure 8–7), telecommunication, and computer science are helping to lessen these fears, but individuals without access to such an experience can harbor technophobia. As competition continues, more emphasis is being placed on reaching buyers who have avoided purchasing a personal computer not because of utility or money but out of fear.

FIGURE 8–7 Making computers "friendly" and appealing to the user is one of the challenges of both computer manufacturers and software developers. The ability of consumers to utilize computers for such everyday tasks as information retrieval, shopping, electronic mail, education, and other services will permit new applications of telecommunication in our society.

Adequate Software

The real power of the computer is not in hardware but in software—the programs that make the machine perform specific functions (Figure 8–8). Software has long lagged behind hardware. Some early entrants into the personal-computer market had excellent machines but did not own sufficient software to make the machines a success. Companies are realizing that without software they cannot survive. Most early software was directed at business and professional people. In the future, software must be developed that appeals to the average consumer. Only then can personal computers begin to achieve the widespread use that radio, television, and the telephone enjoy.

TELETEXT

In many parts of the world a television viewer whose set is equipped with a special converter can turn the channel selector and read an electronic newspaper, learn of new products at the grocery store, check airline schedules, or learn what is playing at the local theater (Figure 8–9). Some experts predict that by the mid-1990s the amount of such electronic textual information we consume will have increased dramatically, changing the way we use television and other media. Transmission of textual information by over-the-air signals is called *teletext*. Transmission by a wired two-way interactive system is called *videotex*. Both terms require explanation.

FIGURE 8–8 Sophisticated software, miniaturization of components, and more powerful processing have resulted in new uses for personal computers. Electronic publishing, the on-screen layout and design of pages prior to printing and publication, has been adapted to everything from small company newsletters to national newspapers.

Transmission System

Teletext is primarily a one-way system for transmitting textual information, most commonly by means of the *vertical blanking interval (VBI)* of a television signal. The VBI is the thick black bar that appears on a television screen when the vertical-hold adjustment is manipulated. Teletext signals can also be transmitted over the entire television channel. FM-radio subcarrier signals, signals transmitted on unused portions of the assigned frequency, can also transmit a teletext signal, and some FM radio stations have experimented in sending teletext. The textual "frames," pages of text which fill the television screen, are transmitted in a rapidly repeated sequence, and the viewer "captures" a specific frame by means of the special converter and handheld key pad attached to the television set.

System Capacity

The capacity of the system is limited by the number of frames that the viewer can wait through before frustration sets in and the system goes unused. At a transmission rate of about five frames per second, a thousand frames can be stored and still make the system appealing to the user.

Graphic displays, though more appealing to the user, take longer to present than simple textual information. As a result, combination textual and graphic systems stressing news, sports, weather, and other popular information have become popular. In designing a system that will be profitable,

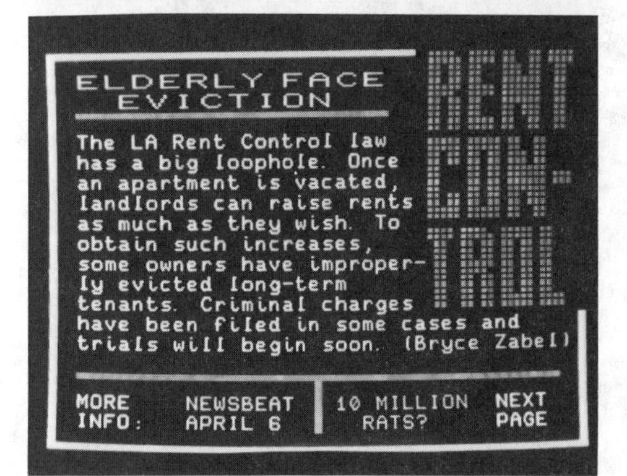

FIGURE 8–9 *Teletext* is primarily a one-way system distributed over the air and through cable systems. When a converter is used on a home television set, the user can access teletext pages by stopping them on-screen. *Videotex* is a wired two-way system. Commercially, teletext has not been overly successful, although it is used as an information service in some parts of the world. These frames are from an early experimental teletext system developed by KCET in Los Angeles.

one must take into account both the technical and content limitations of the system. To be appealing to an advertiser the information must be attractively packaged. To be appealing to a user it must be quickly accessible.[15]

Two-Way Teletext

In two-way teletext the user commands the central teletext storage computer by telephone; the transmission back to the viewer is by standard over-the-air television signals or cable. Using a touch-tone telephone the user dials the telephone number displayed on the teletext page appearing on the TV set. The computer answers the telephone and transmits on a one-shot basis a D-type, or decision page. From the items on the decision page the user can select additional pages by keying in the correct numbers on the telephone.[16]

Touch-tone teletext operates on the theory that a station's teletext computer can store many more pages than the typical teletext cycle can accommodate and still have user appeal. The user wanting a small amount of information not normally being transmitted can telephone the computer and have the information sent over the system.

VIDEOTEX

Unlike teletext, which uses television transmission, videotex is a two-way wired communication system connecting the user with a central computer by telephone or cable (Figure 8–10). Videotex, like teletext, can also be received on a home TV set through the use of an *interactive* terminal or personal computer. The practical capacity of a videotex system is much larger than that of a teletext system. Tens of thousands of pages of text are easily stored by a small videotex system, and larger systems are confined only by the storage capacity of the computer. Although a teletext system could hold an equal amount of information, it would take too long to access it. Unlike teletext, where the user must wait until a frame rolls by and then capture it, the frame is immediately accessed with videotex.

INTERACTIVITY AND USER SATISFACTION OF TELETEXT AND VIDEOTEX

Important to an understanding of how teletext and videotex work is the concept of *interactivity*—the ratio of "user activity to system activity."[17]

The Range of Interactivity

Think of interactivity as two extremes. At one extreme is a one-way cable system that transmits textual news. The teletext concept is not even in use. The user simply tunes to the cable channel and reads the changing frames of news. The user has no control over what appears on the screen. At the other extreme is a two-way interactive videotex system. Perhaps it is used for an instant electronic-mail function in which two people type messages to each other through a central computer.

Both the cable and the videotex system use a central computer. With the cable system, interactivity is zero. In the electronic-mail function of the videotex system it is one to one. Neither is optimal for a general application of videotex or teletext. For ex-

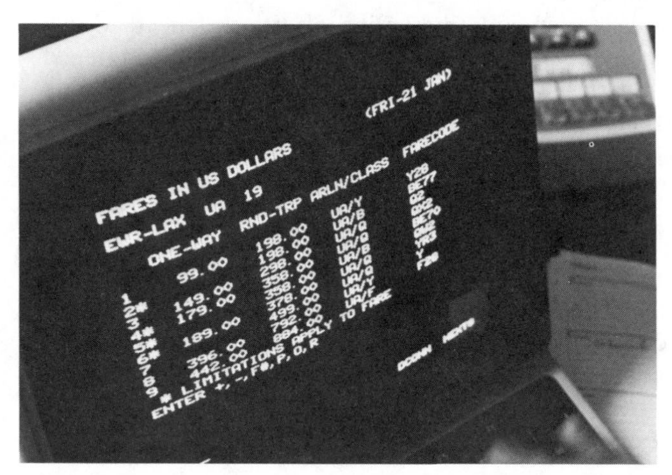

FIGURE 8–10 Videotex is a two-way wired system. Some systems operate with color graphics whereas others display basic textual information. Many data bases are accessed through personal computers; others employ special video display terminals and computer hardware connected to telephone outlets.

ample, if the only use of a teletext system was to send information one way by over-the-air television signals and the user had no control over what was seen on the screen, the system would have little value. On the other hand, if a videotex system could merely access information sent to another user, the telephone or regular mail would be much more economical.

In between is where teletext and videotex systems operate most efficiently. Theoretically, a teletext system containing only a few pages of information can be accessed just as fast and with as much user satisfaction as a videotex system having the same amount of information. When the information increases, however, the interactivity of the system becomes critical.

System Advantages and Disadvantages

Although videotex systems hold more information and can be quickly accessed, not everyone owns a personal computer or the interactive terminal necessary to access the system. Moreover, accessing the system costs money. Charges are similar to economical long-distance telephone rates. With teletext the system can be practically free, the only cost being an original expenditure for a decoder or special television set equipped to receive teletext. Additionally, every TV station and FM radio station could send teletext signals, as can one-way cable systems. Thus, teletext may have more senders and receivers of information than videotex.

An advantage of teletext is that additional decoders can be added to a system at no extra cost to the operator (unless, of course, the operator is buying the decoders). Nor will more television households having decoders affect the responsiveness of the system.

However, the more information that is sent over the system, the more space that is necessary on the electromagnetic spectrum and the more time that is necessary for the system to respond to a user's request. In other words, a user will have to wait longer to capture a page being sent in sequence.

At the same time, remember, videotex has some distinct advantages over teletext. Because the central computer handles the user interaction, much more information can be included in a videotex system without the responsiveness of the system being slowed. True interactive communication between a user and the central computer can take place with videotex.

THE FUTURE OF TELETEXT AND VIDEOTEX

If teletext and videotex are to reach their potential, a potential that some predict will translate into a billion-dollar industry in the 1990s, the industry and the consumers who pay for the services will both have to adapt to the technology.

Consumer Acceptance

Major marketing efforts will need to call attention to the services offered by teletext and videotex and also the utility of these services. Most consumers still look at a television set more as a medium of entertainment than as a medium of information. The public will need to be educated as well as sold on the new services.

Personal computers are also aiding the educational process. Having been exposed to an electronic keyboard and a video screen displaying data, users will be able to understand and accept a television that shows textual information.

Low-Cost Terminals

Any new medium that requires users to spend large sums of money to receive the service begins with a serious disadvantage. Teletext, for example, requires decoders that enable standard television receivers to convert signals carrying textual information. An outlay in the hundreds of dollars can create major consumer resistance. In the early days of teletext such prices were commonplace.

New TV sets, such as those manufactured by Zenith, have built-in teletext decoders. Yet there is no indication that the public will run out and purchase new sets just to receive teletext. For many people the separately purchased decoder will become the gate that opens into teletext. Teletext may come closer to achieving its full potential when TV stations start providing the public with free decoders.

For videotex the problem is much the same as for teletext. Low-cost interactive terminals are necessary if one is to access data banks. Although any owner of a personal computer and a *modem* can access a videotex data bank, there is no indication that large numbers of people will buy a computer just to read an electronic newspaper.

Investment Capital

Another important factor in the success of teletext and videotex will be *investment capital*. Investment capital is money committed to a new venture in anticipation of a large return on the investment, which may, however, hold a sizable risk. Early in its development any new medium is a risk. The greater the risk the more difficult it is to attract investment capital. Thus, the growth of teletext and videotex will be determined by (1) the overall economy, which, when healthy, can provide investment capital; (2) the amount of risk involved; and (3) the potential for profit resulting from that investment. If the economy generates money, if the risk is low enough, and if the profit potential is high enough, then teletext and videotex systems will have a better chance of success.

Advertiser and Subscriber Support

The three factors just discussed—the economy, risk, and profit potential—are important, but getting advertisers and the public to pay for teletext and videotex will determine whether these media succeed. For example, advertisers have been conditioned to spend money on such things as newspapers, the radio station, and television. Moreover, the deregulation policies of the 1980s and the increased number of radio stations and cable systems selling advertising have greatly fragmented the media marketplace. Advertisers understand terms such as full-page ad, sixty-second commercial, and prime-time scheduling.

With teletext a new vocabulary will be necessary. How effective are teletext advertisements versus television advertisements? Can an advertiser accustomed to sponsoring the evening TV news be convinced to invest money in an electronic textual page of advertising that doesn't move and will be seen only when a viewer elects to view the page instead of the entertainment programs of prime-time television?

Videotex, although it may be less dependent on advertising support, faces similar obstacles. A user must pay telephone charges in order to access data banks. Also, because videotex has an unlimited information capacity—as opposed to teletext, which is restricted to the amount of information that can be "rolled" to the public—the chances of any one piece of information being accessed are much less. Therefore, an information supplier must compete with thousands of pages of electronic information.

SUMMARY

High-definition television (HDTV) offers large-screen quantity with almost photographic quality. This developing technology, once it surmounts the technical, economic, and regulatory hurdles, promises to revolutionize the way we watch television. Although HDTV productions are already taking place and experimental systems are in operation, such issues as how HDTV will be distributed, and how fast the public will demand and pay for HDTV, must be resolved before HDTV can realize its full potential.

Multipoint distribution systems (MDS) use microwave technology to transmit TV signals from an omnidirectional master antenna to smaller microwave antennas usually located on multi-unit dwellings such as apartment complexes and hotels. Programming is similar to pay cable, and subscribers pay a monthly fee.

Satellite master-antenna television (SMATV) receives signals from a satellite and then redistributes them to subscribers. Because receive-only earth stations need not be licensed, the system can become operable as soon as it is installed. As long as the system does not connect by cable with dwellings under different ownership, it can operate generally free of regulatory control.

Subscription television (STV) is the use of a scrambled television signal to send programming to subscribers who pay a monthly fee for a descrambler. Most STV systems are leased by television stations who turn over part of their broadcasting day to the STV operator, who obtains programming, markets the service, and places the in-home decoders.

Low-power television (LPTV) operates much like television translator stations except that it can originate programming. Low-power TV has its greatest potential in remote areas not fully served by broadcast or cable stations.

While radio is continuing its traditional role as a broadcast medium, radio technology is being used in a number of other applications. Satellite-aided land-mobile radio permits long-distance, interference-free, two-way radio service valued by such industries as shipping and long-haul trucking. Private-frequency business-news services and new uses of subcarrier frequencies offer radio services to both businesses and households. The utilization of millimeter-wave frequencies is proving promising. Preprogrammed car radios permit the driver to scan stations with particular music formats. Radio stations are adapting to digital-transmission methods in order to meet what is expected to be consumer demand for high-quality sound reproduction, a demand resulting from the introduction of digital audio technology in the recording industry.

Such consumer products as home videotape recorders, videodisc players, electronic still cameras, and video games are offering alternatives to traditional media and providing examples of the application of microelectronics and miniaturization to the field of telecommunication.

Personal computers, through the use of peripheral technology and computer networks, are becoming a new medium of mass communication. Two firms, Apple and IBM, are examples of companies that have become major players in the computer marketplace and are illustrative of the entrepreneurial drive that has helped the personal computer market grow. Widespread acceptance of personal computers will continue to be determined by the industry's ability to identify and capture new markets, to provide adequate distribution channels, to motivate consumers to purchase, and to provide friendly technology and adequate software.

Teletext and videotex continue to be in the experimental stages. Teletext is a one-way information system using the vertical blanking interval of the television signal. Videotex is a two-way system with much greater storage capacity. Such factors as consumer acceptance, low-cost terminals, investment capital, and advertiser and subscriber support will determine the future of teletext and videotex.

OPPORTUNITIES FOR FURTHER LEARNING

AUMENTE, J. *New Electronic Pathways: Videotex, Teletext and Online Databases*. Newbury Park, CA: Sage, 1987.

HACK, D. B. *High-Definition Television (HDTV) in the United States—What Does an "Even Playing Field" Look Like? (With Policy Options)*. Washington, DC: Library of Congress, Congressional Research Service, 1988.

JACKSON, C. L., AND ARNHEIM, L. A. *A High-Fiber Diet for Television? Impact of Future Telephone, Fiber and Regulatory Changes for Broadcasters*. Washington, DC: National Association of Broadcasters, 1988.

LEVY, M.R. *The VCR Age: Home Video and Mass Communication*. Newbury Park, CA: Sage, 1989.

MOSCO, V. *Pushbutton Fantasies: Critical Perspectives on Videotex and Information Technologies*. Norwood, NJ: Ablex, 1982.

NATHAN, R. ASSOCIATES. *Television Manufacturing in the United States: Economic Contributions—Past, Present, and Future.* Chapter IV: "High Definition TV's Potential Economic Impact on Television Manufacturing in the United States." Washington, DC: Electronic Industries Association, 1988.

SIEFERT, M., ed. *The Information Gap: How Computers and Other New Communication Technologies Affect the Social Distribution of Power*. New York: Oxford University Press, 1989.

SIEFERT, M., GERBNER, G., AND FISHER, J. *The Information Gap: How Computers and Other New Communication Technologies Affect the Social Distribution of Power*. Special issue of the *Journal of Communication*. New York: Oxford University Press.

SINGLETON, L. A. *Telecommunications in the Information Age: A Nontechnical Primer on the New Technologies*, 2nd ed., Cambridge, MA: Ballinger, 1986.

WILSON, K. G. *Technologies of Control: The New Interactive Media for the Home*. Madison: University of Wisconsin Press, 1988.

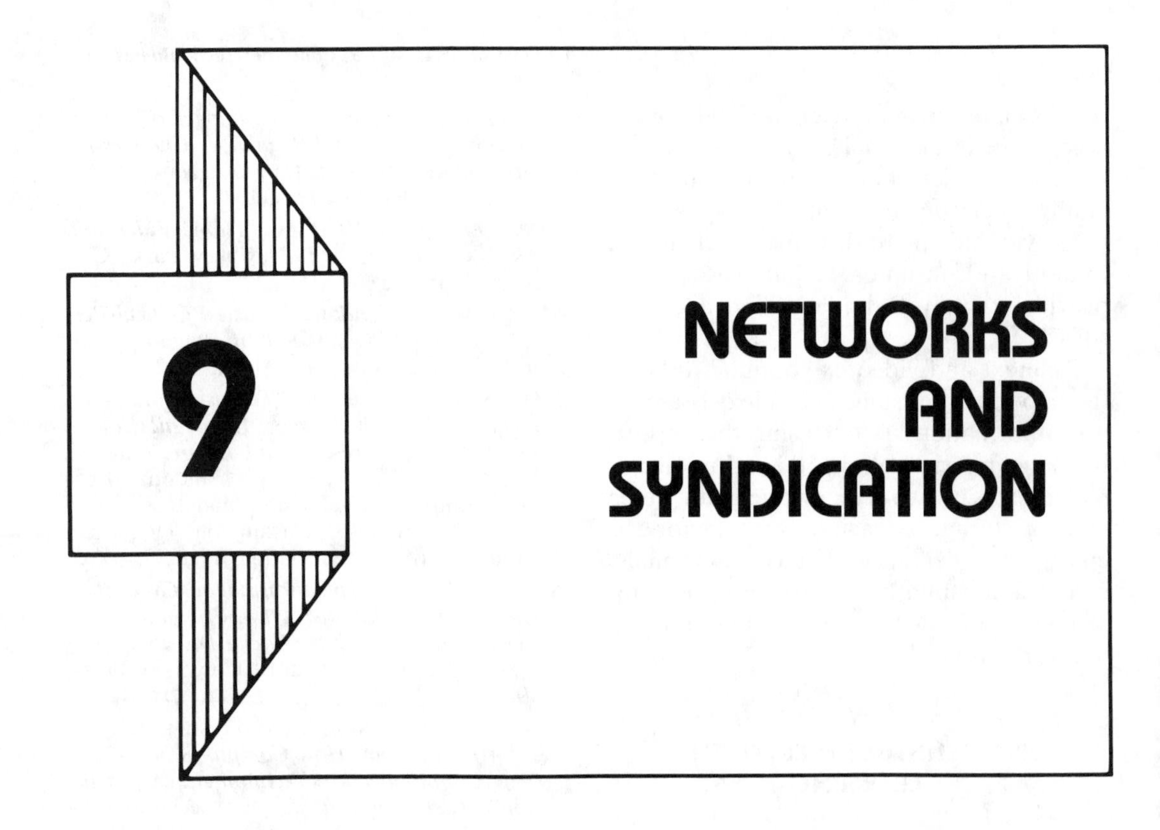

When we think of the term "network," what comes to mind for most of us are television networks such as ABC, CBS, and NBC. In the case of these three commercial networks, they provide the United States with much of its television programming. In Britain, the BBC serves a similar role as does NHK in Japan. In addition to television, many radio networks exist and carry specialized programming ranging from news and information to rock music and even motivation lectures. Cable networks serve individual cable systems with radio and TV programs. Data networks connect individual computers together permitting the exchange of information and the accessing of data banks storing electronic newspapers, books, and other information services.

THE NETWORK CONCEPT

Networks not only provide the public with broadcast programming but also serve the advertiser and the broadcaster. For the advertiser, broadcasting networks provide a medium through which an advertiser can reach large numbers of people at economical rates. Even though the cost of a minute-long commercial on a major network can be tens of thousands of dollars, the cost of reaching the same number of people through another medium, such as direct mail, would be far greater. For the broadcaster, airing network programming is much more economical than producing and airing comparable local programming. In addition, individual stations many times receive compensation from the network for airing its programs.

Acquiring Programs

Within the networks, the news and public affairs units are responsible for informational programming. Supported by major staffs of reporters, producers, directors, and technicians, these units produce daily newscasts seen and heard by millions. They also provide special coverage of such events as elections, inaugurations, press conferences, and space flights. They offer special features and documentaries, which can range from a superficial look at a national fad to an investigative report of political corruption. Documentaries and reports produced in cooperation with network-owned and -operated stations, called *O & Os*, are still another source of network news programming, as are reports contributed by local affiliates.

With entertainment programming the story is different. Here, various independent companies called *production houses* work closely with the networks to supply them with everything from detective thrillers to situation comedies. The ideas for programs come from the network, the production house, writers, or anyone else connected with the creative process. If an idea generates enough interest within a network and production house, a single test program, called a *pilot*, is produced. Both the network and production house may share a financial interest in the program. If the ratings from the pilot program show it to have substantial audience appeal, then the program may move into full-scale production. Usually a complete season of shows is produced only after the network tests the program during a new television season to see whether its popularity will last more than a few weeks. Conversely, if the pilot is disappointing, the program may never get on the air. Other factors determining an air date may include affiliate reaction to a program, scheduling availability, and competition from other shows.

Affiliate Relations and Clearance Ratios

The individual stations, both radio and television, that carry a network's programs are called *affiliates.* How much attention do the networks really pay to their affiliates' wishes? On such major issues as clearing programming (agreeing to air a network program), the networks carefully heed them. The number of affiliates that agree to carry the program is called the *clearance ratio*, and this ratio is a form of direct feedback to the network.

Clearance ratios are critical to networks. If a group of stations decides not to carry a program, then the network's audience suffers, the ratings go down, and the advertising dollar follows. The clearance aspect of the network-affiliate relationship has taken on considerably more importance because of the concern that has arisen over violent and sexually explicit television programming. Many local affiliate stations have simply refused to carry such programs. Others have rescheduled the programming to a later hour when children will not be watching. Although that idea may sound reasonable, to a network it can spell disaster, since the late-night viewing audience is only a fraction of the prime-time audience.

Affiliate Compensation

The arrangements through which affiliates receive compensation for carrying network programming vary. For example, many radio stations pay for network programming because it fills up so much of a station's programming day and the station has the opportunity to sell local advertising, which airs in the network programming. In other cases the audience reached by a station many be large enough and important enough to the network that the network pays the station to carry the network programming. The network receives income from advertising carried by the network.

The compensation arrangements between affiliates and networks is an emerging topic of discussion, especially among ABC, CBS, and NBC television networks. Two factors—the decrease in network television audiences caused by alternative media such as cable and the fact that stations pay for programming from syndicators—have made the networks ask whether they can or should continue to pay stations to carry network programming.[1]

ABC

ABC was formed when NBC was forced to dispose of its dual-network operation. When ABC acquired its own identity in 1943, it began a concentrated effort to compete with its two closest rivals, CBS and NBC.[2] Because of some advantageous breaks, the late 1940s was a profitable time for ABC radio. One of its biggest breaks was singing star Bing Crosby. When Crosby wanted to produce a prerecorded show instead of meeting the demands of weekly radio appearances, ABC gave the innovative idea a try. The success of the program proved that even a prerecorded show could be a hit, and other stars followed Crosby's example. No longer plagued by the image of an unsteady toddler, ABC moved forward from radio into television.

Launching Television Programming

In a special program originating from Broadway's famous Palace Theatre, ABC launched its television ventures on Tuesday, August 10, 1948. ABC's flagship station, Channel 7, WJZ-TV, showed a documentary on the progress of New York City, narrated by Milton Cross. Cameras caught live the action of a parade outside the Palace, street dancing, music from Times Square, an eighty-piece combined police and fire-department band, majorettes, and horse-drawn fire engines and streetcars. Later that evening, viewers watched "Candid Microphone" with Allen Funt, a radio version of the future "Candid Camera." A month later, ABC regional network programming began in Chicago with hookups linking stations in Chicago, Milwaukee, Cleveland, and Toledo. A football game between the old Chicago Cardinals and the Pittsburgh Steelers launched the regional network.

Merger with United Paramount

Not all of ABC's corporate maturity came from television. Needing capital to make inroads against its older competitors, ABC announced plans for a merger with United Paramount Theaters (UPT) in 1951. That year both the ABC and UPT boards of directors approved the merger, and in 1952 the FCC held hearings on it. By early 1953 the merger was complete, and cash reserves of $30 million were added to ABC's bank account. In a shrewd personnel move, Robert T. Weitman, a vice-president at UPT, was placed in charge of ABC talent. Weitman had previously been instrumental in advancing the careers of Frank Sinatra, Danny Kaye, Red Skelton, Betty Hutton, and Perry Como.

The mid-1950s signaled changes in both station operations and corporate structure at ABC. New call letters were assigned to the network's O&O New York station, and WJZ became WABC. Within the company five new divisions were formed: the ABC Radio Network, the ABC Television Network, ABC Owned Radio Stations, ABC Owned Television Stations, and ABC Film Syndication. By the end of the decade ABC had become a formidable opponent of both NBC and CBS.

Edging the Competition: ABC Sports

Unfortunately, the "formidable opponent" status was where ABC television remained. Although profitable and popular, moving out from under the dominance of NBC and CBS was no easy task. It took the combination of a greater number of television stations on the air and popular programming to bring ABC out of the cellar.

The magic formula started to work in the mid-1970s. The network had already managed to excel in one important area, sports programming, with its popular "ABC Wide World of Sports." Then, when weekends seemed saturated with football, ABC introduced "Monday Night Football," originally featuring Howard Cosell, Frank Gifford, and Don Meredith. Olympic coverage added more gold to ABC's pot.

Entertainment and "Roots"

But sports could not manage by itself. Entertainment had to put in its share. ABC launched a *talent raid* that plucked comedian Redd Foxx and "Today" host Barbara Walters from NBC and programming executive Fred Silverman, now with his own production company, from CBS. Walters gained publicity for her reported million-dollar salary, and Silverman for making ABC stock jump upward the day he announced his resignation from CBS. Perhaps ABC's biggest push into dominant prime-time television occurred with its presentation "Roots" (See Figure 3–9 in Chapter 3). First aired in January 1977, this story of black struggle traced through the "roots" of author Alex Haley's family set new records in television viewing. The twelve-hour production included stars such as John Amos, Madge Sinclair, LeVar Burton, Lorne Greene, Ed Asner, and Cicely Tyson.

Faced with increasing competition from cable and other video media, ABC, like other networks, has worked to strategically place itself in a position of leadership in prime-time hours. Two programs, "thirtysomething" (Figure 1–4 in Chapter 1) and "Twin Peaks" (Figure 9–1) helped launch ABC into the 1990s.

Daytime Profits

A strong daytime lineup contributes to ABC's profits. Programs such as "Ryan's Hope," "All My Children," "One Life to Live," "General Hospital," and "The Edge of Night" placed ABC in the top daytime spot for years running and resulted in ABC's eighteen-to-forty-nine-year-old female viewers outnumbering those of NBC and CBS combined. Some of ABC's day-

FIGURE 9–1 The ABC program "Twin Peaks," set in the Pacific Northwest, was one of the programs ABC used for its prime-time lineup to begin the 1990s. The program was directed by film director David Lynch, who was responsible for such works as "The Elephant Man" and "Blue Velvet." An ABC press release called "Twin Peaks" a soap opera for the 1990s and a "disturbing, sometimes darkly comic vision of the ominous unknown lurking beneath the commonplace and the everyday," and containing "seething undercurrents of illicit passion, greed, jealousy, and intrigue." Shown is actor Kyle MacLachlan, who stars as FBI agent Dale Cooper in the series.

time programming consistently captures higher ratings than prime-time programming. The network controls a large percentage of available daytime advertising dollars, which is significant because daytime programming is the cheapest to produce.[3] Moreover, ABC produces a majority of its own soap operas, thereby keeping money inside the network that would normally be spent on production houses.[4]

"Nightline"

In 1980, when Americans were taken hostage in Iran, ABC decided to expand coverage of the hostage crisis by airing late-night news reports on most ABC affiliates after local evening-news programming. The audience that was found to exist for these reports resulted in the decision to air a late-night news program, "ABC News Nightline," which began on March 24, 1980, with a Monday-through-Thursday run anchored by veteran ABC writer, producer, and chief diplomatic correspondent Ted Koppel (Figure 9–2). The experiment proved successful and the program expanded to Friday.

Ten years later in 1990, Koppel again bolstered ABC's news programming when he became the first American network journalist to enter Iraq after it invaded Kuwait.

The Capital Cities/ABC Merger

In 1985, ABC was purchased by Capital Cities Communications, Inc., for $3.5 billion. The merged companies became Capital Cities Communications/ABC Inc., with the ABC network a subsidiary of the parent company. With its origin in Hudson Valley Broadcasting Co., which had been bought in 1954 by famed newscaster Lowell Thomas and his associates, Capital Cities grew through acquisitions, the first of which was WTVD-TV in Durham, North Carolina, which is now an ABC O&O station. News-

FIGURE 9–2 Ted Koppel brought ABC's "Nightline" to a position of preeminence in broadcast journalism. Koppel, a veteran ABC news correspondent, began anchoring the program in 1980. "Nightline" has permitted ABC to provide in-depth coverage of major issues and to improve the network's reputation in news programming.

papers and business publications also added to Capital Cities profits. With bank financing, it executed the ABC merger, which one industry publication called "the minnow swallowing the whale."[5]

CBS

CBS had both the resources and the personnel to be a formidable opponent of ABC. By the late 1940s William Paley had established a track record as an excellent administrator and builder. He was joined at CBS by an exceptional management team.

Klauber, Kesten, and Stanton

The report card of Paley and his management team shows high marks for corporate growth. From net sales of $1.3 million in 1928, the network, exclusive of allied businesses, had climbed to an annual revenue of $1 billion by the late 1970s, which made it the largest single advertising medium in the world.

In addition to Paley, three individuals contributed greatly to CBS's early growth—Ed Klauber, Paul Kesten, and Frank Stanton. Klauber was a newspaper reporter until Paley coaxed him away from the editor's desk of the *New York Times* in 1930. Klauber, hired as Paley's assistant, is credited with shaping the character of early journalism at CBS.

Paul Kesten came from advertising. Recruited as head of sales promotion in 1930, he had earned his stripes with the New York ad agency of Lennen and Mitchell. One of his first decisions was to hire the accounting firm of Price, Waterhouse to "audit" NBC's claim of having the highest radio listenership. In the audit, CBS came out on top.

Dr. Frank Stanton joined CBS in 1935. A psychology professor at Ohio State University, Stanton had an interest in measuring radio listenership. When CBS learned of his work it brought him to New York at a $55 weekly salary and gave him the number-three position in a three-person research department. Stanton continued to measure radio listenership and developed an electronic device that could measure immediate responses to radio programs in a closed laboratory setting. He later left the research department and by 1942 was an administrative vice-president. As much a statesman for the entire broadcasting industry as a CBS executive, Stanton rose to the top of CBS in 1946, giving the company what many said was a sense of character and responsibility.

Trial and Error in Corporate Expansion

CBS did not confine itself to the broadcasting business. Like ABC, it began to apply its profits to acquisitions that in some ways directly supported, yet were different from, network operations. As we read earlier in this book, a lucrative artist-management business gave the network ready access to top talent. However, the FCC questioned the propriety of the network's control of talent, so CBS sold its interests to the Music Corporation of America. The more profitable acquisitions, like those discovered by the other networks, turned out to be radio and TV stations in large markets, which gave the network not only an affiliate station but also a share of the affiliate's profit. CBS also ventured into sports, publishing, and the recording industry. The sports venture was not successful. After CBS bought the New York Yankees baseball team in 1964 the team promptly sank into the doldrums. Attendance dropped, and CBS sold the Yankees in 1973.

Entertainment and News

For many years, until ABC began inching its way up the ratings ladder in the 1970s, CBS dominated television programming with such shows as "I Love Lucy." CBS Television City in Hollywood produced such shows as "Playhouse 90." Led by the long-running "Captain Kangaroo," children's programming also contributed to the network's success. Such adult variety programs as the "Ed Sullivan Show" continually topped the ratings. While ABC was leading the daytime soap-opera race CBS continued to develop its prime-time programming, turning out successful shows such as "Dallas," and "M*A*S*H," which made its successful last run on the network in February 1983.

CBS News also had its share of successes with Edward R. Murrow, who became a legend in broadcast journalism. "Douglas Edwards and the News" and exclusive interviews with such notables as President John Kennedy and Soviet leader Nikita Khrushchev kept CBS News on top. CBS anchorman Walter Cronkite was found by opinion polls to be the most credible person in the United States. Cronkite retired in 1981 and was replaced by Dan Rather. Cronkite had been a familiar part of CBS for twenty years, and many CBS executives feared his departure would cause audiences to change their viewing habits and shift their loyalties. At first it appeared there was indeed some reason for concern, but Dan Rather's reputation for broadcast journalism and some promotional help resulted in CBS holding its own. The highly rated CBS news program "60 Minutes" has also contributed to the success of the network's news operations.

Laurence A. Tisch Buys CBS

In 1986 the network was bought by entrepreneur Laurence A. Tisch, who owned the Loews Corp., an $8 billion conglomerate that Tisch and his brother developed through real estate, insurance, and tobacco companies. That background brought some criticism when Tisch first bought CBS, but the criticism became more pointed after he sold off some of CBS's long-time holdings including CBS records for $2 billion to Japan's SONY Corp. CBS's book publishing subsidiary, Harcourt Brace Jovanovich, and its music publishing and magazine divisions were also sold to permit the company to concentrate more on broadcasting.

Sports Spending Spree for the 1990s

Absent from Olympics coverage for three decades, the CBS network secured the rights to the coverage of the 1992 Winter Olympics from Albertville, France, as a means of building ratings and a stronger image in sports broadcasting. Paying $243 million for the rights, CBS gave some of the official U.S. Olympic sponsors the first opportunity to buy ads in specific categories.[6] Protection for products was also part of the contract. For example, if a car company is a U.S. Olympic sponsor and requests advertising in a specific category of available air time, then another car company would not be able to purchase ads in the same category. The contract language, which ABC cited as a reason for pulling out of the bidding, was introduced partly because in a previous Olympics, Fuji Photo Film, an

Olympic sponsor, was shut out because Eastman Kodak had exclusivity rights for television. The U.S. Olympic Committee also benefited by wording in the CBS contract that restricted the use of the Olympic logo in print advertising by companies that bought broadcast time but were not official sponsors of the Olympics.

CBS also owns the right to the 1994 Winter Olympics in Lillehammer, Norway, which the network purchased for $300 million. That purchase, and the 1992 coverage, signaled to others, including ABC, that CBS intended to be an aggressive contender in sports programming in the 1990s. Baseball and other sports are also part of the CBS coverage. The network has the rights to baseball's League Championship Series and the World Series through 1994.[7]

NBC

The oldest of the three commercial networks is NBC. With its Red and Blue dual-network concept, it gained momentum early and was well entrenched when CBS arrived on the scene.

The David Sarnoff Era

When we look at NBC's past and present, one individual stands out—David Sarnoff. Sarnoff's career bloomed at RCA, NBC's parent company. The shore-bound radio operator during the *Titanic* disaster, Sarnoff came to RCA from the American Marconi Company. His energies at RCA were directed to two areas—developing new broadcast technology and promoting television stars as a means of winning audiences. In the first area he encouraged the development of FM radio and committed millions of dollars to Zworykin so that he could continue his experiments with an electronic television camera. Led by Sarnoff, RCA became a pioneer in color television, developing the system finally approved by the FCC for full-scale production.

Stars, Color, and Innovation

Radio had been a medium of programs, but Sarnoff realized that television made people bigger than life. So people would be where NBC invested its efforts. Major moves were made to attract and sign top talent. Names like Milton Berle and the "Texaco Star Theater" gave America a new night at home with the television set. Sid Caesar's "Your Show of Shows" and Eddie Cantor's "Comedy Hour" contributed to the "people" orientation.

Two other factors helped even more—color television and innovative programming. Following the FCC's approval of RCA's color system, it was only natural for NBC to move ahead and air as many color programs as possible. The first network colorcast was the 1954 Tournament of Roses Parade. Ten years later, NBC was producing almost all of its programs in color.

NBC initiated a series of firsts in programming, many later copied by other networks. With host Dave Garroway, "Today" dawned in 1952, followed two years later by the "Tonight Show" with host Jack Paar, then Steve Allen, and finally Johnny Carson. Years later, David Letterman followed Carson with his "Late Night with David Letterman" program. During the 1980s NBC's long-running "Today" program had been eclipsed by ABC's "Good Morning America." For a while the network was headed by Grant Tinker, who had formally headed M-T-M Productions, the creator of a number of highly successful shows including the long-running sitcom "The Mary Tyler Moore Show." NBC regained a foothold in the ratings with such award-winning programs as "The Cosby Show," "Hill Street Blues," and "L.A. Law."

General Electric Buys RCA

In what became a clean sweep of acquisitions during the 1980s, industrial giant General Electric completed the purchase of NBC's parent company, RCA, in 1986. It brought together two giants of American industry in a $6.3 billion deal. In some ways the companies were similar. Both GE and RCA manufactured television sets, home appliances, and aerospace systems. Both companies engaged in information services including satellite, electronic mail, and data processing.[8] GE was also in broadcasting, although in a small way with one Denver, Colorado, station.

Westwood One Buys the NBC Radio Network

A year after acquiring RCA, Westwood One, a California-based owner of radio networks, purchased NBC Radio for $50 million. Westwood One started in the mid-1970s with a $10,000 investment by Norm Pattiz, a former TV account executive. After the acquisition, many of the NBC operations were combined with Mutual Radio, already a Westwood One subsidiary. The deal brought to Westwood One a number of NBC Radio formats including a young-adult network and a nighttime talk service. The first few years after the acquisition were a bit rough, primarily because of a sluggish market for national radio advertising. Pattiz, however, worked to improve NBC Radio's position by adding affiliates and improving programming.[9]

Fox and Other Television Networks

With the launch of limited prime-time programming in 1987, Fox Broadcasting Co. made a commitment to compete with the three established commercial networks. While early promises of guaranteed ratings had to be scaled back, the network managed to weather some rough starts and remain on the air.[10]

Regional television networks providing limited specialized programming operate with anywhere from a handful of affiliates in a single state to dozens of affiliates covering wide geographic areas. The California Farm Network, the Dakota Giant Network, and the Pacific Mountain Network are three examples.

RADIO NETWORKS

When television reached its golden age, some predicted the demise of radio, especially the radio networks. This simply has not happened. Although they do not provide the same amount of programming they once did, radio networks are still a vital part of radio broadcasting. The major television networks just discussed all developed radio networks. ABC in particular capitalized on the trend toward specialized audiences in radio and divided its radio network into the ABC Contemporary Network, ABC FM Network, ABC Rock Network, ABC Information Network, ABC Entertainment Network, and the ABC Direction Network. Each has its own format, which blends with the format of the affiliate station.

Like television, specialized and regional radio networks provide programming to specialized audiences. For example, "The Wall Street Journal Radio Network" provides affiliates with stock market reports and original business news features as well as stories from the *Wall Street Journal* and other Dow Jones publications (Figure 9–3). Concentrating on affiliates in the top 50 markets, the network pulls affluent listeners who appeal to advertisers. Another example of a specialized radio network is the Winners News Network (WNN), which distributes motivational programming. Regional radio networks include such names as the Maine Radio Network, Florida Network, Texas State Network, California Agri-Radio Network, California Farm Network, Louisiana News Network, and dozens of others.

FIGURE 9–3 The Wall Street Journal Radio Network delivers business news and financial information to radio stations nationwide. Networks such as The Wall Street Journal Radio Network provide an avenue for advertisers who want to reach an up-scale audience with higher than average incomes. The network draws on the information and reputation of the *Wall Street Journal* newspaper.

UPI AND AP RADIO NETWORKS

Although wire-service audio is not a network in the traditional sense, many radio stations affiliate with these services. Local stations, paying a fee to subscribe, can then sell time within the wire-service newscasts to local advertisers.

United Press International (UPI) first launched an audio service in 1956 by establishing state-news telephone feeds in North Carolina and California. By 1960 the wire service had established a New York audio headquarters, and three years later it added a Washington, D.C., audio bureau. The next year an audio bureau in London was opened, and by 1965 coast-to-coast hookups were operable. The 1970s brought hourly newscasts, expanded sports coverage, and experimental satellite transmission. The 1980s saw UPI plunge into financial troubles but by the end of the decade the wire service had reorganized under new owners and UPI Broadcasting Inc. was formed to expand the UPI radio network and provide services to television and cable.

The *Associated Press (AP)* began an audio service in 1974 called *AP Radio*, basing it on the same principle as that of UPI Audio's national news service. Affiliates pay a fee to receive the service and can sell local advertising time within regularly scheduled newscasts. Audio cuts from both services can be incorporated into locally produced newscasts.

ETHNIC, EDUCATIONAL, CABLE, AND MDS NETWORKS

Whereas ABC divides its listeners according to such demographic categories as age, education, and income, other networks have been successful in directing programming to audiences on the basis of race and national origin. The National Black Network (NBN), a radio network formed in 1973 and headquartered in New York, serves affiliates covering black markets and reaching America's black population. News on the network emphasizes events and issues of importance to black Americans.

The Sheridan Broadcasting Network (SBN) is another black-owned network. Sheridan owned WAMO AM/FM in Pittsburgh, Pennsylvania, then in the late 1970s bought the Mutual Black Network, changing its name to The Sheridan Broadcasting Network. Along with special programming such as documentaries and features, SBN also airs regularly scheduled news and sports programs.

Networks also connect stations reaching the Hispanic-American audience (Figure 9–4). With Spanish-language programming and content that reflects this important culture, these networks are gaining an increasing number of listeners as well as advertiser support.

Many noncommercial radio and television stations not only provide instructional programs for in-school use but also originate and transmit programming for state and regional networks. Often these networks are part of a state system of higher

FIGURE 9–4 Highly specialized ethnic networks reach target audiences that tune in to hear programming in their native language. The Telemundo Network broadcasts in Spanish and is an avenue for advertisers who want to reach Hispanics. The growing number of Hispanics living in the United States also results in other media services such as specialized rating services highlighted in Figure 1–3.

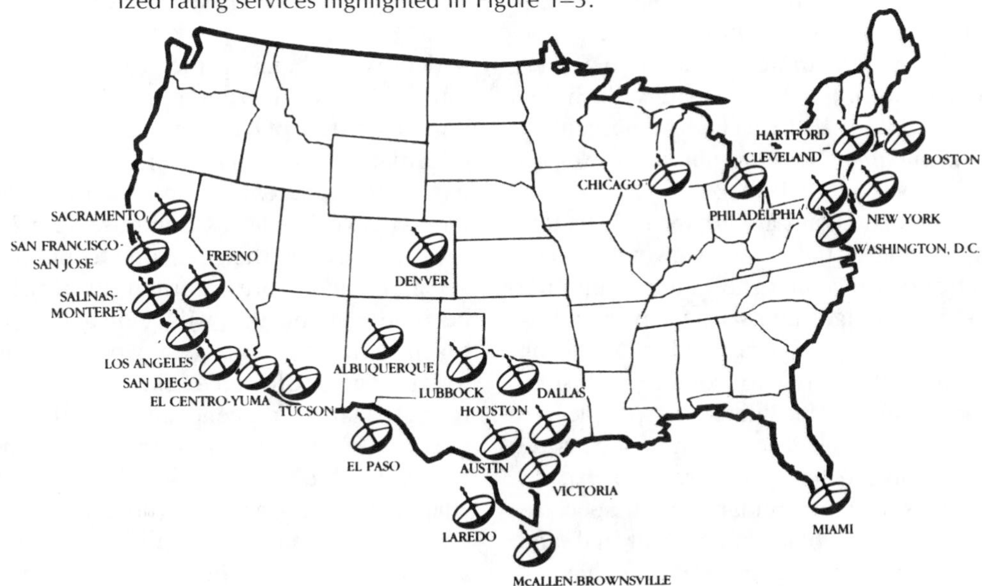

education. The Indiana Higher Education Telecommunication System (IHETS) links both state and private colleges and has two-way television capabilities. A professor can lecture from one campus to students on another, and the students can immediately ask the professor questions through a two-way talk-back system.

Cable television's growth has spurred interest in networks linking cable systems. The technology already exists, in the form of satellite interconnection systems, to offer satellite-distributed programs to cable systems. Multipoint distribution service (MDS)—the use of microwave to distribute television signals over a small region—has enabled such organizations as schools and churches to enter the world of network television.

DECLINING NETWORK AUDIENCES

While the networks venture into cable and pay TV, these same technologies are siphoning away the network audience. The decline in the major networks' share of television viewers has been steadily declining until in 1990 it had reached 60 percent of the *prime-time* viewing audience, down from 92 percent in 1978.[11]

New technologies are not the only factors contributing to the loss in viewership. The practice of cancelling shows and filling the slots with replacements in a never-ending battle for ratings has resulted in television schedules that are, to some, confusing. Some observers speculate that people are not watching television as much because they simply do not know what to expect. Regular shows are replaced with specials, reruns appear in place of new material, and shows are cancelled before they have a chance to build viewer loyalty. Strikes by writers have also hurt audiences because networks traditionally rely on new material to start fall seasons. In addition, as cable systems become more profitable they can improve the quality of their own productions and compete with the networks for new material. *Direct syndication*, the practice of distributing first-run material direct to stations without going through a network, is becoming easier through the use of satellites. Moreover, the programs are higher quality and thus they can generate more income.

Although networks cannot stop the development of alternative technologies, they have made some efforts to stop using reruns. Attempts to develop fresh material, especially during the summer months, may prove helpful, but the trend of declining network viewership appears firm, and networks are adjusting by maximizing profits on the audience that does exist.

DATA NETWORKS

Videotex is changing the traditional definition of network from a broadcasting system to a system of data services that provide information to subscribers. Subscribers have direct access to information data-based networks such as Compuserve, which they tap through their personal computers. As more and more information networks become available they will begin to compete for subscribers just as the networks compete for audiences. Like advertisers who want to reach the largest audience, information suppliers will select among the information networks, for their services will have a better chance of being chosen where more subscribers exist. Many of the same considerations we now see with network radio and television will apply to these new networks. If, for instance, an information network supplies financial or business information, then it will be more appealing to, say, stockbrokers than a network oriented toward consumer goods.

SYNDICATED PROGRAMMING

Thus far we have learned about general radio and television programming, a station's external and internal environment, and how programming is related to station income. We have also discussed a major source of programming, the radio and television networks. Another important source of programming is syndication. *Syndicated programming* is distributed directly to stations, not through a network, although it may have first appeared on a network. With more and more independent stations, the development of cable, and alternative delivery systems such as pay TV and multipoint distribution, the demand for syndicated programming continues to grow.

Breaking the Ice: "Mary Hartman, Mary Hartman"

In order to go into syndication, most programs first started as network material, proved they could attract a loyal following, then made the break with the network and wooed individual stations. The first program to go directly into syndication on a mass national scale and achieve popularity was Norman Lear's somewhat controversial "Mary Hartman, Mary Hartman." Having been turned down by the TV networks, Lear resorted to direct syndication for the show. It drew large audiences, and some bold metropolitan stations even put the program up against their competitors' late-evening news programs, scoring rating points that gave news consultants a headache.

"Mary Hartman, Mary Hartman" started a chain of events that did more than merely attract additional viewers. Local affiliates have become increasingly dissatisfied with the type of programming coming out of network hoppers. With the success of "Mary Hartman, Mary Hartman," station managers realized that there were alternatives to the network distribution system. What is popular in one city may be unpopular in another, and syndication gives managers a freedom of choice that network television does not.

The Deer Hunter and Other Early Syndicated Programs

Another successful syndication effort that caught the eye of the networks was MCA's decision to release the motion picture *The Deer Hunter* directly to independent stations instead of to the networks. Many of these stations ran the movie against the 1980 election-night coverage and scored impressive results. *The Deer Hunter* gave WOR-TV in New York one of the highest ratings in its history and KCOP-TV the highest metered rating ever for a televised movie in Los Angeles. Many stations ran the film with almost all of its violence and strong language intact. Nineteen stations purchased the film immediately before or after its initial TV release.

Even such early favorites as "The Lone Ranger" are seen regularly in syndication. Contemporary artists such as Barbara Mandrell have achieved syndication success with original musical variety programs. Rock-and-roll radio authority Dick Clark has been syndicated on both radio and television. Major novels adapted for television have appeared in syndication through release to independent stations.

Direct Syndication

The success of syndicated programs has lured more syndicators into the marketplace and thereby increased programming variety and kept costs competitive. Direct syndication eliminates the need to "fit" the network audience. For example, if you own a production company that produces a weekly series on skiing techniques, the network probably will not be interested. Although its affiliates in New England and the Northwest would consider the show, stations in the Deep South would very likely preempt the program with something more popular in their local areas. Yet by going into direct syndication you can create your own network of stations. In fact, you might find enough stations in New England and the Northwest that are interested in airing your series to make your venture profitable.

The same principle applies to the expanding enterprise of regional syndication. Perhaps your area hosts a salt-water fishing tournament. A documentary of the tournament would be of interest to other stations located near the ocean. Conversely, generating wide acceptance for the show in Nebraska and Iowa would be difficult. Major events are becoming attractive to syndicators. Tying up rights to special productions can make a long-term syndication package very lucrative.

Many syndicated programs find their way out of the broadcasting station and into an educational institution, corporation, library, or church. A syndicated religious documentary that first airs on television can find additional audiences at a Sunday morning Bible class or at a convention of lay preachers. As you can see, the market possibilities for syndicated material are turning it into a booming business.

SYNDICATED RADIO

Thus far we have been talking about television syndication. But syndication is also alive and well in radio, especially since the networks take up so little of the station's programming schedule. Syndication has found a ready market in the increased number of automated stations. Such stations still program commercials and local newscasts, but much of their remaining programming comes from syndication. Automated stations can choose from a wide variety of syndicated radio programming—anything from music to interviews and talk programs. In addition, both nonautomated and automated stations use syndicated jingles between records as an introduction to newscasts and weather reports, and as a musical background for commercial and public service messages.

Why Syndicate?

Syndication is an economical way to operate a station while providing a competitive sound. But before deciding to use syndicated programming you first should evaluate the competition. Is there a programming need not being met or an audience not being reached by the competing stations? Second, you must decide that when your station does air locally produced programming it will be of the highest quality. Commercials, local newscasts, weather reports, and supplemental entertainment by the local disc jockey must be able to blend professionally with the syndicated music. Poor quality in local production is only accentuated by high-quality syndicated programming.

Formats

A wide variety of syndicated formats are available, but not every syndicator defines the same format in the same way. Thus, managers must preview the syndicated "sounds" to make sure they fit the needs of their market. Most syndicators offer a limited number of formats, keeping to a reasonable minimum the size and range of their own musical libraries. Syndicators specializing in contemporary music might syndicate progressive-rock, country-rock, and soft-rock formats. These formats tend to overlap, and hit songs appearing on more than one chart, such as *Billboard*, can be inserted in more than one syndicated format.

Other syndicated packages can be even more specialized, offering "middle-of-the-road string orchestras" or "upbeat string orchestras." Still another may contain "middle-of-the-road orchestras and bands." To offer these specialties, the syndicator purchases mostly albums and interchanges the different cuts on the albums to fit the different formats.

Talk radio is a specialized format in which a host takes calls from listeners who phone the station. The original program can be taped in one market and aired in another. The host is careful not to make any reference to the city in which the show originates, and the topics do not have a highly localized emphasis. If the show does become localized, local references can be deleted before the program is syndicated.

The Consultant as Syndicator

In television syndication, the syndicator has little involvement with the station except to sell the program. In radio, however, the association is much closer. Radio formats are a more "finely tuned" type of programming than are television formats. First, they last longer, in some cases for the entire twenty-four-hour broadcast schedule. Second, the competition may be much greater. Instead of three or four television stations, the market may consist of twenty or more radio stations. Third, the local radio station directly participates in the programming because of its locally produced and inserted commercials and newscasts.

As a result, the radio syndicator often doubles as a broadcast consultant, recommending how the programs should be utilized. Judging whether the local commercial production is up to par or if the musical background of the commercials matches the syndicated format becomes the consultant's major concern. If the station is going to continue to use syndication, it must see an increase in its audience and ultimately its profits. Moreover, the reputation of the syndicator is at stake. A station at the bottom of the ratings is not good publicity for the syndicator. But a station on top can be a valuable asset in selling other stations. Thus, management's willingness to work closely with the consultant once the syndicated programming has been acquired can determine the difference between success and failure.

THE ECONOMICS OF SYNDICATED PROGRAMMING

Before you run out and rent a television camera, enlist your friends as actors, and find a ski slope or a fishing tournament, it's important to understand the economics of syndication. Although profits can be made with good syndicated material, the investments, the competition, and the gamble are all big business.

The Financial Commitment

Regardless of how popular an event may be or how good a script looks, networks and station managers make commitments only to finished products. More than one show that looked good on paper turned out to be a flop once it was produced. The star of the show might not pull it off, what seems like a good idea to New York program executives may fall flat in Peoria, and it can rain on the day of the ski tournament. Networks and stations have been through it all before. The cost of producing a pilot program in the hope of getting it syndicated can run into many thousands of dollars. Many foreign countries with developing television systems purchased syndicated features, then decided they could produce the programs cheaper themselves. After about two years they went back to buying syndicated features—what had seemed like a $2,000 or $3,000 savings turned into a $50,000 investment that flopped.

Before syndication comes a pilot show. The pilot—which requires writers, producers, directors, talent, sets, and equipment—can cost as much as a quarter of a million dollars for the quality that will appeal to management and meet the competition. Perhaps more than one pilot will be necessary. But even if the pilot looks good, the investment can go down the drain if the audience rejects it. It's a gamble.

For major one-time events there is the cost of securing the rights to the event. If more than one company is bidding, the cost can escalate out of reach. Major sports events are one example. Minor sporting events can't compete with the audience-drawing power of major events, even though the cost of production might be less. A million-dollar investment can mean a million-dollar profit. It can also mean a million dollar loss.

Promoting the Commitment

After key stations have made the commitment to air your program and the program is ready for syndication, you must advertise and promote it. You will need to buy advertisements in trade magazines, produce promotional brochures for a direct-mail campaign aimed at station managers, and exhibit your program at major conventions where program executives gather.

In some cases, you may decide to sell your program at a reduced rate to large-market or prestigious stations so that their acceptance can be publicized in your advertising. Seeing that a major Los Angeles station has purchased your program may reassure a Cheyenne, Wyoming, station that the program will draw an audience. You can create a bandwagon effect: the more stations that buy your program, the more that others will want to buy it.

Promotion can be especially difficult if the program is a single event, if it's a bit unusual, or if the syndicator or production company is untried. Managers understand football and situation comedy. But the Minnesota Canoe Championship or the White Water Raft Races will be tougher to sell. A well-known syndicator can deal with credibility, but a new company must prove its stripes. It must promote not only its product but also its reputation as a company.

SELLING SYNDICATION

The selling of syndicated programming is much like a farmers' market. There are more deals, more contractual arrangements, more variety, and more companies than there are fresh vegetables and home-made pies. Many syndicators sell on a market-to-market basis. Others try for contracts with groups of stations, such as network-owned and network-operated stations. The syndicator who can, for example, say that the ABC-owned and -operated stations have already purchased one of his programs is in a strong position to deal with other stations. This is especially true when the syndicator is marketing pilot programs. If too few stations buy the program, production for the series is cancelled. But by the time it's cancelled the stations have lost the opportunity to buy alternative programming, which might then be in the hands of their competition.

Bidding

The syndicator who wants to let the marketplace determine the selling price may decide to auction a program, an increasingly common practice. On a given day a syndicator will send telegrams to all the station managers in a market, announcing the availability of a program. Each manager is given a certain amount of time to bid on the program, and the highest bid buys the show. Variations include open bidding, where each manager knows at any time what the highest bid is, and sequence bidding, where each manager is told the bid of the previous manager contacted. First practiced in the book-publishing trade, auctioneering has produced record prices for syndicated programs.

Barter Arrangements

Bartering, or trading advertising time on a station for the opportunity to air a program for free, is still used to sell syndicated shows, but even here there are pitfalls. Some large-market managers shy away from bartering, feeling that if the program was of top quality it would have been sold outright. In some cases that's true, but there are many exceptions. For example, major advertisers who want barter time may place a large financial commitment behind the program. A major corporation that has a reputation for sponsoring quality programs can use that reputation to make both a syndicated program and the accompanying public-relations efforts a success. Also, some programs are available only through barter. Many advertising agencies are actively involved in barter because the station supplying the advertising time airs not only the program but also their client's commercial. Other agencies provide barter programs as long as their clients receive "commercial credit" in return. For instance, an ad agency may provide a program series in exchange for 100 minutes of commercial time to be used in whatever way the agency wants.

SUMMARY

Networks give advertisers a means of reaching a mass national audience at economical rates. Network entertainment programs are frequently bought from independent production companies. At the heart of the networks are the affiliate stations. The three major commercial TV networks, ABC, CBS, and NBC, still remain a major force in television distribution.

ABC, formed when NBC sold its Blue Network, quickly acquired an identity and became a formidable competitor of NBC and CBS. CBS gained an early reputation as a leader in broadcast journalism. From little more than $1 million in revenue in 1928, the network progressed to yearly income levels above $1 billion by the mid-1970s. NBC, meanwhile, developed under the RCA umbrella and the guidance of David Sarnoff. NBC made its mark with big-name stars, color, and innovative programming, and continues to make profits. Although the networks themselves are big businesses, all are interconnected with other business ventures. All three of the major networks changed hands in the 1980s. ABC was bought by Capital Cities Communications Inc., and became Capital Cities Communications/ABC Inc. CBS was bought by entrepreneur Laurence A. Tisch. RCA, NBC's parent company, was bought by General Electric, and the NBC radio networks were sold off to Westwood One, a California-based owner of radio networks. Regional and specialized television networks exist and operate with as many as a few affiliates in one state to dozens of affiliates across entire regions.

All three of the major commercial TV networks are involved in radio. ABC, in attempting to satisfy the specialized audience of radio, began splitting into specialized networks in 1968. Today ABC retains the specialized concept with different radio networks geared to the different formats of their affiliate stations. Regional and specialized radio networks also exist and include such enterprises as The Wall Street Journal Radio Network and others.

Both UPI and AP operate radio networks and audio services and provide alternatives for radio stations that may not be affiliated with the commercial networks or for stations that want to supplement their network programming. Other types of radio and television networks include ethnic, educational, and cable networks.

The growth of cable and pay TV and the increased use of VCRs and tape-rental stores have all contributed to a decline in the networks' television audience, which is now at approximately 60 percent of the prime-time television audience.

An alternative to network programming is syndicated programming, which bypasses the major commercial networks and is distributed directly to stations. Syndicated programming is available for both radio and television. Many successful network programs have been released in syndication.

Syndication companies that deal with broadcasters are subject to controls. So that local stations can retain the authority to determine programming for their market, syndication companies are prohibited from specifying such things as broadcast hours, the amount of syndicated programming the station airs, commercial loads, or time devoted to news.

To be successful, syndicated programming must receive a financial commitment and skillful promotion. The sale of syndicated programming includes both bidding and barter arrangements.

OPPORTUNITIES FOR FURTHER LEARNING

BEDELL, S., *Up the Tube: Prime-Time TV and the Silverman Years.* New York: Viking, 1981.

BLUM, R. A., AND LINDHEIM, R. D. *Primetime: Network Television Programming*. Stoneham, MA: Focal Press, 1987.

BOTEIN, M., AND RICE, D. M., *Network Television and the Public Interest.* Lexington, MA.: Heath, 1980.

DORDICK, H. S., BRADLEY, H. G., AND NANUS, B. *The Emerging Network Marketplace.* Norwood, NJ: Ablex, 1981.

SPENCE, J., WITH DILES, D. *Up Close and Personal: The Inside Story of Network Television Sports*. New York: Atheneum, 1988.

10 EDUCATIONAL, PUBLIC, AND BUSINESS TELECOMMUNICATION

At 6:55 A.M., radio and television stations are keying up for another broadcasting day. For some, sign-on came in the predawn darkness with "Today," the "CBS Morning News," or "Good Morning America." In foreign countries, 7:00 A.M. ushers in news, special-feature programming, entertainment, and public affairs.

Sign-on preparations are also bustling at this hour in many school systems. A closed-circuit television system warms up for a "broadcast" of the daily calendar of events, students begin to produce a morning news program, and teachers preview instructional television lessons that they will incorporate into afternoon lectures. At a nearby college, a professor is preparing a lecture that will be aired over a statewide educational television network. In a famous medical school a television camera focuses on the operating table and broadcasts a color picture to interns in an observation room across campus. It is all part of the world of educational broadcasting.

ETV: THE BEGINNINGS

For our own purposes, we shall define *educational television (ETV)* as all noncommercial television programming and commercial programming produced especially for educational purposes, whether or not the program is used for direct classroom instruction.

Although a closed-circuit television system was in use at the State University of Iowa as early as 1932, it was in 1938 that the first over-the-air experimental broadcast for educational purposes took place.[1] Originating in a studio on the third floor of the RCA building in New York City, the program was broadcast from the transmitting tower atop the Empire State Building. A class of 250 students seated in a large auditorium on the sixty-second floor of the RCA building viewed the program, which consisted of an explanation of how television worked. A two-way radio hookup connected the studio with the auditorium. Capturing the flavor of the event, program instructor C.C. Clark asked a student in the auditorium to come to the studio to have a question answered. About ten minutes later, when the student arrived at the studio and appeared on the screen, the group in the auditorium broke into applause.

The Experimental Era

The early 1940s were years of continued experiments with the new medium. The Metropolitan Museum of Art in New York City arranged with CBS to televise its painting collection. Francis Henry Taylor, director of the museum, predicted television would be "just as revolutionary for visual education as radio was for the symphony and the opera."[2] The following year saw such mass-oriented educational television programs as New York's WCBW broadcast of a series of first-aid programs in cooperation with the American Red Cross. Programs of a more informative and educational nature were aired during the height of World War II, including one from Schenectady, New York, on blood plasma.

Early enthusiasm for educational television was sidelined by World War II, but when the war ended the industry began to concentrate on it again. NBC announced the first "permanent" series in educational broadcasting, "Your World Tomorrow."[3] Some of the early program titles in the series included "The Mighty Atom," "Jet Propulsion," and "Huff-Duff, the Radio Detective." The network secured the cooperation of the New York City Board of Education in having students watch the programs in special "viewing rooms." The students then evaluated the programs. This was an early example, though on a very limited scale, of the systematic evaluation of educational programming.

Despite the encouragement from the networks and the willingness of certain school officials, ETV was a long way from widespread acceptance. As late as 1947, the *Journal* of the National Education Association reported the efforts of the State of Virginia to make the transition to what was termed visual education.[4] Although Virginia was known for its pioneering efforts in the field, the report did not mention educational television. The medium had simply not been able to rise above all the movie projectors, slide projectors, charts, models, and posters of the typical classroom. In 1946 two researchers studied the skills and knowledge that elementary teachers needed in order to use audiovisual aids.[5] Out of the forty-two survey items that the study incorporated, which included mechanics, utilization, production, and facilities, none referred to television.

ETV Gains Acceptance

Finally, at the turn of the decade, educational television began to obtain recognition. In 1949, Crosley Broadcasting awarded a fellowship to a Kentucky high-school principal, Russel Helmick, to "carry on research of how education by television can best serve the needs of the general public." The broad charge assigned to Helmick included a review of literature on "radio education."[6]

In 1950, public awareness of the importance and potential of educational television rose when Dr. Earl J. McGrath, U.S. Commissioner of Education, appeared at hearings before the FCC and called for at least one channel in every broadcasting area to be reserved for educational purposes.[7] The FCC responded favorably. When the commission lifted its freeze on new station licenses, it allocated 242 channels for educational use. The first to take advantage of the newly allocated frequencies was the University of Houston, whose station KUHT (Figure 10–1) went on the air in 1953.

Organized Support for ETV

Organized support for educational television came in the early 1950s when the American Council on Education coordinated the formation of the Joint Committee on Educational Television (JCET). The committee brought together seven supporters of ETV, all of them organizations that had originally called upon the FCC to hold hearings on the subject. Financial help was also provided by a $90,000 grant from the Ford Foundation. One of the committee's main goals was to assist educational institutions in establishing stations.

Other financial support for ETV went directly to colleges and universities. Syracuse University received a $150,000 gift earmarked for graduate programs in radio and television. In the fall of 1950 Syracuse announced it was conferring the new degree of Master of Science in Radio and Television.[8]

Pockets of Resistance

Teachers, however, were beginning to object to administrators' requests that they allocate time for television teaching on top of

FIGURE 10–1 KUHT at the University of Houston was the first educational noncommercial TV station to sign on the air after the FCC lifted its freeze and allocated 242 new channels for educational use.

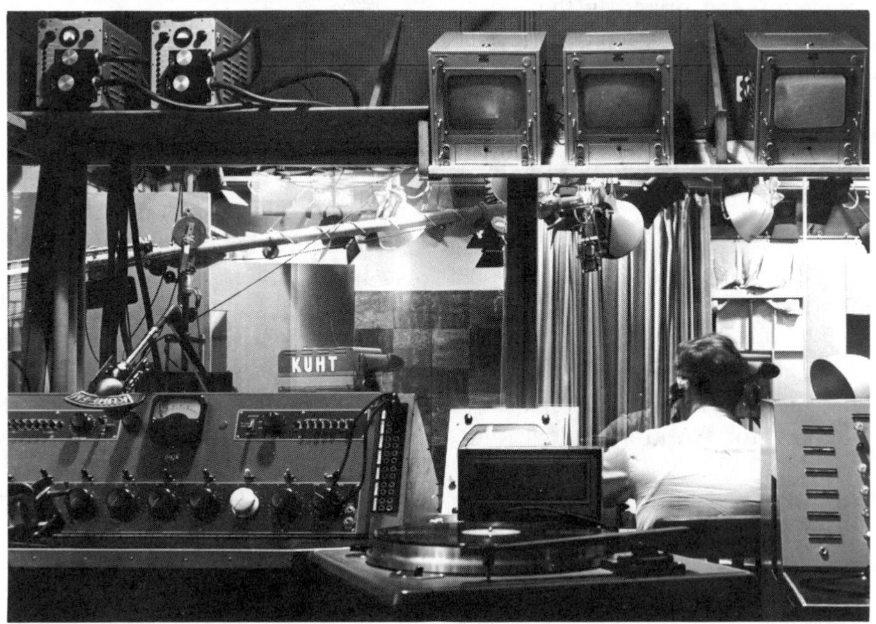

their already crowded classroom schedules. They also wanted to participate from the very beginning in the formulation of instructional television programming. This is now a routine way of developing good instructional programming. The thought of television as a substitute teacher, understandably, received considerable resistance. Classroom teachers were fully aware that the interaction between student and teacher contributed much to the learning process. They appreciated television's particular abilities but wanted to protect their own role in the student-teacher interaction.

Political and Economic Concerns

The public's focus on the economics of educational television also concerned educators. The politicians' and taxpayers' misconception that the medium could eliminate personnel and save tax money was worrisome. Educators realized that they had to reemphasize the importance of the teacher in the classroom. They faced still another economic concern—the financing that educational television would receive. Television studios could easily incur bills in the hundreds of thousands of dollars. Money like that could not only pay several salaries but also provide many of the traditional classroom teaching aids.

AIRBORNE ETV: THE MPATI EXPERIMENTS

One of the most ambitious experiments in early ETV occurred in Indiana through the Midwest Program for Airborne Television Instruction (MPATI).[9] A DC–6 aircraft equipped with a television transmitter and antenna and based at the Purdue University airport would fly in a circle and beam educational television programs to outlying areas of Indiana, Illinois, Kentucky, Michigan, Wisconsin, and Ohio.

The idea had been conceived in 1944 by Westinghouse engineers and was supported by Westinghouse and other organizations including the Ford Foundation. The MPATI transmitter was on the air between 1961 and 1968 and was eventually superseded by satellite delivery of ETV where the antenna in space could relate ETV to entire continents. Nevertheless, MPATI fulfilled an important purpose and much was learned about technology and how teachers and students would use ETV in the future.

EDUCATIONAL VERSUS INSTRUCTIONAL BROADCASTING

As educational television matured and more and more uses of in-school programming were developed, a second term—*instructional television (ITV)*—evolved. Although both terms tend to be used interchangeably, instructional television refers to programming designed specifically for use in the classroom or in some other direct teaching role. Notice we did not use the words *in school* in our definition. Both ETV and ITV are employed beyond the confines of the classroom.

Differentiating between the two terms is also important for economic and political reasons.[10] Part of this criticism of ETV went beyond the classroom to policy issues. Because much of the programming of ETV stations was designed for in-school use, and because major federal funding was emerging, some saw the potential threat of a national school system under federal control. Congressional advocates of ETV were charged with maintaining "a sinister conspiracy directed from the U.S. Office of Education to homogenize the nation's moppets by a standardized curriculum spread from sea to shining sea."[11] The furor caused ETV stations to stop and to consider seriously where their future programming dollars would come from.

During this time, smaller portable videotape equipment arrived in the marketplace. Such equipment permitted many educators to continue tinkering with television, but this time without the threat of outside controls. For school administrations this inexpensive portable television equipment was particularly satisfying since they could buy it with local money and thereby avoid the public criticism caused by large federal expenditures. Although new, the idea of airplanes and satellites beaming federally funded programming to local school systems was sold most easily to both school officials and taxpayers as something "strictly experimental." The portable equipment satisfied those who demanded new technology in the classroom. Instead of ETV for large numbers of students, something new appeared—instructional television used strictly in the classroom, the programs often produced by the teachers themselves.

TRANSITION TO PUBLIC BROADCASTING

In the mid-1960s, as ITV programming switched to local production and distribution, educational television stations, their programming supplemented by locally produced videotape material, began to move toward a greater variety of programs beyond in-school programming. At about the same time, foundations and government agencies were supporting not only in-school production but also the development of state and national systems of noncommercial broadcasting.

In 1962 the Educational TV Facilities Act provided over $30 million over a five-year period for the development of state systems of educational broadcasting. Today many of these systems are integral parts of a national system of noncommercial radio and television stations.

Full-scale planning for what would eventually become a national system of noncommercial radio and television stations began in 1965 with the Carnegie Commission for Educational Television. The commission, whose members were a broad range of industry leaders, was charged with conducting a "broadly conceived study of noncommercial television and to focus attention principally, although not exclusively, on community owned channels and their service to the general public."[12]

THE PUBLIC BROADCASTING ACT OF 1967

The commission's recommendations were the impetus for the passage of the Public Broadcasting Act of 1967, which allocated $38 million to the improvement and construction of facilities for noncommercial radio and television in the United States. Also formed by the act was the *Corporation for Public Broadcasting (CPB)*, a quasi-governmental company that would administer the funds appropriated for *public broadcasting* by Congress. The act authorized the CPB to develop programs, establish a system of interconnection, and assist stations to get on the air.

With the passage of the act, noncommercial radio and TV stations became known as "public" broadcasting stations, signifying their ability to secure income from the public as well as from corporations, foundations, and government agencies.

PUBLIC TELECOMMUNICATIONS FINANCING ACT OF 1978

In 1978, Congress passed the Public Telecommunications Financing Act, which required the CPB, in consultation with interested parties, to update a five-year plan for the development of public telecommunications. The planning funds supported research projects that helped to determine how funding impacted on station operations and how new technologies could be used to improve telecommunication services.[13] Public broadcasting stations that meet certain operating standards are eligible for financial support from the CPB.

Community Advisory Board Requirement

Also part of the Public Telecommunications Financing Act of 1978 was the requirement that each station receiving funds through the CPB establish a community advisory board that would be permitted to review station programming goals. Restricted by law to an advisory capacity only, the community advisory board could be delegated other advisory capacities beyond programming. Today, successful stations receiving money through CPB use these required advisory boards not only as a link with their communities but also as an asset in fund-raising and other station-sponsored activities.

CPB Program Fund

Two years later in 1980 the CPB Program Fund was created to select and fund programs free from outside influences and covering such programming areas as children and the family, minorities, drama, health, and public affairs.[14]

THE INDEPENDENT TELEVISION SERVICE (ITVS)

In 1989, the Independent Television Service (ITVS) began operations under an appropriations amendment to the Public Broadcasting Act.[15] ITVS operates with its own board of directors and is charged with supporting independent public television productions. The ITVS operates without direct oversight from any noncommercial television entity including the CPB Program Fund. A constant problem for many public stations was the lack of upfront monies available to independent producers who wanted to produce programs with a possible eye to distribution on public stations. The ITVS makes such monies available and the end products can be distributed beyond public broadcasting.

FUNDING AND FOCUS IN THE 1990s

Although always subject to Washington political pressures, the Omnibus Budget Reconciliation acts have funded CPB into the 1990s. At the same time, however, the focus of CPB's funding in the 1990s has narrowed in focus to a limited number of higher priority items instead of trying to fund many smaller projects. The top priority of this narrower focus as approved by the CPB board of directors is children's programming followed by news-outreach-public affairs programming, and arts and cultural programming.[16]

NONGOVERNMENT FUNDING OF PUBLIC BROADCASTING

In addition to government funding, many individuals, corporations, and foundations also support public broadcasting. Some of the contributions are quite substantial. For example, Chrysler Corporation contributed $3.9 million to fund a five-hour television prime-time series dealing with education. Other companies fund such popular programs as "Wall Street Week with Louis Rukeyser," and "Washington Week in Review." State Farm Insurance has had a long association with the public broadcasting program "The Woodwright's Shop." At the local level, companies fund productions, which may also be distributed over the public broadcasting network.

Many local stations also conduct their own direct-mail and on-air fund-raising drives to supplement their operating budgets. Such on-air drives interrupt programming, resulting in complaints from the viewing audience. Some stations promise not to fund-raise on the air if enough money is raised through direct-mail solicitations.

Some fund-raising efforts have been radical departures from traditional income-producing activities. For example, a congressionally mandated advertising test resulted in stations in New York, Chicago, Pittsburgh, Miami, New Orleans, Binghamton (New York), Muncie (Indiana), Philadelphia, Erie (Pennsylvania), and Louisville participating.[17]

THE PUBLIC BROADCASTING SERVICE

To help meet the goals of the Public Broadcasting Act of 1967, the CPB joined with many of the licensees of noncommercial television stations in the United States to form the Public Broadcasting Service (PBS) in 1970. Today the PBS is the primary distribution system for programs aired on public broadcasting stations. These programs are distributed through a multichannel satellite system that enables individual stations to choose programs that best suit their viewers. Although stations determine their own schedules and broadcast PBS–distributed programs at various times, the vast majority of the stations carry a core of programs fed by PBS.[18]

PBS is similar to commercial television networks in that it distributes programs. Beyond that, however, most of the similarity ends. PBS is in many ways more sensitive to its affiliates, for the affiliates represent the public, which at least in theory owns the PBS stations through its contributions and tax dollars. Part of this sensitivity is generated by the PBS board of directors. The board represents the general public and station managers, and it gains insights into the success and failure of its system. The board's communication with stations and the public makes it a fairly broad-based indicator on which to base decisions.

THE SCOPE OF PUBLIC TELEVISION PROGRAMMING

Public television has become a system of broadcasting whose wide variety of programs range from traditional in-school instructional programs to programming having a wide national and even international appeal. "Sesame Street" (Figure 10–2) has become an internationally popular program and the characters have resulted in licensing fees—fees paid for the rights to use or associate with Sesame Street characters—in

FIGURE 10–2 Big Bird of the popular PBS series "Sesame Street." This popular children's program very early showed the potential of educational and public television for mass audiences beyond the classroom.

the millions of dollars.[19] "Wall Street Week with Louis Rukeyser," "Wild America," "This Old House," and the "MacNeil/Lehrer News Hour" have had long-running associations with public broadcasting. WNET–TV's "The Adams Chronicles" depicted America's Adams family of presidential fame and its role in history from 1750 to 1900. Some programs such as "The Adams Chronicles" were accompanied by teacher and curriculum guides, which are especially helpful when the program is used in the classroom. Regional programs have also reached a broad-based audience. "Mr. Rogers" (Figure 10–3), which started out as a regional program on WQED–TV in Pittsburgh, became popular nationally.

Many colleges are taking advantage of the outreach programs that employ ITV to reach adults who may not want to take the trouble to come to campus or, because of the distance and time, simply cannot come. The colleges list the television courses as part of their regular schedules, enabling students to enroll as they would for any other course.

FIGURE 10–3 Fred Rogers of "Mr. Rogers' Neighborhood." The program began as a regional show broadcast over TV station WQED in Pittsburgh. Later it became a national favorite of preschool-age children. Rogers' slow, quiet presentations and gentle manner are successful in holding the attention of youngsters and are in sharp contrast to the fast-paced presentation of "Sesame Street." Rogers has achieved recognition as one of the pioneer educators in the field of public television and has been, along with the program bearing his name, the recipient of numerous awards.

NATIONAL PUBLIC RADIO

Radio also benefited from the Public Broadcasting Act, and in 1971 many noncommercial radio stations became members of *National Public Radio (NPR)*, the radio equivalent of PBS. NPR differs from PBS, however, in that it also produces programs, whereas PBS's chief responsibility is distribution. NPR affiliates also produce programs, which are often syndicated by NPR and made available to member stations. One of the more famous programs aired on NPR is the daily newsmagazine program "All Things Considered," acclaimed by both educators and the public for its informative, in-depth coverage of news and public affairs. NPR's growth has been closely aligned with the overall growth of noncommercial radio.

Fiscal and Management Problems

For a brief period of time in 1983, NPR found itself on shaky financial ground when an audit turned up a debt of approximately $9 million. The audit led to the resignation of some of the NPR administrative staff, not over charges of any wrongdoing, but in an atmosphere of alleged mismanagement. The result was a rift between NPR and many local NPR affiliates, who had to reconsider both their relationship with the network and their financial structure in the wake of possible funding shortfalls.

National Program Production and Acquisition Grants

Partly because of NPR financial and management problems, the financial structure was changed in the mid–1980s to send CPB monies direct to public radio stations, which in turn could choose to affiliate with NPR by paying an annual membership fee. The monies, distributed through a National Program Production and Acquisition Grant (NPPAG) to individual stations, can also be used for other purposes such as paying for independently produced radio productions. While many stations use the NPPAG monies for NPR dues, the key to the fiscal restructuring was to free stations to use the money as they see fit and make NPR accountable to the affiliates.

AMERICAN PUBLIC RADIO

American Public Radio (APR) was formed in 1982 and today is somewhat of a competitor of NPR. APR was originally conceived to assist public radio stations in producing and distributing their own programs. While that philosophy remains, the organization has grown to the point where it has seen some national success with programs such as "A Prairie Home Companion," produced by Minnesota Public Radio. The program dealt with a fictional Midwestern town of Lake Wobegon and starred Garrison Keillor. The program eventually halted production, but the tapes continue to be sold, but above all it launched APR into national success and made it a viable force in public radio programming (Figure 10–4).

More recently APR has broadened its programming to include radio news from such sources as *The Christian Science Monitor* and business programming produced by CBS News. Other news and informational programs come from such sources as the Canadian Broadcasting Corp. and the BBC World News Service. Perhaps most significant is the fact that APR has not hesitated to schedule such programming in the time slots traditionally reserved for NPR's "All Things Considered."

PUBLIC RADIO AS INSTRUCTIONAL RADIO

Along with its entertainment and informational programming, public radio continues to be a viable educational force with a rich tradition in instructional programming. WHA (Figure 10–5) radio began at the University of Wisconsin in 1919 and still serves a wide spectrum of listeners through direct instructional programming. By the late 1950s, students in Wisconsin schools had been listening to instructional radio for more than thirty years.

Other states have also found radio important to their instructional missions. FM stations located at strategic points in a state are programmed as part of a network so that schools over a wide region will be within earshot of the broadcasts. The broadcasts are directed toward elementary grades through high school as supplements to regular classes. To facilitate reception, states may offer special radio receivers (Figure 10–6) pretuned to the educational stations. Actual airings are scattered throughout the school day, and supplementary teaching materials are available for teachers who wish to plan lessons around the broadcasts.

BUSINESS TELEVISION

At a time when networks are losing audiences and new technologies are impacting on traditional broadcasting, one area of

FIGURE 10–4
Garrison Keillor and his popular radio program "Prairie Home Companion" became one of public radio's most popular shows. The program was set in an imaginary town in Minnesota called Lake Wobegon. Homespun humor brought radio drama and entertainment to millions of listeners in the 1970s and 1980s. The American Public Radio Network distributed the program, which made Keillor a media star. He has been featured on the covers of such magazines as *Time* and *The Saturday Evening Post*.

telecommunication is growing in both size and importance. This nonbroadcast aspect of telecommunication, called by various names such as "business television," "corporate video," "nonbroadcast-television," "corporate telecommunication," involves the application of video technologies to business and industry (Figure 10–7). Companies are spending upwards of $5.5 billion annually on video production equipment and those expenditures are increasing by 20 percent annually.[20] By business, we also mean nonprofit sectors of the economy as well as government uses.

The next time you walk into an automobile dealership, a department store, or a shopping mall, be observant and locate some of the video monitors that may be showing product information. At the automobile dealership a videotape showing the latest road-handling features of a new model can be viewed in the showroom and

FIGURE 10–5 WHA radio in Madison, Wisconsin, went on the air in 1919 and is considered the first noncommercial public radio station in the United States. This painting, which is part of a larger mural displayed at the University of Wisconsin at Madison, shows the early studios of WHA. The station became a source of news, agricultural information, and education for its listeners in the upper Midwest.

used to convince the customer to take a test drive. At the shopping mall a department store uses a video monitor playing upbeat music and showing a beach scene to help sell swimwear.

In a nearby corporate office an executive uses a personal computer interfaced with a voice, video, and data network to discuss a new product with a business associate thousands of miles away. High-resolution graphics display the product while the latest sales information can be accessed and displayed next to the product.

All three examples illustrate the uses of video and telecommunication beyond the realm of standard commercial, educational, or public broadcasting. In addition, such uses as video teleconferencing, interactive video, and other applications proliferate.

In-House News and Information Programming

Most of us think of television as the network prime-time programs or daily newscasts broadcast in our area. These programs are the result of decisions on what stories to use, how to edit them, what the audience wants and what it should have, what graphics to use, which audio cuts to include, and many more matters. Those same decisions are also made every day in places far from the network and newsrooms. They are made at corporations where TV production crews and corporate newscasters are preparing the daily newscast that will be sent to employees at the downstate plant or through corporate networks to international offices (Figure 10–8).

Applications of Corporate Newscasts

One example of a daily corporate newscast is the Federal Express program "FedEx Overnight," which is sent daily to more than 900 Federal Express locations in the United States and Canada.[21] More than 300 programs annually are distributed via satellite over the Federal Express business network. Plans for the international Federal Express television network grew out of the compa-

FIGURE 10–6 Instructional radio is used as a classroom supplement in many elementary schools. Some school systems are linked together as part of statewide instructional radio networks. Teachers use the educational radio broadcasts to supplement in-class personal instruction.

ny's success with earlier "state-of-the-company" briefings first distributed over a land-based network. With the satellite network, programs originate at the company headquarters in Memphis and are scrambled for security purposes. Receiving locations are equipped with a video receiving monitor and a VRC, which can be automatically activated from the Memphis location to record programs when no one is available at the receiving site.

To diversified companies spread over wide areas, corporate news programming is especially valuable. An oil company may consist of oil exploration, refineries, and gas stations, as well as the corporate office. How is it possible to link the people and activities of these varied enterprises? The main instrument for many companies has been, and still is, the corporate magazine or newsletter. Filled with pictures and articles about the corporation's activities, it is sent to all employees. Different parts of the company have their own reporters or stringers who contribute to the magazine. Now, although continuing the corporate magazine, corporations are turning to television. A gas-station owner wins a community award; an oil-rig worker is promoted; a pipeline crew starts a new project. The corporate news cameras catch it all and it appears in the corporate television newscast.

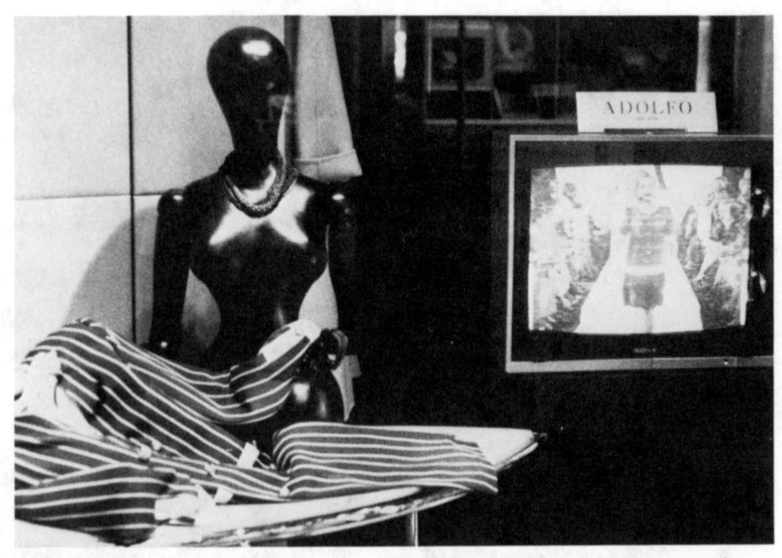

FIGURE 10–7 Nonbroadcast uses of television are found everywhere from retail sales, to corporate training programs, to in-house business news programs. In a department store window in Zurich, Switzerland, this TV monitor shows a continuous videotaped program highlighting a designer's latest fashions, which includes the dress on display. The videotaped program showing in the store window also aids in attracting the attention of shoppers, and it adds additional elements of motion and color to the window display.

Content of Corporate News Programming

The content of a corporate newscast is not unlike that of other traditional TV newscasts, namely the programming of timely information of interest to the audience. In a corporation the news may include how the company performed that particular day in filling company orders. Comparisons may be made between different plants or warehouses to see which ones were the most efficient and then leaders are signaled out for special recognition. The president of the firm may make brief remarks on some issue of importance to employees, such as stock options or new company health benefits. Even news bulletins are used for alerting employees to particularly important information. By no means are the newscasts dull. Sophisticated production techniques, much like network television, can be seen in business television.

CORPORATE NETWORKS

The companies we have been discussing use their own corporate telecommunication networks to link together offices, plants, dealerships, and others who will benefit from the information transmitted from a central location to employees and, in some cases, the news media. New product information can also be sent to potential customers. For example, IBM operates its own Field Television Network (FTN). FTN programs are broadcast live to more than 350 IBM locations throughout the United

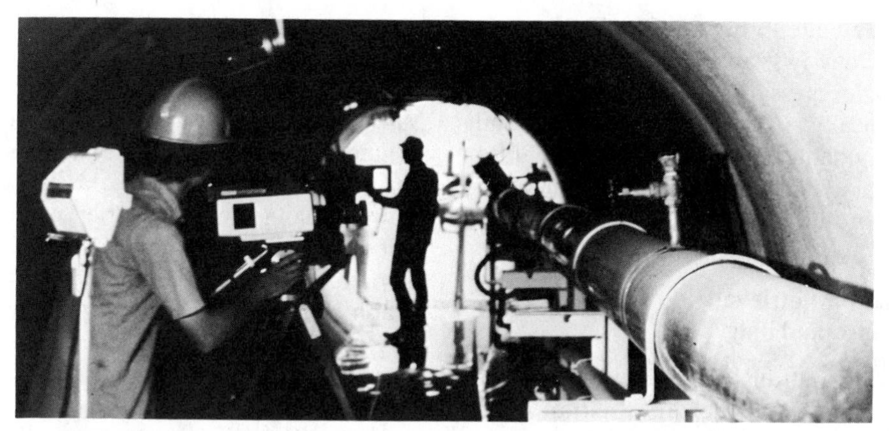

FIGURE 10–8 Television can be used as an effective corporate training tool and can go into areas that are too dangerous for the untrained employee. Here a portable TV camera and recorder are used to record a welding demonstration at a steel company.

States. An interactive telephone system permits viewers at the various locations to call questions back to the presenters at the main studio. Customers and users can interact with specialists who are authorities on IBM equipment. The FTN, which serves as a marketing and information tool for IBM, broadcasts on business days and during business hours.

CORPORATE VIDEOTEX

More and more offices are finding that a personal computer equipped for videotex is as important a component of an office communication system as is the telephone. That same computer may be connected with personal computers at the desks of other employees, and all may feed into a central mainframe computer with a large storage capacity. These corporate videotex systems are equipped with graphic capability that permits data to appear as pictorial representations. Printers can make a hard copy of anything seen on the desk-top display terminal.

Interoffice Communication

A corporate videotex system permits electronic mail to be delivered between personal computers as easily as it can be delivered between post offices. Instead of writing a letter or memo and having it typed, placed in an envelope, and delayed while traveling through the interoffice mail system, an employee can type the letter into the videotex system and send it instantaneously to the intended receiver. Either the sender or receiver can then electronically discard the letter or store it in the computer for future reference.

Employees who want to electronically mail a letter outside the company can easily do so by interfacing personal computers at one location with those at a distant location. A branch office across town or around the world can be connected through telephone lines and microwave links. Security against tampering is achieved by access codes assigned both to individuals and to the system.

If the memo is intended for anyone in the company it is keyed into the corporate electronic bulletin board. Employees starting their day first access the bulletin board to check for messages that have been posted electronically. An alert mechanism built into the videotex system activates a light or beeper alarm when priority information is waiting on the board.

Information Retrieval and Electronic Filing

Retrieving a letter is one form of information retrieval, but other information is also retrievable, everything from the latest inventory to airline schedules. Any information entered in the central data bank or accessible elsewhere can be called up on each videotex terminal. The graphic capabilities of the system enable employees not only to know what's in inventory but also to look at the screen and see a computer-displayed illustration of each item.

An executive's assistant is planning a trip for his boss. From the videotex terminal he calls up a map of the United States that features color-coded weather zones and temperatures. He then accesses the airline schedule and keys-in the flight information, the corporate account number, and his boss's name. Within minutes the trip is planned. The executive knows what plane to catch and what to wear when she arrives. All of the information, all of the planning and decision making, was done from the videotex terminal located on the assistant's desk. The same system that checked the weather and made the plane reservations can also show a diagram of the plane's seating arrangements, the dinner menu, and a computer-generated picture of the food served on the flight.

If the executive wants to consult with one of the sales managers before leaving for her trip, she telephones and asks the manager to check her videotex terminal for the sales projections for the coming month. Viewed by both the executive and the sales manager on their respective videotex terminals, the sales projections appear as computer-generated bar graphs. As the two of them confer about the monthly quota, they discuss how a change in inventories would affect sales. They key-in the pertinent information, and the screen displays a new bar graph showing different projections.

Both the executive and the sales manager like the new projections and decide to change inventories. So that they can refer back to their discussion later, they electronically file the projections. Without the computer-based videotex system the same information would not have been available or simply would have taken too much time to retrieve to be practical. Companies in which it is practical to retrieve such information may very well put their competitors out of business. The old adage that time is money takes on new importance in our computer society.

In addition to the information in our example, such things as telephone numbers, stock quotations, credit authorizations, accounting information, financial notices, job openings, and reference guides are among the thousands of pages of information that can be accessed through a videotex system.

Retail-Sales Support

Earlier in this chapter we discussed how a television program can facilitate showroom sales. Videotex can be used for similar functions. For example, a retail sales clerk can take orders at a central catalogue location and call up on a videotex screen a description of the catalogue item, its price, and a computer-generated picture of the item. The customer and the clerk can both view the article to verify its specifications and make sure the order is correctly entered.

At another videotex terminal the customer searches through an electronic catalogue for the item he wants. There on the screen he sees a picture of the item and its price and specifications. He also spots a notice on the screen that that particular item is on display and on sale in the hardware department. If he wishes, he can go there and see the merchandise firsthand. The same videotex system that showed him the merchandise can take his order. By keying-in his credit-card number he can charge the item to his account and then pick it up at the warehouse or have it delivered to his home.

Employee Training

The same employee who in the past sat at her desk with a training manual can now sit at a videotex terminal and use the electronic manual accessible through the videotex system. If she needs additional information she can check other pages right on the screen. Illustrations appear along with instructions. If the employee wants to jump ahead to the advanced section of the manual she simply keys-in the correct information and there on the screen appear the instructions and graphics at an advanced level. The employee's office is not piled up with volumes of seldom used books, paperwork, manuals, printouts, and other space-consuming items. All of this clutter is stored neatly in a computer accessible by any employee with a videotex terminal.

SUMMARY

Educational and public broadcasting stations serve both the public and educational institutions that incorporate radio and television in their educational mission. From a station that beams an educational program via satellite across the globe to a professor using ETV in the classroom, educational broadcasting plays a major role in teaching and learning.

ETV began in 1932 at what was then the State University of Iowa. In 1938, RCA beamed a program from the third floor of the RCA building, via a transmitting tower atop the Empire State Building, to the sixty-second floor of the RCA building. A class of 250 students viewed a program showing how television worked.

Other experiments took place in the 1940s, including a broadcast of the paintings in the Metropolitan Museum of Art, first-aid demonstrations, and information and training programs during World War II. By the 1950s educational stations were on the air, including KUHT, the first ETV station, broadcasting from the University of Houston. Organized support for ETV came with the formation of the Joint Committee on Educational Television (JCET). Some pockets of teacher resistance occurred but were gradually overcome as planners realized television would complement, but not replace the classroom teacher.

The Midwest Program for Airborne Television Instruction (MPATI) proved one of the more interesting experimental systems and employed an airborne antenna in a DC–6 aircraft that flew over the Midwest beaming educational television programs to schools in Midwestern states. Satellites in space replaced the MPATI concept. MPATI did, however, move educational television forward and showed the uses that instructional television (ITV) held for the nation's educational system.

Gradually, educational television became public television with a much broader audience and the support of Congress, which supported it with legislation such as the 1962 Educational TV Facilities Act, the Public Broadcasting Act of 1967, the Public

Telecommunications Financing Act of 1978, and other legislation that firmly established a system of public radio and television in the United States. At the head of this federal commitment is the Corporation for Public Broadcasting (CPB), which distributes funds to individual public broadcasting stations. The Public Broadcasting Service (PBS) for television and both National Public Radio (NPR) and American Public Radio (APR) distribute programs to affiliate stations. The Independent Television Service (ITVS) makes grants to independent producers and operates without direct oversight from any noncommercial entity including the CPB Program Fund.

Programming on public television varies greatly from locally produced programs, which may be of local or regional interest, to national programming for specialized audiences such as "Wall Street Week with Louis Rukeyser," "Washington Week in Review," and others.

National Public Radio and American Public Radio are two major sources of programs for public radio stations. Qualified stations receive funds directly from the CPB and can select which programs to buy and air. Radio also serves as an instructional medium. With roots such as those dating back to WHA at the University of Wisconsin, educational radio today serves many individual school districts with in-school programming.

Non-government funding for public broadcasting comes from such sources as individuals, foundations, and corporations. Many stations engage in direct over-the-air fund-raising as well as direct-mail fund-raising efforts.

The application of video technologies to business results in more than $5.5 billion in equipment expenditures annually with these expenditures increasing at the rate of 20 percent per year. More companies use television for in-house communication than there are commercial television stations in the United States.

Corporate newscasts are one of the uses of business television. For example, Federal Express beams its daily news program to more than 900 locations in the United States and Canada. The content of in-house business news programs ranges from state-of-the-company addresses by chief executives to stock information, order processing, and news and features for employees. Corporate policy, executive reports, and the latest regulations affecting a company are subjects readily found in corporate newscasts.

Business networks also support companies in sales and marketing. IBM's Field Television Network broadcasts live programs where IBM customers can interact with experts through telephone hookups to learn new applications of IBM products.

Corporate videotex permits graphics and data to be exchanged within the same building or across the country.

OPPORTUNITIES FOR FURTHER LEARNING

BERG, L. V., AND TRUJILLO, N. *Organizational Life on Television.* Norwood, NJ: Ablex, 1989.

DOUGLAS, S. U. *Labor's New Voice: Unions and the Mass Media.* Norwood, NJ: Ablex, 1986.

FULK, J., AND STEINFIELD, C., eds. *Organizations and Communication Technology.* Newbury Park, CA: Sage, 1990.

PALMER, E. L. *Television and America's Children.* New York: Oxford University, 1988.

SCHILLER, H. I. *Information and the Crisis Economy.* Norwood, NJ: Ablex, 1984.

SMITH, T. J., III. *The Vanishing Economy: Television Coverage of Economic Affairs, 1982–1987.* Washington, DC: The Media Institute, 1988.

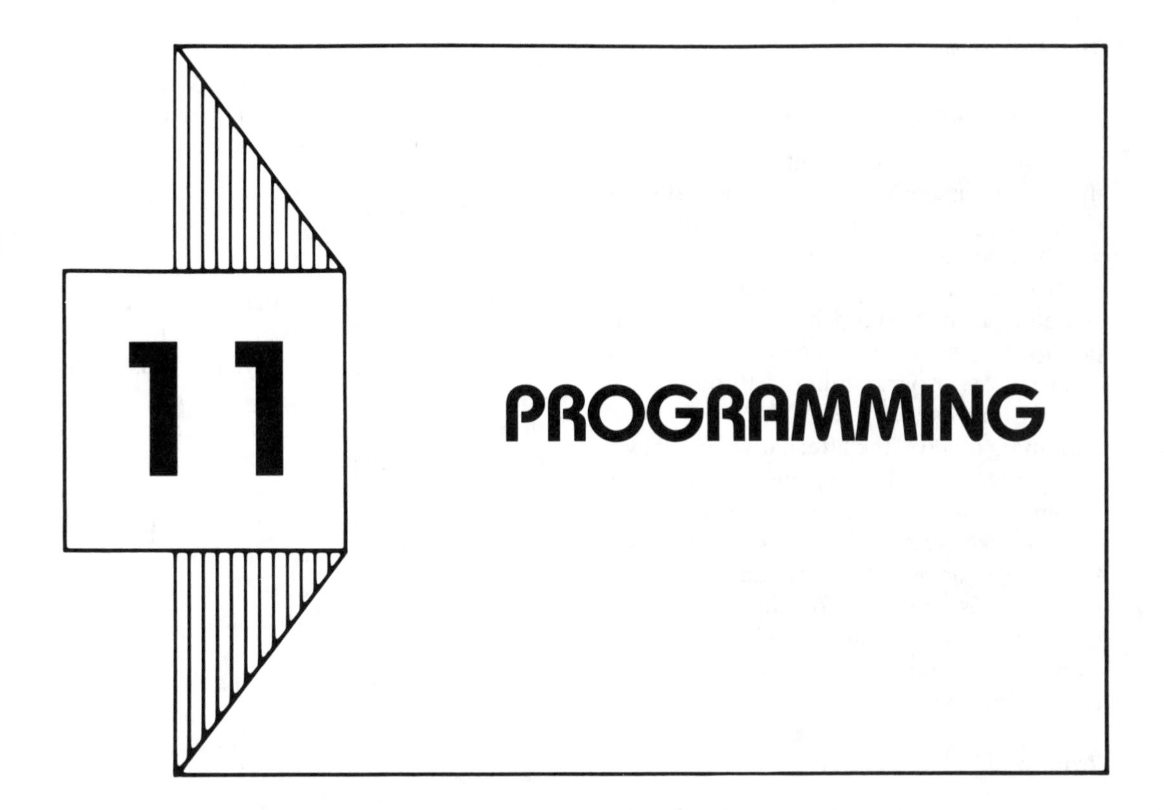

We often take for granted the programming available to us on radio and television. Over-the-air broadcasting is mostly free. Programming ranges all the way from a network television spectacular to a local radio station's swap-shop program. We may watch an art exhibit from Italy, a news report from Iran, or a bicycle race from Indiana. Despite the seemingly effortless way in which programming reaches our living room, behind the scenes are talented strategists who must combine creative decisions with economics, technical reality, and political constraints. The purpose of this chapter is to look more deeply into the field of radio and television programming and production and to learn more about the decision making that results in what we as consumers hear and see on radio and television.

THE CONTEXT OF BROADCAST PROGRAMMING

Programming is the product of broadcasting.[1] Just as a store sells goods or a law firm sells advice, broadcasting sells programming. Just as store owners set prices for their goods and lawyers set fees for their services, broadcasters set rates for the commercials that will share time with programming. Even public broadcasting solicits contributions on the basis of the type of programming it can offer. But if programming on either commercial or public broadcasting should be irresponsible or not meet the public's needs, then, like the lawyer who gives bad advice or the store owner who sells inferior goods, broadcasting will be out of business.

The Marketplace

But not all broadcast programming is produced and disseminated in a democratic society. Where it is not, the marketplace in which it operates and the government that controls it will have a profound effect on the end product. Keep in mind that competition in our society is a great determinant of broadcast programming. Although such competition can evoke outcries over programming quality, the alternative to our system could be total government control, as is common in authoritarian countries. Thus, even though we may not like everything we hear or see on broadcast media, if the programming is preferred by the majority of the viewing or listening public, and if the competitive marketplace will support it, then it will likely stay on the air.

Profit and Public Service

Stations operate to reach as many people as possible in a limited audience environment. In other words, with a limited number of people available to tune to radio or television the station reaching the largest number of people has a financial advantage in being able to charge more for advertising. On the other hand, that station must also operate in the public interest, and that may entail programming decisions that serve the public but are not profitable. For example, to avoid covering a local disaster and thus not be of value to people who need emergency information simply so regular programming and commercials will not be preempted is irresponsible. The critical question for commercial broadcasters, rather, is this: How can we effectively program a station to serve the needs of the public while making a profit? If this dual responsibility can become the foundation of broadcast programming, then program planners and broadcast management can work cooperatively with the public.

UNDERSTANDING RADIO FORMATS

Radio programming evolved from the theatrical elements of radio drama to the music and news formats of today. Decreasing its reliance on network programming, radio became a specialist in locally produced programming. It even specialized in different formats in order to compete not only with other stations but also with television.

Today radio enjoys an almost endless number of formats and combinations of formats, each designed to reach specific audiences. One of the earliest specialized formats was Top-40 radio, which developed in the 1950s and concentrated on rock-and-roll. Top-40 radio has now become more of a middle-of-the-road format, still catering to rock-and-roll fans but mild in tempo compared with the progressive and acid-rock formats that developed in the late 1960s.

Music plays a major role in the specialized formats of many stations. Still other stations cater to the needs of various population groups — blacks, Hispanics, speakers of other foreign languages, and religious groups. Public and educational broadcasting stations fill an additional programming niche.

For *program directors* in competitive markets, trying to reach a specific audience has become a very challenging task. This is due to the vast number of stations, closely overlapping formats, considerable variance in station power, and a wide variety of on-air personalities. Reaching an audience has become a select art that combines musical tastes, demographics, an understanding of popular music, and the skill to put them all together.

PROGRAMMING STRATEGIES IN COMPETITIVE MARKETS

General managers of large-market stations usually delegate programming responsibilities to a specialist, a program director, while retaining overall station responsibilities.

Analyzing the Competition

As the program director for a large-market station, you would be concerned not only with the needs of your audience but with competing stations. Which stations are the leaders? What do they program? In addition, you need to realize that long-cherished listening and viewing habits are very hard to break, and that the reputation of being the leading station and having seniority in the community is a powerful advantage. A television news image developed over decades of dominance with established anchors is tough to beat, although by no means impossible. Rights to certain programming can also be a big factor. When a competing TV station owns syndication rights to a group of popular programs that are being programmed in key time slots, securing programs to compete with these market leaders can be difficult and expensive. Similarly in radio, twenty years of the same radio personality, of programs heard in the same time slots, and of familiar newscasters can be tough to beat.

Adjusting to Programming and Formats

Assuming that no station dominates the market and that sharp programming decisions can be made, careful planning can significantly improve your station's position. In radio, a good way to start is to examine the other stations' formats. You may discover a format that is not covered by the competition. Another possibility is formats that seem to be already covered may lend themselves to alteration. Even though a competing station may be programming a tight playlist (few songs but each song repeated often) with the top twenty hard-rock songs, the market may have room for a soft-rock format—more mellow rock songs with slower tempo.

In television, analyzing syndicated programming and scheduling is critical, especially lead-ins to news programming and local access periods, since these are the only areas where programming changes can be made. Network programming is fixed and for the most part remains firm, although a manager is under no legal obligation to carry specific network programs. Independent stations have much more programming flexibility, although not being a network affiliate can be a disadvantage, and if the more profitable stations own all the rights to the good programming, buying rights to strong competitive programming can be difficult.

Radio Personalities

With radio, a strategy is to examine the personalities in your area. You may find that although a popular personality gives a station an identity, when that personality signs off the air the other personalities do not retain listeners. At this point you can make one of two decisions. You can develop and promote a personality at your own station and then place that person on the air head-on with the competing personality. Or you can schedule your personality's show so that it avoids such a confrontation. If you have the money, you can sometimes hire the popular personality away from the competing station. This practice is risky, however, especially if the personality is linked with a certain format. Remember, popularity may have been achieved because of a combination of personality and music. Having one without the other may simply prompt the listeners to turn the dial.

Bringing in a personality who has been a big hit in another city may also spell disaster. Listeners develop habits and tastes based on a variety of factors. Perhaps the previous popularity was achieved by a series of successful promotions. But the listeners in the new city have not been exposed to the same promotions, so they simply view the newcomer as a rank amateur and give their loyalty to another station. A famous Chicago station once brought a top disc jockey out of the South to run a popular morning show. Although his southern humor had made the announcer popular in his home state, he immediately turned off Chicago listeners who weren't attuned to southern speech patterns.

You may discover that your market simply has no dominant personality, and that by adding and promoting one of your own you can bring your station a loyal following. If you're successful at this, you will probably find your idea becoming very expensive as other stations try to hire the person away. If and when the person leaves, the popularity vacuum may be filled by another station. But it's all part of large-market programming.

Jingle Package and Production Aids

In addition to considering personalities and formats, you will also want to obtain a good set of *jingles*. Jingles are a set of short musical recordings, all designed around a common musical theme and usually related to the station's call letters. Consider an upbeat combination of musical notes behind the call letters WABC. Each letter of the call has a musical identity, as does the set of call letters as a whole. The same musical identity is woven into a musical background for a thirty-second commercial or public-service announcement, a ten-second musical background for a station identification, or a promotional announcement. The station's news programming will also use the same musical tones. A good jingle package, although costing many thousands of dollars, is not only a good investment but a must in large-market programming.

Television Personalities

Television personalities may also be well known in a market and help promote and hold the station's dominance. This is especially true for anchor people who may have been in the market a long time. Since news programming can be very popular, stations go to great lengths to hold popular anchors and keep them under contract. Noncompetitive clauses are common in contracts and prevent anchors from working at a competing station for a period of time after they leave the station where they are employed. Sometimes a competing anchor can be lured from a competing station, although it is usually expensive and risky since the image of the anchor's previous station follows the person to the new station. Thus, it is necessary to publicize the anchor by using the new call letters, something that can confuse the public.

TELEVISION NETWORK PROGRAMMING

Beyond the strategies for competitive markets are programming decisions involving a station's network programming.

Television vs. Radio

To understand how network programming fits into television's picture one must first understand that television programming differs from radio programming in two ways. First, much of radio programming is music, although news, talk, and information stations do dominate in some markets. Television, on the other hand, includes everything from cartoon shows to coverage of major news events. Second, although independent stations and cable are certainly factors to contend with, the real competition in any market is usually found among the affiliates of the three major commercial networks. Networks play a powerful role in determining the position of their affiliates in local ratings. The investor wanting to buy a television station or the program director programming it inherits the merits and demerits of the network, and each has only a limited amount of flexibility in instituting programming changes.

The National Audience

For the network, the audience is national, and the programs that make the networks are those that appeal to the largest segment of the population. After all, the network is in the business of convincing advertisers that buying its commercial time will enable them to reach the largest national audience. Any program that weakens that audience base has little chance for survival. The result has been what critics call *programming to the lowest common denominator*, or trying to reach the largest mass audience.

If television advertisers wish to reach a more specialized audience, they select different types of programming rather than changing stations or networks the way a radio advertiser can. Television has not yet become a specialized medium, although cable, the rise of independents, and new technologies are starting to change this (Figure 11–1). In the meantime, local stations that want an alternative to network programming are finding an increasing variety of syndicated programs available. Affiliates are also exercising their right to have a say in network programming. The networks, facing competition from syndication, independent stations, cable, and stations implementing their own satellite distribution system, are taking the time to listen.

Affiliate Decisions

Despite the powerful position of the network, it could not exist without its affiliates. Unless its affiliates agree to carry its programming, the network will have no market for its advertisers. If a network program does not receive affiliate clearance, its national ratings can suffer. Advertising dollars aren't spent on network programs that show up poorly in the ratings. In a sense, a station imposes economic sanctions on a network when it refuses to carry a network program.

Instead of refusing to clear network shows, some affiliates record the programs and air them at a different time. A violent or sexually explicit program may be rescheduled late at night. The show is still broadcast, but the new time slot may reach a much smaller audience than it would have reached during prime time. The audience may also have a different makeup, perhaps a different income level. The program shown opposite the rescheduled program may seriously cut into its viewership; if the program had appeared in its regularly scheduled time slot, it would have faced weak competition. Even more important, there is no way to measure this new audience successfully, for each station rescheduling the program may air it at a different time. So a rescheduled program can spell

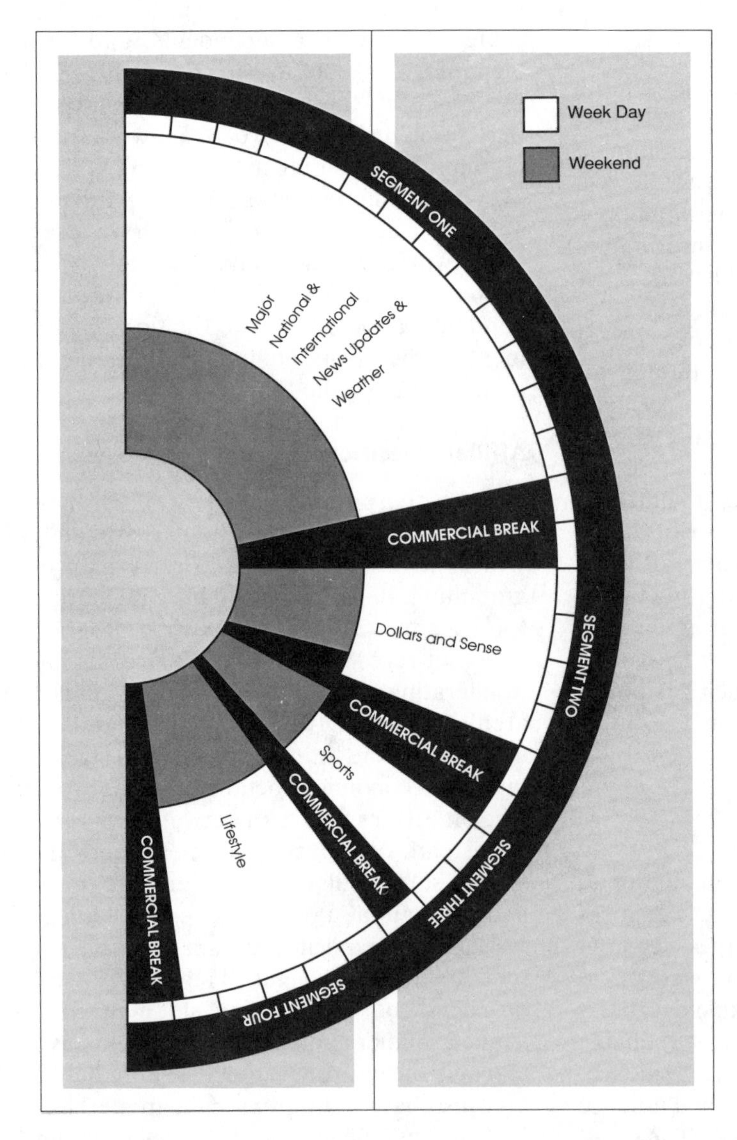

FIGURE 11–1 CNN Headline News is one example of specialized programming delivered primarily through cable systems. CNN Headline News reaches an up-scale audience interested in news and information programming. In addition, local cable systems that carry CNN Headline News can sell local advertising within CNN Headline News programming, thus providing smaller businesses with access to this more specialized service. It also provides the businesses with a way to reach consumers who have higher income and education levels. An increasing number of cable television systems, independent TV stations that provide alternative programming to the networks, and satellite channels are making television a much more specialized medium.

the same economic disaster as a preempted one.

THE STATION'S COMPETITIVE ENVIRONMENT: EXTERNAL FACTORS

Now that we have acquired an overview of radio and television programming, we will examine the competitive environment in which a station operates (Figure 11–2). Many of the concepts we have just discussed apply to this environment. The competitive environment is made up of both internal and external factors. The key to effective station operations is to integrate them so that they produce the most favorable competitive climate. External factors are largely beyond the control of station management. Internal factors, on the other hand, are

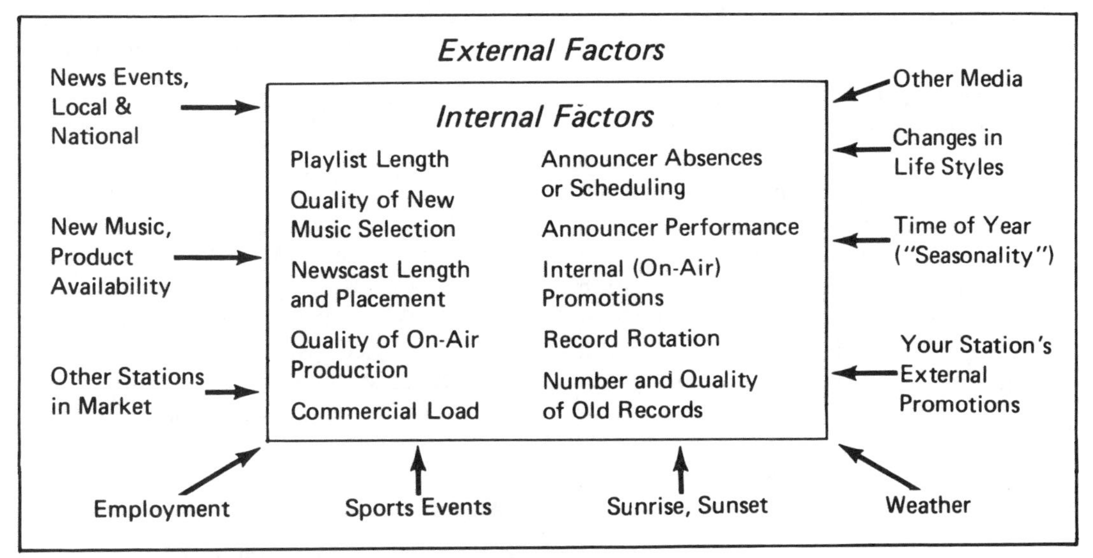

FIGURE 11–2 Both internal factors, those under the control of the station, and external factors, which are for the most part not under the station's control, can influence decisions that program directors make in radio and television. For example, a major local news event can increase the number of listeners and viewers to local radio and TV stations. Similarly, a hit record or popular recording artist whose songs fit the format of a radio station can increase the number of listeners to that station.

under the complete control of station management.

News Events

Every news director has finished a day saying, "It's been a slow news day." In other words, very little has happened in the community. This doesn't mean every broadcast journalist is interested in sensationalism, but if nothing happens, nothing can be communicated to the public. News programming tends to draw more listeners or viewers on a day when things are happening than on days when very little is happening. If the mayor announces an investigation of alleged corruption in the police department, if a hurricane wipes out a portion of the city, if an earthquake strikes, if a plane crashes, if a new school bond passes—it all makes news. Essentially, a station and its news programming are at the mercy of news events.

New Music Affecting Radio

In radio, while the news department is spending time covering events, the programming department is concerned with the latest music or syndicated features. The availability of recorded music determines what a station has available to attract listeners. If a hit song becomes popular in the same format the station programs, then the station benefits. Awareness of new trends in recorded music and the ability to play selections that are emerging in popularity can give a station a significant advantage.

Other Stations in the Market

The number of stations in a market can be one of the biggest factors in determining success. Many small markets have stations that are extremely successful, partly because they face no competition. Stations in these markets are not overly worried about programming since there is no competition to take away the advertising dollar. But consider the station competing against a full complement of network television stations and dozens of radio stations. Finding a niche is tough, and wooing a particular audience with a particular format can be a formidable task.

Employment

Employment conditions can affect a station in two ways. First, the level of employment in a community will be related to the overall economy. If people are working, they are buying. If they are buying, the merchants are making money and in turn can afford to advertise. Second, if there's a large employment pool, the station can find ample talent. Typically, good talent is scarcer in smaller markets than in larger ones.

Sports Events

How can sports events affect a station? Coverage of both the local high-school sports schedule and—if a market includes a college town—the college sports schedule is an excellent way to attract advertisers. And, naturally, a winning team can attract a larger audience and consequently more advertisers than can a losing team.

Radio Sunrise-Sunset Restrictions

The type of license a radio station is issued can affect its competitive environment. Such considerations apply to radio where stations may be required to change their power or the direction of their signal after sunset because of the different way radio waves travel.

External Promotions

An external factor a station can control is its external promotions. Billboards, community-sponsored events, and such special activities as Easter-egg hunts or Christmas-gift drives all contribute to a station's visibility. Most stations in highly competitive markets utilize external promotions to their fullest.

Seasonality

For any advertising medium, the season of the year can change the competitive environment. Christmas is a period of heavy advertising as stores prepare for the onslaught of shoppers. Primaries and general elections bring politicians flocking to buy campaign advertising. August is perfect for back-to-school specials, and gardening shops are ideal spring sponsors.

Changes in Lifestyle

Closely tied to the season of the year are changes in lifestyle. The opening of a new factory can signal the arrival of a "blue-collar prime time" and create an opportunity for an all-night format in radio or special late-night television programming that keeps company with the assembly line. Smart station managers keep close tabs on the lifestyles of their communities.

Other Media

Just as other stations in a market are important external factors, so are other media. The most common competitor of broadcasting is the local newspaper, and in some communities the economic rivalry is fierce. City magazines or weekly newspapers for shoppers also contribute to the competition. Station managers must be alert to the advertising that local businesses buy in nonbroadcast media. Why does an advertiser prefer the local newspaper to radio or television? What can the station do to lure money away from print advertising? These are questions management must answer in order to survive in the competitive environment. All of the factors we have discussed fluctuate from market to market. The key to broadcasting's success as a business lies in the station manager's ability to understand how each of these external factors relates to a given community and how the station can adapt to it.

INTERNAL FACTORS AFFECTING PROGRAMMING DECISIONS

A station's ability to adapt to the competitive environment is partly determined by internal factors.

Radio Station Playlist Length

In radio, *playlist length* is the number of different songs played with regularity. Some stations have rather long playlists, which means that in any given week they will play a large number of songs, but each one only once or twice. Stations with short playlists usually play twenty or so songs with regularity. A short playlist can catch a large number of listeners but will not necessarily hold them for any length of time. After all, hearing the same songs over and over can become monotonous. These stations aren't too concerned with the "dial hoppers." Rather, their objective is to reach a large number of different listeners. On the other hand, a station with a long playlist and more diversified programming may not capture as many different listeners, but those they do have are more loyal and will listen to the station longer. It then becomes the job of each station to convince the advertiser that it is the better buy.

Record Rotation in Radio

By *record rotation* we mean how often records are changed on the playlist and how frequently they are played on the air. Stations with longer playlists, naturally, have a less frequent rotation.

Quality of New-Music Selection in Radio

In many markets, good program directors are hard to find. Such a person must not only be able to supervise people, make out schedules, and publish a playlist but also intimately understand the music business. In demand is the program director who understands trends in music before they reach the listener, and who has an ear for interpreting this trend by selecting for air play those few songs, from the hundreds released, that will appeal to a wide segment of the listening audience. Talent such as this is rare. The ability to choose music well can strengthen the competitive position of any station.

Newscasts

News is an important commodity to sponsors, and in multistation markets where each station places heavy emphasis on local news coverage, the competition can be keen.

Placement of newscasts is important in both radio and television. Most stations have a regular newscast time slot with which the listeners or viewers and sponsors can identify. For radio, the decision must be made on whether the placement of the newscast will be part of the programming strategy or even if news will be aired at all.

With television, however, local news coverage contributes heavily to a station's overall image. Determining when network programs will air in relation to local newscasts, whether the newscasts will be one-half hour, one hour, or even longer—all play a part.

Quality of On-Air Production

Almost anyone can sit in front of a radio control board and play records, talk over the microphone, and punch a few buttons to play jingles. The skilled radio announcer, however, works those controls like an artist. Every word, every song, and every jingle is finely tuned to the second. For the listener, such skill creates a sound that massages rather than irritates. The professional announcer who is creative and can understand on-air production stands out in the competitive environment.

Similar talent is necessary in local television production. Since locally produced programming of all types is compared with the sophisticated production standards of networks, sloppy local production reflects negatively on the station and can hurt both its image and its ratings. A TV station with an image of sloppy local production may have a tough time competing, regardless of how sophisticated its network or syndicated programming is.

Commercial Load

Some radio stations have been known to sign on the air and operate for months without commercials just so they can bring themselves up in the *ratings* and then use the data on their increased listenership to sell advertising. Other stations pump their commercial load up so high that little time is left for music. The balance between these extremes, in both radio and television, is where the station's optimum income and listenership meet. A heavy commercial load will lose listeners. A light commercial load will lose money.

Announcers and Anchors

The human factor is what makes any station run. Good announcers and anchors identify with their audience, and knowing this, good program directors try to match the two as closely as possible. For example, a radio station's Top-40 morning disc jockey may be a complete disaster at midnight, and a soft-talking, easy-listening personality may put listeners to sleep at morning drive time. For a station to be successful, the match between audience and on-air personality must be a good one. And once that match is made, loyalty grows between the two.

Similarly, in television the anchor is the local station's showplace to the public. If one station has professional, mature-looking, experienced, and competent anchors, then stations that lack that professionalism will suffer.

Internal On-Air Promotions

Promotions—contests, exhibits, call-ins, giveaways—especially when well planned and carried out, can generate more listeners or viewers, and more income for a station. In recent years, broadcast promotion has been recognized as a vital element in a station's overall performance. Many stations are forming in-house promotion departments, either to take the place of outside advertising agencies or to work in conjunction with them.

SUMMARY

Programming is the product of broadcasting. It can be compared to the goods and services sold by other businesses. Even in public broadcasting, which solicits contributions for programming, programs are the "goods" that people want to continue to hear and to view.

In radio, the format is the heart of programming. Formats vary among stations and range from foreign-language broadcasts to hard-rock music. Specialized formats are the key to modern radio programming. They reach specialized audiences and, accordingly, permit advertisers to target their commercials and messages. The person responsible for programming a station is the program director, who combines an awareness of musical tastes, an understanding of demographics, and the ability to put them together.

A number of programming strategies exist in competitive markets. It is first necessary to analyze the competition and determine how other stations are programming for the market. Assuming that no station dominates the market, it is necessary for the program director to adjust to the formats that will fit into the competitive marketplace. The radio personality will play a large part in this adjustment. Jingle packages complement air personalities and help create unified sound for the station.

Programming strategies also exist for TV programming. Some of these strategies are dictated by the network, which controls the majority of programming for network-affiliated stations. But the goals of affiliates must be taken into consideration by the networks, which after all depend on affiliates to carry network programming to a mass audience.

A station's competitive environment consists of a number of external and internal factors. External factors include news events, new music, other stations in the market, employment, sports events, sunrise-sunset limits on broadcasting, external promotions, seasonality, changes in lifestyle, and other media. Internal factors include playlist length, record rotation, availability of new-music selection, newscasts, quality of on-air production, commercial load, and internal on-air promotions.

OPPORTUNITIES FOR FURTHER LEARNING

ALLOWAY, L., AND OTHERS. *Modern Dreams: The Rise and Fall and Rise of Pop*. Cambridge, MA: MIT Press, 1989.

ANG, I. *Watching Dallas: Soap Opera and the Melodramatic Imagination*. New York: Methuen, 1986.

ARMER, A. A. *Writing the Screenplay: TV and Film*. Belmont, CA: Wadsworth, 1988.

HOWARD, H. H. *Radio and TV Programming*. Ames: Iowa State University Press, 1983.

KEITH, M. C. *Radio Programming: Consultancy and Formatics*. Stoneham, MA: Focal Press, 1987.

KEITH, M. C., AND KRAUSE, J. M. *The Radio Station*. Boston: Focal Press, 1986.

KLATELL, D. A., AND MARCUS, N. *Sports for Sale: Television, Money and the Fans*. New York: Oxford University Press, 1988.

LEVIN, M. B. *Talk Radio and the American Dream*. Lexington, MA: Lexington Books, 1987.

LULL, J. *Popular Music and Communication*. Newbury Park, CA: Sage, 1987.

MATELSKI, M. J. *Broadcast Programming and Promotions Worktext*. Stoneham, MA: Focal Press, 1989.

SANDERS, M., AND ROCK, M. *Waiting for Prime Time: The Women of Television News*. Urbana: University of Illinois Press, 1988.

TUROW, J. *Playing Doctor: Television, Storytelling, and Medical Power*. New York: Oxford University Press, 1989.

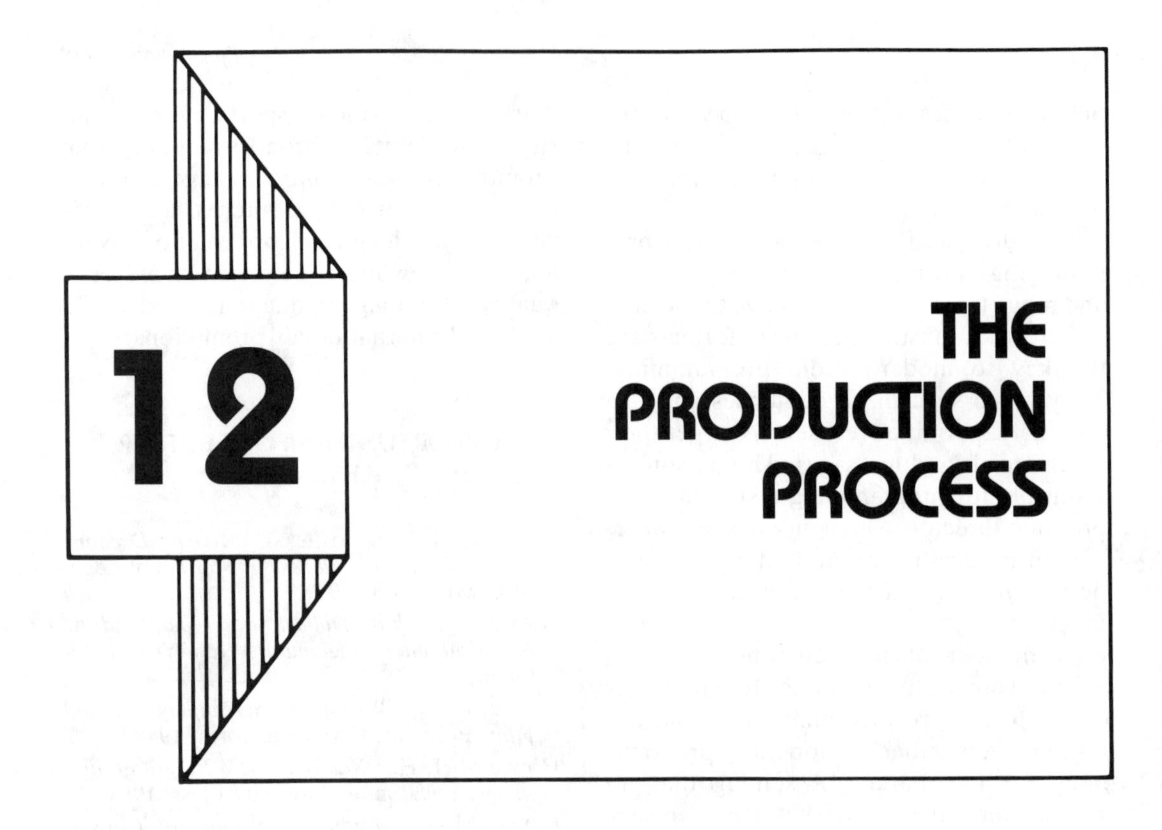

In broadcasting and other technologies such as cable, VCRs, and videodiscs, we are witnessing an explosion of new programming as the public's appetite for new artistic forms continues to increase. In a local video store we can rent new releases, many of which have not yet been aired on the media. On cable and satellite systems the variety of programming ranges from nature programs to ballet. Where once only television stations produced original programming, today small independent production companies are turning out programs for schools, businesses, and other outlets. An evening televison newscast may carry a story, the accompanying video of which was shot by an amateur photographer using a home video camera.

Behind the programming decisions are individuals who possess artistic talents such as writers and directors, and technical talents such as engineers. The combination of these talents produces what we see and hear on radio and television.

THE PRODUCTION PROCESS

The sounds we hear on the radio or the programs we view on television involve a complex process of art and technology blended together by skilled behind-the-scenes and on-air people. A better understanding of this process is possible by examining both the radio and television studio and then the production process.

THE RADIO STUDIO

Although diversity exists among radio stations, most stations contain the basic elements of a control room, production studio, and an on-air studio that can also function as a news studio. Radio stations operate in permanent structures that range from modern buildings with the finest equipment to hand-me-down facilities in a small trailer that could be towed by an automobile. The size and sophistication of the facilities are not as important as the talent of the people who use them. The potential of even the best equipment is lost when human skills are missing. Talent, on the other hand, can work wonders in a studio equipped with very basic equipment.

MICROPHONES

Microphones are classified primarily by types, patterns, and style.

Types

The major types of microphones are carbon, dynamic, condenser, and ceramic.

Carbon microphones, although rarely used in studios because they produce a hollow "dispatcher" sound, contain carbon granules that vibrate with the sound of the voice, causing changes in the electrical energy that reaches the transmitter.

Dynamic microphones, which along with condenser microphones are the most commonly used studio microphones, have a diaphragm that collects the sound and compresses a coil that varies the electrical current.

Condenser microphones use their own supplementary power supply. Condenser microphones are used where high-quality musical reproduction is desired.

Ribbon microphones use a thin metallic strip, hence the term "ribbon," which vibrates and varies the electrical current. Over time the ribbon becomes stretched and makes the voice sound deeper, which is why some announcers still like to use a ribbon microphone.

Ceramic microphones are low-cost devices found on many portable tape recorders.

Pickup Patterns

Pattern refers to the direction the sound enters the microphone. Each pattern has its own variations. An *omnidirectional* pattern picks up sound in all directions. *Directional*, also called unidirectional, microphones pick up sound from one direction. Subpatterns of the directional microphone are *cardioid*, which picks up sounds in a heart-shaped pattern, and the *shotgun,* which picks up sounds in a narrow area directly in front of the microphone. A *bidirectional* pattern picks up sounds from two opposite directions.

Applying these terms to an actual working situation, we would use the omnidirectional microphone to do a news report where we wanted some background sound. A directional pattern would work well when we are announcing an event, inside or outside the studio, and did not want background noise to interfere. A shotgun microphone would be used at a sports event where we wanted to pick up the official's call from the sidelines of a football field.

Styles

Style refers to the actual use of a microphone. For example, a *lapel* microphone is small and inconspicuous and can clip onto a jacket lapel or necktie. *Boom* microphones are connected to a large horizontal rod, the boom, which is attached to a wheeled tripod and can be moved around the studio. *Hand-held* microphones are a favorite of broadcast journalists. A *parabolic* microphone is frequently used to cover a sports event and is located inside a dish-shaped reflector that captures sound much like an antenna dish captures microwaves. A *utility* microphone is located on a fixed flexible shaft, such as that which would be installed on a fixed console in a radio studio. *Head-word* microphones are used by play-by-play announcers and extend from the head-set earphones. Each type of microphone we have just mentioned is of a specific type and pattern.

RECORDING AND PLAYBACK EQUIPMENT

Radio has benefited greatly from advances in the technology of recorded sound, first in the development of records, then magnetic tape, and now digital recording. Commercials, jingle packages, and various types of programming have been greatly enhanced by these developments in recording technology. Whether using analog recording—which reproduces through variations in current, the sound waves which are "analogous" to the original sound source—or digital recording, which uses the numerical binary code of a computer language—the equipment found in today's radio studio is basically the same. Reel-to-reel recorders, cassette tapes and cartridge tape recorders, compact disc players—which are exclusively digital—and control consoles make up the core of the radio studio.

Compact Disc Players

The development of digital recording and playback technology has resulted in the widespread use of compact discs. Using a laser beam to read the computer language of the recorded sound, the compact disc player eliminates the distortion that can occur with vinyl records or magnetic tape. With digital recording, the sound is sampled tens of thousands of times per second. These different samplings are actually different frequencies of sound, which are then converted into the binary numbers of computer language. Keep in mind that although the compact disc produces clear, noise-free sound in the studio, all signals that leave the station transmitter are subject to radio wave interference.

Tape Recorders

The reel-to-reel recorders have different numbers of tracks, each track devoted to a different audio channel. In large sophisticated recording studios 32-track machines are not uncommon, but in the typical radio studio, two-track stereo machines usually serve most of the needs of the typical radio station and many one-track machines are still rendering service. On the reel-to-reel machine the tape travels from a feed reel to a take-up reel and flows over both an erase and recording head.

Cartridge tape recorders have for years been the indispensable workhorse of radio. Commercials, audio cuts for newscasts, jingles—all find their way onto the cartridge tape. The cartridge itself contains a continuous loop of tape that automatically stops at the point where it starts. This is because an electronic cue is placed on the tape at the point where the tape begins to move when the machine is started in the "record" position. The capacity of the cartridge is determined by how much tape is wound on the loop, but 10-second to 60-second cartridges are the most common.

Cassette recorders also play a major role in station operations. Technology has made the quality of the audio cassette equal to network standards, and digital technology promises to make its application even more widespread. The small cassette recorder is ideal for radio news production. Many times the recording from the cassette tape is transferred to reel-to-reel tape for editing and then to cartridge tape for airplay.

Turntables

While the future belongs to digital technology such as the compact disc, the inventory of records at radio stations, accompanied by the fact that many songs are not available on tape or disc, means that radio studios still have their share of turntables. Today's turntables contain delicately balanced tone arms and direct-drive motors operated by special electrical circuitry that keeps speed variations to a minimum and permits as high a level of reproduction quality as is mechanically possible.

THE RADIO CONTROL AREA

The sounds filtered through microphones, reproduced on tape recorders, or played back from a compact disc do not go directly to the station's transmitter.

Instead, the sounds pass through a control console, also called a *control board* or *audio control console*, where sounds are monitored, altered, and mixed together into the sounds we the listeners hear. Control consoles vary from those that control a small number of audio channels or "inputs" to those that can handle more than thirty sound sources. Studios found in the recording industry handle many more channels. The inputs are controlled by a slide bar called a "slider" or a knob called a "pot." In addition to these and other controls, the control console also contains meters, which monitor program levels.

After the sound leaves the control console it travels to the station's transmitter where it is sent to the antenna and then by radio waves to receivers. Other monitoring steps are also built into the process to assure the signal is modulated correctly and that the transmitter does not exceed its power limits as prescribed by law.

RADIO PRODUCTION

Some people feel that radio is a more creative medium than television. That may seem strange when we consider the visual elements of television, but radio does force us to use our imagination and an imagination is limitless. Radio is certainly a medium of the mind and the mental image we create is determined by the sounds that reach us.

The Mental Image

Imagine in your mind the scene described in the following lines:

> *A scarlet valley of fragrant roses shimmered in the caress of a soft breeze descending from lilac-splashed hills.*

We would find it a difficult assignment if we were part of a TV crew and were assigned to go out and find such a valley. Yet with radio we can imagine the valley in our mind, even, perhaps, smell the roses and lilacs. It is this mental imagery that radio is capable of and which the skilled producer tries to create.

In a radio documentary about a mountain garden in Idaho, the producer walked with us up a dirt road into the old mining town of Warren. We heard the barking of dogs, first distant and then gradually nearby, just as you do when you are in Idaho and walk up the same mountain road.[1]

A commercial for the Fuller Paint Company wanted to emphasize the intensity of the color yellow. Think of yellow. What images come to mind? Now visualize the images from these lines of the Fuller Paint Company commercial:

> *. . .Yellow is a way of life. Ask any taxi driver about yellow. Or a banana salesman. . . Dandelions, a dozen; a pound of melted butter, lemon drops and a drop of lemon, and one canary that sings a yellow song.*

If we viewed these same images on television we might even lose their impact because instead of concentrating on the color *yellow* we would have our concentration broken by the actual sight of a downtown taxi caught in the snarl of rush hour traffic, or the lemon drops that would have to be in a container that sits on a table with a colored background or positioned in front of a window and an accessory such as a plant next to the container. Get the picture? The television commercial may actually clutter the pure, intense, radiant mental images of *yellow*.

Radio Production in Daily Programming

For most people in radio, the majority of radio production centers around the daily programming of a local radio station. The disc jockey prepares a program based on such programming considerations as how the lead-in to a song can be used to introduce the song, be heard under the weather forecast, or be just the right length over which the time and temperature can be announced. The station's format is the basic foundation upon which the station production process evolves. Most stations keep this in mind when producing the station's daily programming. The jingle package that captures the overall station image, the type of music that accompanies commercials, the lead-ins to news programming, and the quality of news production, are all daily issues in radio production.

The salespeople, copywriter, and others involved in the production of commercials work together to produce the visual images we just discussed. Part of their success is their ability to mix words and sounds that complement each other. If these individuals are skilled and experienced they will not add sound just for the sake of accompaniment, and will not add words just to jam as much information into commercial copy as possible. They will understand the art of radio production and appreciate the unique qualities of the medium.

THE TELEVISION STUDIO

Of the components we have just discussed, such things as microphones, tape recorders, and control consoles are also found in a television studio. But to these are added components that provide the video portion of the signal. Keep in mind that in addition to equipment, the average television station also makes a sizable investment in sets and props for its own locally originated programming such as news shows. Sets and scenery may also exist for other types of programming a local station may produce such as interview shows, children's programs, and others.

Lighting

Lighting a television studio is not unlike lighting a theatre set. In television, the camera takes the place of the human eye, and lighting aids the camera in being able to see things as much like the human eye as possible. Lighting is used to highlight a set and give it a three-dimensional effect so it is seen by the TV camera in much the same way that the human eye would view it.

In television two primary types of lights are employed—floodlights, which are also referred to as "scoops," and spotlights. Floodlights light an area with diffused light and spotlights illuminate specific areas.

Lights serve three basic functions: *key lighting, back lighting,* and *fill lighting. Key* lights illuminate subjects from the front, *back* lights from the back, and *fill* lights function to fill in shadows. Lights are controlled using dimmers, which alter brightness by varying the amount of electricity going to the light.

Cameras and Lenses

Over the years cameras and lenses have changed drastically. In the 1936 Berlin Olympic Games the TV camera and camera lens were experimental and so large it took two people to operate them and two more to operate the power supplies. Later a rotating turret with different lenses attached permitted the camera operator to select between different lenses. Still later the zoom lens was perfected, which gave the camera much more versatility. Wide-angle and other shots are possible with the zoom lens, which can be operated manually on smaller cameras frequently used for electronic news gathering (ENG) work or may be motor driven, as is common on most larger TV studio cameras.

Cameras are positioned on a mounting head, which permits the camera to move in a horizontal motion, said to "pan," and a vertical motion said to "tilt." When the director speaks to the camera operator the director gives such commands as "pan right," "pan left," "tilt up," "tilt down." Cameras are mounted on tripods, which are three-legged supports, or on pedestals, which operate on the same three-legged support system as a tripod but contain a large flexible shaft that permits the camera to be easily raised or lowered.

TELEVISION CONTROL AREA

What the audio console is to the radio studio the video switcher (Figure 12–1) is to the television studio.

Video Switcher

The video switcher gives the director the option of choosing different camera shots, mixing audio, superimposing special effects, and a wide assortment of other creative functions that result in the program becoming an on-air reality. The basic components of a video switcher consist of rows of switches called "banks" or "buses." Different buses control different functions. For example, one bank controls a studio camera, another a videotape input, another special effects. By using a "fader bar" the different banks can be selected and the director can move from one camera to another, from a camera to a videotape, from a camera to special effects, and a wide variety of other choices depending on the complexity of the video switcher.

Other Video Control Equipment

In addition to the video switcher there are monitors that tell the director which program is being aired and what video is available from other program sources. A "slide chain" and "film chain" are used to project slides and film into a camera, which is then fed into the video switcher. *Character generators* superimpose words over the picture. Typically, they are used in such applications as showing credits or the name of an anchorperson or field reporter. *Special effects*

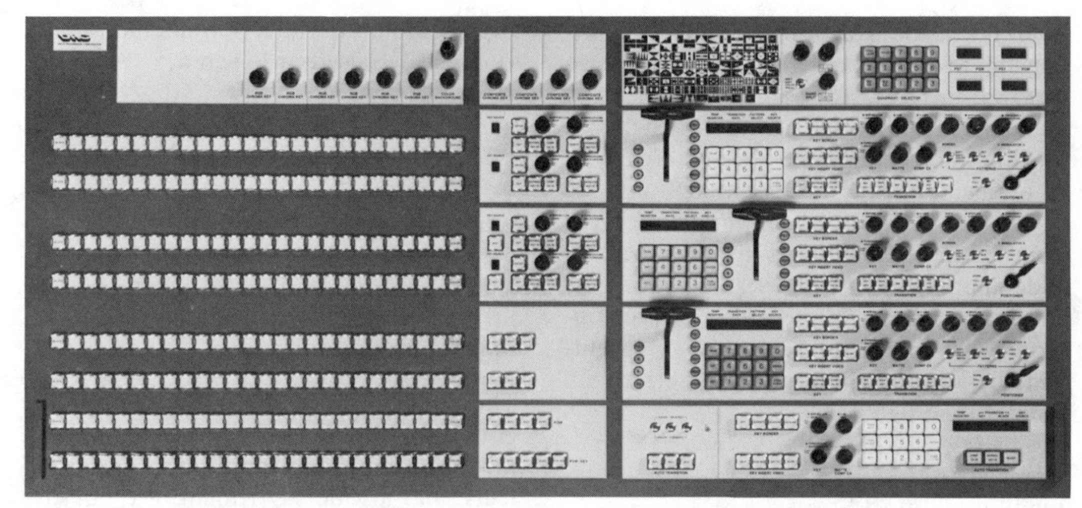

FIGURE 12–1 The control console is a technical extension of the director or producer. Rows of "banks" or "buses" (the square buttons on the left side of the console) can access different program sources, such as individual studio cameras. The lever, called a "fader bar," permits the person controlling the program to select from one row of banks over another. Special-effects controls are located in the upper-right-center of the control console.

generators permit the picture to be maneuvered and manipulated in many different video compositions. Videotape recorders using video cassettes handle most of the station's recording needs, although film is still employed on a very limited basis in some stations. An interstudio communication system using headsets permits the director to communicate with the camera operators and engineers.

TELEVISION PRODUCTION

With television production many of the same decisions take place as in radio, but with the addition of video. Producers, writers, directors, and technical crews all come together to produce a finished product that may find its way into a local schoolroom or appear on an international network.

From Idea to Script

The idea conceived in the mind of someone is transposed into a workable script by a scriptwriter who must understand the relationship of dialogue to setting. We can draw from our discussion of aesthetics earlier in this chapter and imagine, for example, two couples linked in a romantic dialogue on screen. The couple can be placed on a moonlit beach or on a factory assembly line. One setting may enhance the romantic atmosphere by complementing the dialogue with a romantic atmosphere of the beach. The second setting may enhance the romantic atmosphere by contrasting it with the harsh environment of the assembly line. The script will determine which setting will appear in the program and even then the director's artistic judgments may alter the original script.

The actors may also play a part in the

final script's development. For example, if you are writing a situation comedy, knowing how different actors handle specific dialogue will partially determine who says what to whom. Keeping all of this in mind results in a preliminary script, which is then revised, sometimes many times, into a final script. A change in a word here or a line added or deleted there can make a substantial change in the theme and mood of the production. Such changes can also determine whether the program flows smoothly or bogs down in unnecessary dialogue.

Changes can occur for a variety of reasons. The script may be too long. It may need to be trimmed to fit into a given time frame, but trimmed without losing its appeal. A scene between two people may drag on until all the humor is sapped from the scene, much like a joke that is too long before its punch line. The actors may feel uncomfortable with the dialogue and claim it simply doesn't work. At that point the script is revised again.

The same principles apply for both dramatic shows and commercials. In commercials a storyboard is drawn that provides still-frame drawings of potential on-camera action. The script of the commercial is combined with the pictures, resulting in a series of drawings that can give the director and actors an idea of the sequence in which the commercial will be shot.

Producing the Program

After the script or storyboard is ready, the production process begins. If shot outdoors, the position of the cameras in relation to the sun is taken into consideration. If indoors, the director must determine how the set will be lighted with key, back, and fill lights.

Camera shots are taken into consideration and the script again becomes a part of the production process. In a dramatic production the camera shots are added to the script. For example, the script may say "MS of Donald and Sally." MS means "medium shot" and tells us that the camera will frame Donald and Sally about midpoint between a close-up shot (CU) and a long-shot (LS). An extreme long-shot (XLS) would place Donald and Sally in the distance. Other positions might include a tight close-up (TCU), which would frame just the heads of Donald and Sally. An extreme close-up (XCU) would show even less of Donald and Sally.

The script, the designated camera shots, the camera movements discussed earlier in the chapter—all combine with the creative talents of the director, actors, and others to produce the final program.

TELEPRODUCTION

The application of computers to the TV production process has resulted in what is termed "teleproduction." However, the application of computers to the production process is not new. In 1967, Ampex interfaced a computer with videotape recorders to create a video editor during the CBS coverage of the 1968 Grenoble Olympics.[2] Since that time the television industry has made increased use of computers in the production and editing process.

Standardized Time Codes

Through the use of standardized time codes, equipment manufacturers have been able to build computer editing into videotape technology permitting preselected commands to be used for electronic editing. Other applications of computer technology include software that can develop storyboards, word-process scripts, generate computer graphics, and other support for both administrative and creative functions. The development of the personal computer has also revolutionized the work of the graphic artist.

The Video Paintbrush

Software now available makes the computer keyboard and a "mouse" device become the artist's brush and pencil. Adding animation capabilities makes this marriage of video and computer software even more spectacular, and interfacing desktop computer-generated graphics with a video input device makes one person an artist, director, and video editor.

SUMMARY

At the heart of a station's programming are aesthetic decisions and the production process, both of which determine the quality of the programs received by listeners and viewers. The production process begins in the studio, which at a radio station can range from a complex engineering masterpiece to a few pieces of equipment in a small room.

Within the studio are microphones that are classified by types, pickup patterns, and styles. Recording and playback equipment also make up the radio studio. The compact disc player has become standard equipment in most radio stations. Tape equipment such as reel-to-reel and cassette recorders are also standard equipment. Although not used as much as in the past, stations retain turntables to play vinyl recordings.

The heart of the radio control area is the audio control console. Using a series of "pots," which are actually control knobs, or bars, called faders, the various sound inputs are combined and altered before being sent to the station transmitter.

Radio production involves the skillful mastery of equipment to produce sounds that create visual images. Sometimes referred to as the "theatre of the mind," radio is restricted only by the limits of our imagination.

Some of the same equipment found in radio studios is also found in TV studios. Television expands on the production techniques of radio by adding video to the production process.

The TV studio includes cameras that capture the images through the skillful use of camera movements. The images are combined at a video control console.

The actual elements of the production process involve developing a script, deciding what camera shots are to be used, and then combining these elements into a final program.

Application of computer technology to the production process has brought the concept of teleproduction into the creative process, and the use of personal computers for desktop videographics has resulted in new skills and new opportunities for graphic artists. Interfacing personal computers with videodisc machines has resulted in interactive instructional television.

OPPORTUNITIES FOR FURTHER LEARNING

Aldridge, H., and Liggett, L. *Audio/Video Production: Theory and Practice*. Englewood Cliffs, NJ: Prentice-Hall, 1990.

Alten, S. R., *Audio in Media,* 2nd ed. Belmont, CA: Wadsworth, 1986.

Burns, G., and Thompson, R., eds. *Television Studies: Textual Analysis*. New York: Praeger, 1989.

Chambers, E. *Producing TV Movies*. Englewood Cliffs, NJ: Prentice-Hall, 1986.

Cohen, A. A. *The Television News Interview*. Newbury Park, CA: Sage, 1987.

Fensch, T. *The Sports Writing Handbook*. Hillsdale, NJ: Erlbaum, 1988.

Killenberg, G., and Anderson, R. *Before the Story: Interviewing and Communication Skills for Journalists*. New York: St. Martin's Press, 1989.

Oringel, R. S. *Audo Control Handbook for Radio and Television Broadcasting*, 6th ed. Boston: Focal Press, 1989.

Shook, F. S. *Television Field Production and Reporting*. New York: Longman, 1989.

Shook, F. S., and Lattimore, D. *The Broadcast News Process*, 3rd ed. Englewood, CO: Morton, 1987.

Wulfemeyer, K. T. *Beginning Broadcast Newswriting: A Self-Instructional Learning Experience*. Ames: Iowa State University Press, 1984.

York, I. *The Technique of Television News*, 2nd ed. Stoneham, MA: Focal Press, 1987.

13 INTERNATIONAL DOMESTIC SERVICES

To confine the study of broadcasting and telecommunication to one country is to restrict oneself to a very narrow view of both the world and the broadcast media. The international flow of broadcast programming and direct-broadcast satellites beaming signals across national boundaries demand an international perspective of these electronic media.

We read earlier in this book that broadcasting developed simultaneously in many different parts of the world, especially after World War I. The different political, economic, and social conditions in which broadcasting operates are as varied as the countries themselves. Not every country permits commercial advertising, has a free press, or allows private ownership of broadcasting. Each country has its own system serving both domestic and international audiences.

We will not examine every country, nor will we view each country from the same perspective. What we will do on our world tour is acquaint ourselves with many different elements of international broadcasting.

In addition, although we will concentrate on broadcasting, satellite, and cable systems in individual nations, many countries are hosts to other services besides their own domestic and international services. For example, both the American and the Canadian Armed Forces Radio networks as well as BBC relays and Radio Free Europe/Radio Liberty stations operate in other parts of the world. The United States Information Agency (USIA) controls RIAS Berlin radio and airs German-language programs. The Cable News Network (CNN) is based in the United States, but operates CNN International with foreign bureaus.

THE SCOPE OF INTERNATIONAL BROADCASTING AND TELECOMMUNICATION

As we visit different countries, remember that the United States represents only one model of broadcasting. Our commercial radio and television stations are supported by advertising, and our public broadcasting is backed by government funds, the public, corporations, and institutions. In other countries, advertising also supports some systems of broadcasting, as do public contributions. We can also find systems totally supported by the government. We can find systems supported in part by a tax paid on radio and television receivers. We can find systems supported by license fees that the public pays in order to listen to radio or watch television. In some countries, different methods of financial support are found side by side.

Not every country has the freedom of expression found in some democratic societies. Moreover, the content of radio and TV programming is determined by many things, such as the ratings, the government, and advertising. Some countries operate very large systems, and some operate very small ones. In some parts of the world we can watch television only a few hours a day. In other parts of the world we can watch television beamed in from many different countries. This diversity, the opportunity to compare and contrast different broadcasting systems, is what makes international broadcasting and telecommunication fascinating topics.

CANADA

Canada's place in the history of international telecommunication began when Marconi selected Newfoundland to test his transatlantic wireless. Those dots and dashes started a chain of events that eventually gave birth to the Canadian Marconi Company. Today Canada is the home of a broadcasting system that stretches from Eskimo villages in the north to the American border in the south.

Canada is a country in which the evening news can be delivered by satellite, yet in which television can be a strange phenomenon to an inhabitant of the northern tundra. Canada is also a country whose pride and loyalty are reflected in everything from broadcast regulations to television programming.

Protecting Cultural Integrity

Canadian broadcasting is regulated by the Canadian Radio-television and Telecommunications Commission. Canada's broadcasting policy is stated in the Broadcasting Act of 1968:

> (a) Broadcasting undertakings in Canada make use of radio frequencies that are public property, and such undertakings constitute a single system, herein referred to as the Canadian broadcasting system, comprising public and private elements;
>
> (b) the Canadian broadcasting system should be effectively owned and controlled by Canadians so as to safeguard, enrich and strengthen the cultural, political, social and economic fabric of Canada;

(c) all persons licensed to carry on broadcasting undertakings have a responsibility for programs they broadcast, but the right to freedom of expression and the right of persons to receive programs, subject only to generally applicable statutes and regulations, is unquestioned;

(d) the programming provided by the Canadian broadcasting system should be varied and comprehensive and should provide reasonable, balanced opportunity for the expression of differing views on matters of public concern, and the programming provided by each broadcaster should be of high standard, using predominantly Canadian creative and other resources.[1]

Canadian Broadcasting Corporation

Comparable in many ways to the PBS in the United States, the Canadian Broadcasting Corporation (CBC) is Canada's government-owned public broadcasting service. Although the CBC reports to Parliament through a designated cabinet minister, the responsibility for its programs and policies lies with its own directors and officers. CBC is financed by public funds and by advertising.

Private Commercial Broadcasting

The private sector has a well-established system of commercial broadcasting stations serving all of Canada. The main commercial television network is the CTV Television Network, which is owned by broadcasters. Inaugurated in 1961, CTV reaches about 95 percent of Canadian television households and transmits over sixty-six hours of programming per week.

In addition to the CTV and CBC networks, the Canadian Association of Broadcasters has a program-exchange service for its members. Despite its huge land mass, Canada has managed a coordinated policy of broadcast development that uses the latest technology to reach the diverse Canadian population.

Broadcasting in Canadian Provinces

Individual Canadian provinces also have well-developed broadcasting systems. For example, Radio Quebec was formed in 1968 and answers to the Quebec National Assembly through the Minister of Culture and Communications. A network of TV stations bring Radio Quebec television programs to all of Quebec province under a programming mandate that states programming must:

favor the exercise of the right to education; promote the citizens' access to economic and social well-being and to their cultural heritage; encourage the exercise of the rights to freedom of expression and to information.[2]

Broadcast schedules are categorized into "current events and public affairs; social and public service programs, general cultural programs and instructional programs."[3]

In Ontario, TV Ontario is an educational television production and programming entity that has been in operation since 1970. Programs produced in English and French are distributed over local television stations and sold internationally. Many of its programs, especially those oriented toward young children, are of especially high quality and capture a large and loyal television audience. For example, the program "Polka Dot Door" (Figure 13–1) is a half-hour variety program for young children based on the premise that children learn through play. Another popular TV Ontario program is "Today's Special" with a light-hearted approach to developing language skills.

FIGURE 13–1 On the set of TVOntario's "Polka Dot Door," an internationally distributed program in French and English. The program is based on the premise that children learn through play activities. Characters on the program, including real people and fantasy characters such as Marigold a rag doll, and a teddy bear appropriately named Bear, lead young viewers into activities that stimulate learning experiences.

MEXICO

The regulation of broadcasting in Mexico is divided among different government agencies with some overseeing programming and others operations. Along with Mexico's government stations and government-supported TV networks, privately owned radio and TV networks criss-cross the country. A federation of coordinated television channels called Televisa serves most of Mexico. To some degree the channels represent an attempt to reach more specialized audiences in Mexico. For example, one "channel" or "network" covers most of Mexico and is received by a predominantly middle-class viewing audience. Its programming includes state-produced programs for young children, such as a Mexican version of "Sesame Street," soap operas, the Mexican version of "Today," weekend sports events, and family entertainment programs. Another channel covers metropolitan Mexico City and its environs. Aimed at the mass urban public, it features specials on Mexico City's neighborhoods. Evening programs feature films, amateur hours, and variety artists.

Still other channels are geared to younger, better-educated Mexicans. For example, one channel reaches about half the population of the country via a series of repeater stations, which receive and retransmit the signal. The target audience is the youthful middle class, including university students. Programs on current issues are featured, as are American, British, and Japanese programs. Approximately thirty hours of national productions are also seen. Another channel reaches about 10 million viewers in the urban valley of Mexico with many cultural programs. Academic groups and people with differing political beliefs make up its target audience. Much of Mexico's programming consists of performances by recognized artists, award-winning international films, and international productions.

THE UNITED KINGDOM

Although North America enjoyed the fruits of Marconi's labor, Europe saw the seeds of his achievements in communication germinate much earlier. Perhaps nowhere was the impact of the new medium felt more than in the United Kingdom, whose naval vessels, merchant fleet, Post Office Department, high-powered stations linking continents, and British Marconi Company all used it from its inception. Two systems of broadcasting have grown out of the wireless in England: The BBC and the Independent Broadcasting Authority (IBA). The BBC is noncommercial; the IBA is commercial.

The British Broadcasting Corporation (BBC): Organization

The BBC started in 1922 as the British Broadcasting Company and was made a nonprofit, public corporation by royal charter in 1927. Today it is an independent broadcasting organization, although it receives its budget for overseas broadcasting from Parliament. The fees it collects from licenses are also determined by Parliament. Directed by a board of governors appointed by the Queen, the corporation operates under the advice of a series of advisory boards. Over a hundred countries use BBC productions, and on the average 500 programs are aired weekly in various parts of the world. The BBC has a policy of not relinquishing editorial control over any of its programs and does not tailor its programs to any specific region.

BBC Radio

The BBC Radio programming is disseminated throughout the United Kingdom by four domestic networks. BBC Radio 1 and BBC Radio 2 are the popular formats. BBC Radio 1 is more progressive, with pop and rock-and-roll music formats. BBC Radio 2 attracts a more general audience, with sports, light music, and information programming. Together they capture about 80 percent of Britain's listening audience. BBC Radio 3, on the other hand, programs more classical music as well as dramatic and cultural programs. Live concerts, both in Britain and in other countries, are emphasized. Masterpieces of world theatre and discussions of scientific and philosophical subjects round out BBC Radio 3's programs. BBC Radio 4 is devoted to speeches, news and information programming, dramatic entertainment, and current events. "Today," "The World at One," "PM Reports," "The World in Focus," and "The World Tonight" are typical of its extended magazine-type news programs. Phone-in programs, panel games, plays, and readings are also heard on BBC Radio 4. Together these four national networks complement the local radio stations, which serve small geographic areas.

BBC Television

Although radio continues to be the foundation of the BBC, television does not take a back seat. Experimental television was launched in 1936. Suspended during World War II, it went back on the air in 1946. Today, two BBC television networks serve nearly the entire United Kingdom. Over 80 percent of the programs are produced by the BBC; the remainder come from independent producers and other countries. As in the domestic radio system, television is financed by license fees collected from owners of receiving sets.

The Independent Broadcasting Authority (IBA)

The IBA operates Independent Television (ITV) and Independent Local Radio (ILR).[4] A commercial broadcasting system, IBA was created by Parliament in 1954 to broadcast side by side with the BBC. Its sole income is from advertising. Commercials are aired between programs, not in the middle of them. IBA's structure was amended by the Independent Broadcasting Authority Act of 1973. IBA monitors the quality of the programs carried by its radio and TV stations.

In addition, the IBA supervises a developing system of local radio outlets (Figure 13–2) and fifteen television production companies. Operating much like American television stations, the production companies serve different regions of the United Kingdom. The production companies, although supporting themselves by regional advertising, receive their assignments from the IBA, which also operates their transmitters and receives a percentage of their income in return.

Like the BBC, the IBA is free to sell its programs to other countries. At home, it has the authority to assign time to such specific programs as education, news, religious broadcasts, and documentaries. The nature and the amount of advertising are controlled by the IBA in keeping with the mandate of the 1973 legislation. Other guidelines are provided by the IBA's Code of Advertising Standards and Practice. Television advertising is limited to an average of six minutes an hour, and radio advertising approximately nine minutes each hour. Like the BBC, the IBA is advised by a group of quasi-citizen, quasi-governmental committees on advertising, medicine, religion, education, local radio, and other operations.

IBA's Channel 4

The Broadcasting Act of 1980 added a new channel to British television, Channel 4. Channel 4 is operated under the authority of the IBA and is specifically designed to provide an alternative to British television programming seen on ITV stations. Its revenue comes from the IBA in a shared arrangement with ITV. In effect, Channel 4 and ITV share revenue generated by both channels. Channel 4's charge under the 1980 Act is to provide programming with "innovation in form and content."

Channel 4 does not produce its own programs but instead chooses them from other suppliers both in and out of the United Kingdom. Emphasis on educational and international programming brings programs that are not unlike American continuing-education programs and foreign programming addressing multinational audiences in Britain. Subjects such as psychology, history, gardening, sex, and retirement are part of Channel 4's programming as are foreign films in their original language presentations.

The cultural heritage of the Irish, Cypriots, Poles, and other nationalities living in the United Kingdom are reflected in the mission of Channel 4. The service operates under a Channel 4 board picked by the IBA. In addition, a Welsh Fourth Channel Authority oversees a Welsh-language television service in Wales.

Oracle: IBA Teletext

The IBA also operates a teletext service called Oracle on Channel 4.[5] Approximately 5 million teletext receiving sets are

FIGURE 13–2 Independent Local Radio can be heard throughout the United Kingdom. It operates under the auspices of the Independent Broadcasting Authority created by Parliament in 1954.

operable in the United Kingdom and reach approximately 4 million people daily. Regional TV guides, weather information, news, and police reports are some of the information available through the Oracle teletext service.

The Broadcasting Bill and Commercial Competition

British broadcasting is expected to undergo some revolutionary changes in the early

1990s as Parliament passes legislation designed to free up programming and commercial constraints and make British commercial broadcasting more competitive. Although dreaded by many British broadcasters, the new laws are designed to permit foreigners with some British ties the opportunity to own television stations in the United Kingdom by competitively bidding for commercial television licenses that expire December 31, 1992. It means the biggest restructuring of British broadcasting in more than thirty-five years.[6] At the same time, however, it will place the United Kingdom in a strong position to become a major player in the economic changes that will occur in Europe when trade barriers among European nations are lifted in 1992.

REGIONAL DIVERSITY IN THE NORDIC COUNTRIES

In northwest Europe lies the land of the midnight sun—the Nordic countries of Iceland, Sweden, Denmark, Finland, and Norway.

Iceland

In Iceland, until 1986 broadcasting was controlled by the Icelandic State Broadcasting Service, which still derives its income from license fees and advertising. Since the Broadcasting Act went into effect in 1986, private radio and television stations are permitted, and are competing with the State Broadcasting Service. Many homes have videotape recorders (VCRs) as well as satellite dishes, which make it possible for residents to receive foreign stations.

The Broadcasting Act governs programming and requires the state service to promote the Icelandic language, Iceland's history, and its cultural heritage. A general news service is required, as are the expression of diverse views on public issues. A Broadcasting Station Cultural Fund operates to promote national program production and is financed by a 10 percent tax on advertising revenues.

The state's broadcasting network and a private television station constitute television broadcasting in Iceland. The private station operates both free and pay-TV programming services. In addition to subscription fees for pay-TV, the private station carries advertising.

Sweden

Satellite receiving dishes, VCRs, and to a limited extent cable television, all offer television options for the Swedish viewer. Television programming from other Scandinavian countries is also seen in Sweden. The Swedish Television Company operates two national television networks. One broadcasts programs produced primarily in Stockholm, the capital and largest city. The other network airs programs from the outlying regions. The networks are funded by license fees paid by people who own television sets.

Radio in Sweden includes the external service, Radio Sweden, broadcasting to four continents through its short-wave service. The Swedish National Radio Company is a non-commercial public broadcasting service, operating three national domestic radio program services. Like television, receiver license fees support the services. The Swedish Local Radio Company is also non-commercial and broadcasts on a fourth network to local audiences with specialized programming.

A somewhat unusual community radio service also operates in Sweden. Any non-commercial organization which isn't involved in broadcasting can get a temporary license for low-power operation using a transmitter provided by the Swedish Telecommunications Administration.

Denmark

In Denmark, a public corporation called Danmarks Radio (DR) is financed by receiver license fees. Three national radio networks and nine regional radio networks operate in Denmark, all under Danmarks Radio. As is the case in Sweden, many local stations operate on low-power and are licensed to organizations wanting to reach highly localized audiences.

The first television station in Denmark began operation in 1951. Today, two national networks are in operation. One is financed by license fees, the other by license fees and advertising. In addition to the two national television networks, local television stations, many operated by labor movement organizations, operate throughout Denmark. A pay-TV channel has had limited success and a growing number of households are connected to cable systems.

Finland

In Finland, radio and television are controlled by the Finnish Broadcasting Company (YLE).[7] Four primary radio networks are operated by YLE, and a number of local radio stations with commercial programming are also in operation. The Foreign Service directs radio programs primarily to North America, Europe, and Finns at sea.

Television broadcasting in Finland began in 1955 at the Technical University. Today, four television networks are in operation under the auspices of YLE. Two networks are seen nationwide and include programs from YLE, as well as programs from a commercial company, Mainos-TV (MTV) (Figure 13–3), which rents time and sells advertising on the YLE networks. A third channel is jointly owned and operated by YLE, MTV, and Nokia, a Finnish electronics company. The fourth network relays programs from Swedish television networks.

Finland cable television systems offer an increasingly wide range of programming. The English language Sky Channel, French TV-5, the Soviet channel Ghorizont, and the music television channel, (also abbreviated MTV), familiar to American cable viewers can be seen.

Norway

Norway's radio and television remain under the control of the independent Norwegian Broadcasting Corporation (NRK) (Figure 13–4). Financed by receiver license fees, the national NRK network has wide support, especially among older viewers. Younger viewers are split between NRK programs and satellite channels. NRK maintains strong programming control within Norway, although managing a degree of artistic freedom. NRK rules, however, require programs which strengthen Norwegian culture. Local television channels operate through cable systems and Swedish television has a significant following in Norway.

BROADCASTING IN GERMANY

The reunification of Germany is expected to have profound changes in the broadcasting system.

The structure of broadcasting in the Federal Republic of Germany (West Germany) is grounded in a legal system that grew out of World War II. Western Allies were concerned that the broadcasting system be free from state control and consequently set up "public" corporations in zones of occupation.

FIGURE 13–3 MTV Entertainment, a division of Finland's commercial television company, MTV (not to be confused with the music video cable channel), produces programs for MTV. Pictured here are the stars from the highly acclaimed offering "Trouble at the Post Office," which was part of the "Strange Stories" series produced by MTV Entertainment. The program was voted MTV Program of the Year and was recognized in other international program awards competition.

Public Stations: Organization and Programming

The Association of Public Broadcasting Corporations (abbreviated *ARD* in German) in the Federal Republic of Germany oversees the public stations, which are comprised of independent broadcasting corporations operating nine regional radio services: Bayerischer Rundfunk (BR), Hessischer Rundfunk (HR), Norddeutscher Rundfunk (NDR), Radio Bremen (RB), Saarlandischer Rundfunk (SR), Sender Freies Berlin (SFB), Suddeutscher Rundfunk (SDR), Sudwestfunk (SWF), and Westdeutscher Rundfunk (WDR). Joint radio networks with specialized programming operate between many of the regional radio services.

Radio broadcasting in West Germany under the auspices of the ARD is primarily directed at audiences in specific West German states. Each state regulates its own

FIGURE 13–4 One of the transmitting towers of the Norwegian Broadcasting Corporation. Because of Norway's geographic location, some transmitters are susceptible to conditions of heavy snow and ice; thus, the antennas are constructed with a large base to withstand the heavy loads caused by such adverse weather conditions. Towers supported by wires would break under the weight of ice and snow.

broadcasting system with federal jurisdiction existing for some technical facilities and external services. Each of the corporations is responsible for the day-to-day operations of radio and television. Organized groups such as churches and labor organizations provide feedback on programming through various broadcasting councils. An administrative council oversees the actual business and financial aspects of the various corporations.

Financing Public Broadcasting

The public system in West Germany is financed primarily by license fees determined by each German state and supplemented to a small extent by advertising revenues. Over the years the license fees have increased because the number of households and people using radio and television has leveled off while costs have increased.

Increases in license fees, a hotly contested political issue in West Germany, were last instituted in the 1980s but are still less than the typical newspaper subscription. The more profitable and populace German states support some of the smaller states where revenues are less. A central collection office collects the fees. Regional broadcasting stations carry advertising but in limited amounts. No advertising is permitted on Sundays and holidays.

Commercial Broadcasting

A number of commercial radio stations also operate in West Germany. Bundesverband Kabel Und Satelit E.V. is the trade association for private radio and television operators. Private commercial networks serve West Germany as do numerous local stations.

Television in West Germany includes regional programs as well as national programming and satellite distribution of programs for a wider European audience. The nine regional radio stations also operate television stations, and each shares in the production of programs that support the ARD television network.

Some programs are produced individually by the regional stations while other ARD network programs, such as some national news and sports programs, are a joint effort by all stations. In addition to ARD stations, Zweites Deutsche Fernsehen (ZDF) television operates a nationwide TV service but no radio service. ZDF was formed in 1961 as an alternative network to the ARD.

Television advertising in West Germany, like radio advertising in West Germany, is restricted. Along with the ban that prohibits advertising on Sundays and holidays, advertising is also banned during other times, except between 6:00 P.M. and 8:00 P.M. when commercials are grouped together and only separated by some short cartoon features.

East Germany's State Broadcasting Committee

East German radio is controlled by the State Broadcasting Committee of the Council of Ministers. Four principal domestic radio networks are in operation: Radio DDR, Stimme Der DRR, Berliner Rundfunk, and Jugenradio DT 64. In addition, Radio DDR-Feriewelle broadcasts from May through September and is directed to vacationers on the Baltic coast. Radio DDR-Messewelle broadcasts only during the Leipzig trade fair.[8] Television is controlled by the State Television Committee of the Council of Ministers. Broadcast income is obtained from receiver-license fees.

East German television consists of the state-operated G.D.R. Television.[9] VHF and UHF stations are located in a number of East German cities, and the network operates channels called "1st Program" and "2nd Program," each with different formats. Program schedules have been limited to mostly weekend and evening hours. News correspondents report from major capitals around the world. The primary news program "Here and Now," called "Aktuelle Kamera" (Figure 13–5) on G.D.R. Television, reports international and domestic news. Although drama and feature presentations are included in G.D.R. Television programming, the East German economy, still based on emerging agriculture and industry, results in these same themes in programming.[10]

Films and pre-recorded programs from other countries are synchronized in German, and G.D.R. Television has cooperative arrangements with television and film companies in about seventy countries. As part of the International Programme Exchange broadcasts from other countries are seen on G.D.R. Television. Among entertainment and music programs from outside the country are the New Year concerts from Vienna and performances from Prague Spring. Rock stars from abroad can be seen in the program "Music Box."

G.D.R. Television supports its own ballet company, primarily oriented toward modern dance. The ballet has been the

FIGURE 13–5 In East Germany, where broadcasting has been controlled by the State Broadcasting Committee, the major news program, "Aktuelle Kamera" ("Here and Now"), reports on domestic and international events. The reunification of Germany is expected to change the structure of broadcasting, which in the past has had sharp distinctions between East and West Germany. The two countries' use of a single currency, more entrepreneurial-type enterprises, and the overall effect of lessening trade barriers elsewhere in Europe will continue to influence the structure and programming of German broadcasting.

foundation of many G.D.R. Television productions and represents the television service with guest performances. Sports is a staple of programming, and about 1,500 sports programs are produced annually with coverage of many international competitions carried from outside East Germany.

THE USSR

On an expansive land mass of 8,600,000 square miles in Eastern Europe, sits the Union of Soviet Socialist Republics (USSR). The Soviet Union's radio and television system reaches both its own population and, through television satellite relays and the broadcasts of Radio Moscow, a global audience.

Control and Operations

Soviet broadcasting, although in transition, has historically been controlled by an arm of the state, the State Committee for Television and Radio under the USSR Council of Ministers.[11] Income has been derived from the state budget and from the sale of programs, announcements, and public concerts. Satellites relay programming to the USSR's Pacific coast, Central Asia, and to a limited extent, other areas of Europe. Central operations (Figure 13–6) are housed in the Moscow TV Center Tower Building, which has full production facilities for everything from small-studio to large-auditorium productions. Color television uses the Soviet-French SECAM system.

FIGURE 13–6 Soviet television falls under the control of the USSR Council of Ministers. Central operations are located in Moscow. State-controlled television networks, aided by satellite relays, help bring TV programs to the large expanse of the Soviet Union. An external satellite channel, Ghorizont, also distributes Soviet television outside the USSR. Under *perestroika,* cooperative production arrangements are taking place with American firms such as Turner Broadcasting, and commercial satellites launched by American companies can use Soviet rockets.

Soviet Broadcasting in Transition

A mostly authoritarian state, the USSR's broadcast policies have traditionally reflected the central government. However, with the democratization of Eastern Europe, the emergence of a more capitalistic economy, and a more free and open society, Soviet broadcasting is also beginning to change. For example, in the late 1980s the Soviet news agency Tass began offering television stations outside the Soviet Union the Tass news feeds.[12] The impetus for the beginning of major change, however, occurred in the early 1990s.

Broadcast Reforms under Gorbachev

As part of the changes affecting the Soviet Union under Soviet President Mikhail S. Gorbachev, broadcasting which came under control of the USSR Council of Ministers, was opened up to less central government control by Gorbachev's own presidential decree issued in July, 1990. In addition, the July, 1990 decree ordered that broadcast news reports, which had traditionally reflected the control and ideologies of the central government's propaganda line, should be completely "impartial."

As a check against upheaval and to help control operational and labor problems, the original Gorbachev order retained the ability to nullify unilateral changes in local broadcast outlets. The long-term effects of the Gorbachev order remained to be seen. Critics pointed out that a free-press, free-speech approach to news coverage was only as impartial as the people responsible for writing and delivering the news. On that account, the deep-seated ideologies and political beliefs of an intrenched broadcast bureaucracy which had long been in control, would not be easily changed.

Political Aspects of Gorbachev's Broadcast Reforms

By opening up control of broadcasting to more local influence, the views of dissidents which had been opposed to the policies of the central government of the past, could now be aired more openly. Many of Gorbachev's opponents had strong ties to the central government. The presidential decree also signaled that the broadcast media would become a more important part of the democratization of the Soviet Union.

In any free society where government leaders are elected, opposing views contribute to an informed electorate. Local control of the Soviet broadcast media, requiring impartial reporting of news in localities of the USSR where government officials are elected, meant that an open discussion of issues would now provide greater credibility for the "process" of democracy. With the central government controlling the broadcast outlets, the electorate's faith in the credibility of a democratic process, driven by public opinion, had understandably been lacking.

Political observers noted that Gorbachev himself had been the key figure in the reform of the Soviet political and economic system. Regardless of how much criticism the new open-media policies might bring upon himself, Gorbachev was seen as benefiting from being the one who instituted the atmosphere of free expression in the first place, and therefore could benefit from a more open discussion of issues, regardless of how critical they were of Gorbachev's own policies.

In theory, by monitoring public opinion as expressed in the media, astute locally elected politicians (and local elections were part of Gorbachev's reforms), would be able to adjust government reform in the Soviet Union to the tolerance level of the public, long before exceeding that level and being voted out of office.

Economic Aspects of Gorbachev's Broadcast Reforms

For political reform to succeed, economic reform must follow. A more capitalistic free-enterprise system demands a conduit for advertising, a means through which suppliers can create an awareness and desire for consumption. In the Soviet Union, much of the economic reform must come from the local grass-roots level. In addition, where more freedom and open political expression exist, there is less concern over the political content in all types of programming. With political freedom comes creative freedom; thus, local stations can program more to the needs and viewing preferences of local audiences, therefore becoming a more effective vehicle for broadcast advertising. The small local merchant whose future success depends on local consumption patterns, may in the future find broadcast advertising a necessary and productive marketing tool. Broadcasting, however, remains primarily government-controlled and government-owned. Whether future reforms go so far as to permit private ownership of broadcasting on a large scale is speculation.

The Future of Soviet Broadcast Reform

Although Gorbachev's presidential decree carried considerable force, many hurdles remain. Being a government-controlled broadcasting system, and delegating authority to local government, means the Soviet Parliament must contend with local elected officials who now enjoy a stake in the operation and governance of local broadcasting stations.

Moreover, while programming generated from Moscow has been the primary source of news and entertainment programming, the new found independence of local Soviet republics, which continue to want more control of local government and less central control, may very well result in the need for more local programming that local governments perceive is more sensitive to their local audiences.

In addition, when local governments change, questions will arise as to who is in charge. Such concerns were already being felt in some television centers in the Soviet Union when Gorbachev announced his original presidential decree in 1990. Moreover, because broadcast signals transcend local boundaries, and because countries must negotiate the allocation and use of frequencies with international bodies such as the International Telecommunication Union, some central control of radio and television is necessary in any country. Resolving these issues of central versus local control, while still maintaining stability, will be a future challenge for the Soviet Parliament, local governments, and the operators of the Soviet Union's television centers spread over a huge land mass that is not always easy to govern.

The Structure of Soviet Radio and Television

The organization of Soviet television, although in transition, consists of national television "channels" relayed to different parts of the Soviet Union through 117 television centers. Two channels, Moscow 1 and Moscow 2, serve all of the Soviet Union. Moscow and the immediate area are served by another channel. Still another channel carries drama, music, film, and literary programs. As part of the television relay system, programs are repeated in different parts of the Soviet Union on other channels to accommodate viewers in different time zones. Typical programs seen on Soviet television are "Time," a half-hour news program; "The 13 Chairs Tavern," a musical satire; and "Come on, Boys," an audience-participation show for boxers, wrestlers, weightlifters, and motorcycle racers. Programs are broadcast in different languages of the various Soviet nationalities.

Soviet radio programming is also heard in different languages. The Home Service of Radio Moscow has four networks. Channel I, like its television counterpart, is heard throughout the country and broadcasts news, commentary, drama, music, and children's programming. News and music are heard on Channel II, the former every half hour. Literary programs and drama are presented on Channel III, and Channel IV concentrates on FM musical broadcasting. FM stereo can be heard in twenty-six major Soviet cities. There are three different channels of cable radio, which reach about four hundred Soviet communities.

Programming Changes with Perestroika

How the Soviets view perestroika, the era of restructuring that began under Gorbachev, is captured in a quote from an issue of *Soviet Life* which talks about Soviet television. When perestroika began:

> television was one of the first mass media to react to the new way of thinking. News and information became more complete and were not "edited for the public at large." International telebridges gave people the opportunity to hear the "other side" and to be heard themselves. Live broadcasts enabled TV viewers to corner officials at all levels with direct questions to which they demanded straight answers. The anchor people suddenly became human, unpursing their lips and speaking to the audience in normal voices. Pop singers and rock musicians whose punk clothes and unorthodox songs had been condemned by aesthetic policymakers from the Ministry of Culture and puritanical citizens became regulars on television. No more complaints are made about boring programs.[13]

Although the validity of the statement, "No more complaints are made about boring programs," may be questioned, the *Soviet Life* description does help capture the revised programming philosophy of Soviet television in the era of perestroika.

FRANCE'S DIVERSIFIED SYSTEM

France has one of the more diversified broadcasting systems which ranges from government control to citizen-access channels.[14] The system consists of independent program "societies," which are much like separate broadcasting companies and networks. Support societies, such as those responsible for program production, also exist.

Radio France

A dominant force in radio broadcasting is Radio France, which operates domestic and foreign broadcasting services. Local and regional stations comprise the Home Service of Radio France (Figure 13–7). Radio France International also operates a foreign

FIGURE 13–7 A major force in French broadcasting is Radio France, which operates stations and networks inside France through the Radio France Home Service. The Radio France International Service also operates a domestic service for foreign workers living in France. Radio France International broadcasts to other parts of the world including French possessions overseas, the Middle East, Southeast Asia, and North and South America.

and home service. The foreign service broadcasts in French to such areas as Europe, Africa, the Middle East, Southeast Asia, and North and South America. Other languages are also heard over the foreign service of Radio France International. The home service of Radio France International broadcasts within France to foreign workers of various national origins. More than 1600 private commercial stations operate in France and are affiliated with various private commercial networks. Such radio networks as KISS FM, Radio Nostalgie, and Skyrock are heard throughout France.

French Television

In television, three major networks (societies) are Antenna 2 (A2), France Regions 3, and TF1, which can be seen in most French communities. Three private networks are Canal+, La Cinq, and M6. La Cinq and M6 operate UHF stations with a variety of programming. On M6 (Figure 13–8) such musical features as "Boulevard Des Clips," "Hit Hit Hit Hourra," and "Zap 6" can be seen. Magazine features such as "Turbo," detailing the expensive cars of Europe, and "Adventure," on individual sports such as mountain climbing are featured. American programs are also seen, and such familiar names as "Wonder Woman," "Miami Vice," and "Kojak," are available.

Canal+ Pay TV

Canal+ is a subscription network, and has made a substantial impact on French broadcasting. It matured quickly, signing on the air as the first private television channel in France in 1984. Canal+ quickly rose to become one of the world leaders in direct-sales pay television. Special decoders are necessary to receive the signal, with the exception of mealtimes when programs are free. Long-term subscriber contracts, automatic bank transfers to pay subscription fees, and high renewals have given Canal+ a solid financial base. Special customer services, such as golf tournaments organized for subscribers and discounts at sporting events, all contribute to subscriber loyalty. About half the programming consists of films and sports with a sizable percentage of Canal+'s income devoted to production of new television features.

The French telephone company, France Telecom, has a well-developed information network called Minitel which is discussed in the next chapter.

FIGURE 13–8 M6, one of many television channels available in France, airs a wide range of popular programming ranging from adventure series to magazines and musicals. Shown here is a scene from the program "6 eme Avenue."

THE PEOPLE'S REPUBLIC OF CHINA: MAINLAND CHINA

Radio broadcasting in China through the Central People's Broadcasting Service (CPBS) began in 1926 and is now an outgrowth of the Central People's Broadcasting Station, which is relayed to most of the country as well as Taiwan through microwave, shortwave, and FM stations.[15]

The Central People's Broadcasting Service

The CPBS and also television broadcasting through China Central Television (CCT), is under the control of the Ministry of Radio, Film and Television. Local radio and television stations are under the "direct leadership" of the Radio and Television Bureaus of the provinces, autonomous regions, and municipalities. The CPBS serves as a network of programming made up of news, educational, entertainment, and service programs.

Of programming percentages, *news* accounts for about 19 percent of the air time and includes programs produced by CPBS as well as local stations. *Educational* programs, accounting for about 26 percent of airtime, include "Study Sessions," "Reading Appreciation," "Tales of History," and others. *Entertainment*, which accounts for about 46 percent of programming, includes "music, operas, literature, ballad singing, and radio dramas. . . ." About 9 percent of the programs the CPBS classifies as *service* include "Mailbag, Information Service, Foreign Exchange Rates, Weather Forecasts, Radio Callisthenics, Programme Announcements, and Advertisements."[16]

In addition to CPBS, more than 200 stations operate in provinces, municipalities, and autonomous regions. Networks to relay programs for these stations, as well as CPBS, operate in more than 2,500 local areas.

China Central Television

China Central Television (CCTV), formerly Peking Television, traces its roots to the late 1950s. More than thirty years later there are more than 200 television stations in China operating on more than 200 channels and relayed to the country through more than 12,000 relay and translator stations.

Microwave, cable, and satellite links further aid in disseminating programs to China's mass population and into Tibet. CCTV operates in full color. Among the channels under the control of CCTV is one that broadcasts nationwide and includes a full schedule including such programs as "Television University" containing lectures shown in both morning and afternoon. Another channel also shown nationwide specializes in economic news along with some entertainment and sports programming.

An experimental channel began operating in the Beijing area in the late 1980s and carries educational and sports programming. Promoting more openness between itself and other countries of the world, CCTV has introduced foreign programming to China and operates television exchanges with approximately fifty nations, even coproducing some programs. Production facilities for CCTV have increased with the construction of new facilities and twenty studios that became operable in the 1990s. Research continues on the application of such technologies as teletext and high-definition television.

REPUBLIC OF CHINA: TAIWAN

Much of broadcasting in the Republic of China falls under the Broadcasting Corporation of China (BCC), which was originally founded in 1928 as the Central Radio Station. After a period of expansion the Central Radio Station Administration was established.[17] In 1947, the Central Radio Station Administration became a private entity and was renamed the Broadcasting Corporation of China, which operates under contract to the government.[18] In 1965, the BCC began accepting advertising.

The Broadcasting Corporation of China

Today the Broadcasting Corporation of China (BCC) is organized around an organization that includes such units as domestic broadcasting, news, engineering, overseas broadcasts, and other services typical of an international commercial broadcasting organization. A flagship station operates in Taipei and a series of remote transmitters, relay stations, and microwave links carry the BCC networks throughout the country. FM broadcasting arrived in the Republic of China in 1968 and AM stereo in 1987. Both AM and FM networks air programs ranging from music to children's programs and cultural programming. Cooperative broadcasting arrangements exist between the Broadcasting Corporation of China and government and military stations.

Radio: A Medium in Transition

Some of the BCC stations are highly specialized. For example, one station broadcasts nothing but traffic reports. Another station caters to the agricultural community and airs farm programming and weather reports in two Chinese dialects.[19] A "labor station" is located in southern Taiwan and broadcasts to factory workers. Other radio stations in the Republic of China, including those of the BCC, are becoming more and more specialized in the 1990s. Part of this specialization is the result of changes in programming caused by a loss of listeners, which reached new lows in the early 1980s. Prior to that time, much of radio consisted of Chinese soap operas, not unlike the era of soaps and radio drama in American radio of the 1930 and 1940s.

Changes in Radio Programming and Marketing

With the diversification of television programming in the Republic of China, much of it popular programming that captured larger and larger audiences, radio, especially in competitive urban areas, had to adapt. With models of American marketing and programming, Top-40 and other contemporary-type formats are developing. Other changes taking place include much more community-service programming and activities. For example, the English-language station ICRT (International Community Radio Taipei) airs a young stars program to identify musical talent.

Disc jockeys from the BCC are making more and more public appearances. The BCC also sponsors such activities as the BCC Chinese Classical Music Demonstration Group, Chinese Art Song Evening, the BCC's Children's Chorus, other cultural and community activities and events, and publications such as *Broadcasting Monthly*.[20] The Walkman and similar small radio receivers have also contributed to the growth in radio listenership.

Television in the Republic of China

Television in the Republic of China operates under three competing companies: China Television Company (CTS), Chinese

Television System (CTV), and Taiwan Television Enterprise Ltd. (TTV), each of which represents a network of stations serving the country. A look at TTV gives us one picture of television in the Republic of China. Taiwan Television Enterprises Ltd. (TTV) began in 1962 as the first commercial TV station in the Republic of China. The TTV is especially strong in its prime-time television news program. Major television news programs appear at other times as well. Divisions of Reporting, Production, and Foreign News support the TTV news organization. Correspondents in such cities as San Francisco, Los Angeles, New York, Hong Kong, and elsewhere feed international stories to TTV and supplement news feeds such as CNN and others.

Programming Philosophies in an Era of Liberalization

The liberalization of the Republic of China in the late 1980s resulted in more emphasis on international news, especially that of mainland China. Other programming has stressed the cultural heritage of China, to some extent part of the early tradition of the TTV network. Exactly how traditional this programming philosophy is can be seen in a report on TTV programming, stating the following aims: "to propagate the riches of Chinese traditional culture, to promote social ethics, to enlighten children's intellects, to help women's learning, and to reinforce the scientific and technological education."[21]

Quality Programming Control

This philosophy, complemented by strong efforts at quality control, is not forgotten by the other two networks. In what may seem rigid compared to the artistic free marketplace that exists in some other countries, CTV (the Chinese Television System) uses what it calls "prerequisite check-up." If a program is found to be improper in content, or to have defective pictures, sound, or lighting, mandatory remedial actions [are] taken "before the program airs."[22] At CTS (China Television Company) a development statement further emphasizes the programming philosophy with the goal of presenting "the brighter side of society, focusing on the promotion of truth and ethics, so as to advance social harmony, and fulfill the social obligation of TV programming."[23]

AUSTRALIA

The Australian Broadcasting Commission (ABC) operates the government radio system, which is called the *National System*. A private commercial system also operates in Australia, as does a public broadcasting system.

Radio

Responsible to Parliament, the ABC controls transmitters scattered throughout the continent. The National System consists of two major networks. One services local and regional areas and the other is nationwide. License fees, supplemented by government subsidies, help finance the system. Programming has emphasized cultural entertainment and classical music. Concerts by world-recognized musicians have been popular. Since these concerts are often held in large Australian cities, an admission charge helps defray their cost.

Australia also has a comprehensive commercial broadcasting system under an association called the Federation of Australian Broadcasters. Its service areas are more limited, but it can sell advertising throughout

Australia. Commercial stations, partly because they must depend on local advertisers, orient their programming to their own locale. As in the United States, commercial stations compete directly with each other as well as with the government stations. A voluntary code of ethics helps to control the content of Australian broadcasting and acts as a buffer to increased government control. In Australia, public stations, similar to those in the United States, are supported by subscription, universities, and organizations.

Television

As with radio, there are two systems of television in Australia: the government-supported system and the private commercial system. Government stations come under the ABC jurisdiction; commercial stations are part of the Federation of Commercial Television Stations (FACTS). Drama, public-interest programs, and sports make up the majority of government television programs.

JAPANESE BROADCASTING

Japan operates a dual broadcasting system consisting of a commercial and a government-public system. Commercial stations are supported by advertising; the public system, called NHK for Nippon Hoso Kyokai, is supported primarily by license fees and supplements its income in a small way by the sale of programs and interest paid on its funds. In addition, the Nippon Telegraph and Telephone Corporation (NTT) operates a sophisticated telephone and information service company with international interests.

Nippon Hoso Kyokai (NHK): General Philosophy

NHK started as the Tokyo Broadcasting Station in 1925 and today has grown to one of the most technologically advanced and respected broadcasting systems in the world. More than 30 million people pay receiving fees for the privilege of listening to and watching NHK programs (Figure 13–9). Its philosophy of broadcasting is articulated in this credo: "The fact that NHK is maintained and operated under this receiving fee system means that the very existence of NHK depends on the trust given it by the entire nation."[24] In some countries such platitudes would be suspect, but in Japan there is validity to the claim of impartiality.

NHK's Television and Radio Services

Two television channels are operated by NHK, General and Educational. Direct broadcast satellite systems, one for high-definition television, beam TV signals to the country. Three radio services—two medium-wave called Radio 1 and Radio 2, and an FM service—are operated by NHK, and both television and radio services are networked through stations to Japan's population.

Of the radio services, Radio 1 is the oldest and broadcasts mostly live programming, somewhat like request radio and talk radio formats, covering issues of daily concern to the populace. NHK described Radio 1 as the "medium closest to the audience."[25]

Radio 2, which originally aired programs for in-school use now handled by television, broadcasts programs about the Japanese language and music. The FM service concentrates mostly on musical programs. NHK's General Television Service broad-

FIGURE 13–9 NHK is a major force in television programming in Japan. Both domestic and satellite systems, as well as HDTV delivered through direct-broadcast satellites, operate in Japan. License fees are NHK's primary source of income and are collected throughout Japan by direct payment and by electronic funds transfer.

casts specials and features much like other systems worldwide but with a heavy emphasis on news, which is reflected in its "NHK News Today" program, an 80-minute prime time news show.

NHK's Contract Fee System

In Japan, license fees are collected from individuals based on categories of receiving contracts. One fee contract covers reception of color television, "the color contract," whereas the "ordinary contract" licenses black-and-white television receivers. Fees are billed every two months and can be paid through a bank account transfer or to a fee collector, where the charge is slightly higher. Exemptions for fees are possible for the needy and for the handicapped.

The contract fee system is administered through more than ninety offices throughout Japan. Efforts are being made to increase the use of the account-transfer method of payment and for advance payments. Not unlike cable systems in other parts of the world, the key to the success of the contract fee system is to make payment as easy and inexpensive as possible, and to collect as much advance payments as possible.

Commercial Broadcasting

Following the end of World War II, the democratization of Japan, and the introduction of television technology, the 1950 Broadcast Law recognized private commercial broadcasting as a competitor of NHK. NHK and the commercial companies operate thousands of radio and television stations covering even Japan's mountainous terrain. The public has a wide choice of programming. Dozens of companies are engaged in commercial radio and television programming in Japan. Four major VHF commercial TV stations operate in Tokyo and have cooperative networking arrangements with other stations throughout the world.

With Japan's interest in high technology and with a well-supported government broadcasting system competing with a commercial system, the future for broadcasting in Japan is particularly bright.

JAPAN'S NIPPON TELEGRAPH AND TELEPHONE CORPORATION (NTT)

The Nippon Telegraph and Telephone Corporation (NTT) calls itself the world's largest company.[26] It provides telephone, telegraph, leased circuits, digital network, and other services to its customers. Like telephone companies in such countries as the United States, NTT is a private company in a regulated industry. Also, like telephone companies in the United States, NTT consists of a number of subsidiaries, some of which are cooperative ventures with other nations. For example, the NTT subsidiary Nippon Information and Communication Corporation is a joint venture of NTT and IBM Japan, Ltd., and it combines IBM's data-processing expertise with the NTT telecommunications capabilities.

Consumer Electronics

How close some of NTT's domestic telecommunication services come to traditional broadcasting is evidenced by NTT's Off-Talk service, which permits a telephone terminal with a speaker to operate as a cable radio and allows the consumer to select from four different news-radio stations. Customers can also obtain a home telephone terminal with a video monitor and receive videotex graphics and other information from a remote data base.

Digital and ISDN Services

Many more services will be made possible as Japan's telephone system continues the transition from analog to digital. Accompanying this change is the increased use of the Integrated Services Digital Network (ISDN). This service, especially as it becomes the world standard, will permit a wide range of telecommunications services to be exchanged simultaneously over the same system and between countries. By the early 1990s in Japan, a commercial ISDN service was operational in Tokyo, Osaka, and Nagoya.

International ties are maintained through such ventures as the sale of NTT technologies to other countries and consulting services to international users of NTT technology. Three examples are a telephone directory service operated as a joint venture with NTT and ITT; a research company operated in the United States as a joint NTT venture; and a joint venture with a Canadian company involved in developing a rechargeable lithium battery.[27]

THE MIDDLE EAST

Broadcasting in the Middle East means radio and television stations programming in different languages, to different cultures, and to audiences with different political and religious beliefs. The region, which is oil-rich but contains extremes of both wealth and poverty, is faced with frequent political and social change. This change often determines who owns and controls radio and television stations, and what gets aired on those stations.[28]

Iraq, Kuwait, and Iran

When Iraq invaded Kuwait in 1990, the Kuwait Broadcasting Service fell to Iraq. At stake were approximately a half dozen VHF and UHF television stations in Kuwait, and more than a dozen Kuwait radio stations, which beamed signals to the Middle East as well as Africa, Europe, and North America. After the invasion, the government radio and television system of Iraq became the instrument of electronic mass communication for both countries. Iraq's Baghdad Television network, relayed on international television networks, became a worldwide conveyor of information. Iraq's government spokesmen and TV news readers (Figure 13–10) used Baghdad Television to make their official policy statements to a waiting audience of broadcast journalists covering the crisis in the Middle East.

At the time of the invasion, the government's Baghdad Television oversaw the operation of all television stations in Iraq, some with limited amounts of commercial programming. Radio stations operated by the Broadcasting Service of the Republic of Iraq transmitted programs in various languages to both domestic and foreign audiences. In addition to Iraq-controlled radio news, radio programming ranged from "The Voice of the Masses" to "FM Radio Baghdad" which aired popular music programs.

Iran, Iraq's neighbor to the East, con-

FIGURE 13–10 A TV news reader on Baghdad Television reads a policy statement from the government of Iraq shortly after Iraq's invasion of Kuwait.

trols radio through an organization called Islamic Republic of Iran Broadcasting. Television falls under a government organization called Islamic Republic of Iran Television. Television stations and repeaters are located throughout Iran, as are radio stations. An external radio service operated by the government of Iran broadcasts to North America, Europe, and Africa.

The Mediterranean Region

Israel, Lebanon, Jordan, Syria, Turkey, a small part of Egypt, and the island of Cyprus, are located in the region of the Middle East close to the Mediterranean Sea and to the north of the Red Sea.

In Israel, where broadcasting comes under the Israel Broadcasting Authority, there are domestic, external, commercial and non-commercial stations operating. Multilingual radio programs for immigrants can be heard in Israel, as can programs in Hebrew, French, English, and Arabic. Programs in other languages are heard on the external radio service. In addition, Israel Educational Television operates VHF and UHF educational television stations throughout Israel.

In Lebanon, much of radio falls under Radio Lebanon, while television comes under Tele-Liban; both organizations are government-controlled. In addition, the Lebanon Broadcasting Corporation operates Christian and commercial television stations. It is radio in Lebanon, however, that reflects the political and religious diversity of the Middle East. Radio stations called by such names as the Voice of National Resistance, Radio of Islam, Radio Free Lebanon, Radio of the Lebanese Forces, and Voice of Hope, operate in Lebanon. Commercial radio stations also operate in Lebanon.

Both Jordan and Syria have government-operated broadcasting systems. Radio and television programs in Jordan are primarily broadcast in Arabic, with some additional programs in French, Hebrew, and English. Broadcasting in Jordan comes under the control of Jordan Radio & Television. Syrian radio and television is under the Syrian government's Broadcasting and Television Organization, headquartered in Damascus. About a dozen stations in the 560–1480 kHz range operate in Syria, and VHF television stations and television repeaters cover the country.

In Turkey, broadcasting is operated by the Turkish Radio-Television Corporation. An extensive network of FM radio as well as television stations operate throughout Turkey, and a foreign radio service covers much of Asia, Africa, Europe, and North America.

A part of Egypt extends into the Middle East and is bordered by Jordan. Much of Egypt, which is considered part of Africa, has a government-controlled broadcasting system operating under the Egyptian Radio and TV Union, headquartered in Cairo. Egyptian broadcasting is one of the most extensive in the region, with a well-developed radio and television network that includes extensive domestic programming in Arabic.

In addition to the countries just mentioned, radio and television stations are found on the island of Cyprus, located off the coast of Turkey. The Cyprus Broadcasting Corporation, an independent agency, operates radio and television stations which broadcast in Greek, Turkish, and other languages. Special foreign language radio programs directed at tourists can be heard during the summer months. The commercial station, operated by Radio Monte Carlo Middle East, and radio stations operated by the British Forces Broadcasting Service, are located in Cyprus.

Saudi Arabia and the Indian Ocean Region

To the east of Egypt across the Red Sea sits Saudi Arabia, where radio comes under the government's Broadcasting Service of the Kingdom of Saudi Arabia, headquartered at the Ministry of Information in Riyadh. In Saudi Arabia, bordered on the east by the Persian Gulf, television comes under the control of Saudi Arabian Television, also a unit of government. An extensive radio and television network covers Saudi Arabia, much of it programming in Arabic, although some external broadcasts are in the African languages of Somali and Swahili. The Arabian American Oil Company also operates radio and television stations for its employees in Saudi Arabia.

On the western and southern borders of Saudi Arabia sit independent states, all with government-controlled broadcasting systems. In this region are Qatar and the United Arab Emirates, both of which border the Persian Gulf. The region also contains Oman, the People's Democratic Republic of Yemen, and the Yemen Arab Republic, all of which border the southern part of Saudi Arabia and lie next to the Indian Ocean. Of these nations, the United Arab Emirates have the most diversified broadcasting system with different operating companies.

The Significance of Broadcasting in the Middle East

The significance of broadcasting in the Middle East rests with its role as a strategic political force in a region where religion, politics, and frequently war, are a significant part of life. With so much of the region's broadcasting under government control, the ruler who controls radio and television also controls the channels of information. When it is necessary to reach a mass audience, to stir political passions, or to issue a call to arms, government-controlled broadcasting is a preferred medium of persuasion. Moreover, in a part of the world where kings, sheiks, exalted rulers, dictators, and religious leaders are the norm, radio and television can mean the difference between social stability or upheaval. The mere broadcasting of masses of individuals in protest or support of a leader or his policy or doctrine, is a powerful persuasive force, both inside the country and internationally. Although the news and information which is often generated from these government-controlled stations is less than objective, it is many times the only information available for Western journalists. The news and information we receive must be tempered by the fact that the pictures and reports are many times a product of news management and censorship, and Western journalists are faced with adding their own commentary in an attempt to balance, and add objectivity to reports which may be biased and controversial.

SUMMARY

The growing interdependence of nations makes it imperative that we understand something about international broadcasting systems. This chapter examined some of the domestic broadcasting systems.

Canada and Mexico offer comparisons to the United States. Laws governing Canadian broadcasting stress the protection of cultural integrity. The Canadian Broadcasting Corporation is the Canadian government's broadcasting arm. Individual provinces also operate broadcasting systems. Both government and private commercial stations are found throughout Canada. Both government and private commercial broadcasting stations operate in Mexico.

The British Broadcasting Corporation (BBC) and the Independent Broadcasting Authority (IBA) are two principal broadcasting organizations in the United Kingdom. The BBC began in 1922 and became a public corporation by royal charter in 1927. BBC Radio is sent over four domestic networks. BBC television operates on two networks. The IBA operates independent radio and television stations which compete for listeners and viewers with the BBC. Oracle is the IBA teletext service.

Regional diversity exists in the Nordic countries of Iceland, Sweden, Denmark, Finland, and Norway. All of the Nordic countries have increased use of VCRs and some have well-established cable systems. A variety of organization and regulatory frameworks exist, ranging from the tightly controlled government system of Norway to the more open and competitive framework of Sweden.

The reunification of East and West Germany is expected to have profound changes in the broadcasting systems. The government's Association of Public Broadcasting Corporations operates nine independent radio services. Commercial stations operate in the West, although advertising is restricted to certain times of the day. The State Broadcasting Committee of the Council of Ministers controls broadcasting in the East.

In the Soviet Union, Radio Moscow is a primary force in both domestic and international broadcasting. Four television networks, as well as satellite systems cover the Soviet Union.

The French have a very diversified system which includes Radio France's extensive network of stations, in addition to many private commercial radio stations. Three primary television networks, A2, FR3, and M6 serve France. A profitable pay-TV service, Canal+, also operates in France.

In mainland China, The Central People's Broadcasting Service is an outgrowth of the Central People's Station which began operating in 1926. Television is under the control of China Central Television (CCTV) which operates more than 200 stations and covers the country through more than 12,000 translator stations.

In Taiwan, the Republic of China, broadcasting falls primarily under the Broadcasting Corporation of China (BCC). Units of domestic broadcasting, news, engineering, overseas, and others are part of the BCC's organization. Radio broadcasting is more specialized, much of it programmed and marketed like American radio stations. Television operates under three competing companies which are the China Television Company, Chinese Television System, and Taiwan Television Enterprise Ltd.

Private, government, and public broadcasting stations operate in Australia. The government stations are under the direction of the Australian Broadcasting Commission (ABC), and commercial stations operate through the Federation of Australian Broadcasters.

Japan's broadcasting system is both government-supported and commercial. Known around the world for its technology and programming is Nippon Hoso Kyokai (NHK), the government broadcasting arm. A high-definition, direct-broadcast-satellite system also operates under NHK.

Broadcasting in the Middle East reflects the tremendous diversity of the region. Most of the countries have government-controlled broadcasting. Radio and television stations are found throughout the region, and many countries engage in external broadcasting, some to international audiences in Europe, Africa, and North America. Others concentrate their external broadcasts on other countries in the Middle East. Radio and television stations play a big part in conveying the governments' policies and directives.

OPPORTUNITIES FOR FURTHER LEARNING

ASANTE, A. K., AND GUDYKUNST, W. B., eds. *Handbook of International and Intercultural Communication*. Newbury Park, CA: Sage, 1989.

BISHOP, R.L. *Qi Lai! Mobilizing One Billion Chinese: The Chinese Communication System*. Ames: Iowa State University Press, 1989.

BROWNE, D.R. *Comparing Broadcast Systems: The Experiences of Six Industrial Nations*. Ames: Iowa State University Press, 1989.

CHANG, W.H. *Mass Media in China*. Ames: Iowa State University Press, 1989.

DINH, T. V. *Communication and Diplomacy in a Changing World*. Norwood, NJ: Ablex, 1987.

FOX, E. *Media and Politics in Latin America*. Newbury Park, CA: Sage, 1988.

GANLEY, G. D., AND GANLEY, O. H. *Global Political Fallout: The First Decade of the VCR, 1976–1985*. Norwood, NJ: Ablex, 1987.

HACHTEN, W.A., WITH HACHTEN, H. *The World News Prism: Changing Media, Changing Ideologies*. 2nd ed. Ames: Iowa State University Press, 1987.

HARTMAN, P. AND OTHERS. *The Mass Media and Village Live*. Newbury Park, CA: Sage, 1990.

JUSSAWALLA, M., LAMBERTON, D. L., AND KARUNARANTE, N. D. *The Cost of Thinking: Information Economies of Ten Pacific Countries*. Norwood, NJ: Ablex, 1988.

LENT, J.A. *Mass Communication in the Caribbean*. Ames: Iowa State University Press, 1990.

LULL, J.L. *World Families Watch Television*. Newbury Park, CA: Sage, 1988.

PICARD, R. G. *The Ravens of Odin: The Press in the Nordic Nations*. Ames: Iowa State University Press, 1988.

ROGERS, E. M., AND BALLE, F. *The Media Revolution in America and in Western Europe*. Norwood, NJ: Ablex, 1985.

SENNITT, ANDREW G., ed., *World Radio TV Handbook*. New York: Billboard, 1989.

14 INTERNATIONAL EXTERNAL, AND TELECOMMUNICATION SERVICES

Many countries operate external and overseas broadcasting systems designed to reach audiences in other parts of the world. Some of the more familiar names in overseas services include Radio Canada International, the BBC, and Radio Moscow, among others. Most offer a wide variety of foreign-language programs that reach people of diverse cultures and political perspectives in other regions of the globe. This chapter examines these external overseas services in addition to some of the emerging satellite systems.

RADIO CANADA INTERNATIONAL

Radio Canada International (RCI) traces its roots back to its first broadcast in 1945 when the Canadian Prime Minister announced the new service would "serve both a national and international purpose; it will also bring Canada into closer contact with other countries."[1] The early years of the service were marked by some tough economic constraints until priorities were established and the service gained a more stable financial footing. Some early foreign-language broadcasts were started and then discontinued, but Radio Canada International continued to grow in other areas, including agreements with the BBC to relay programs to Eastern Europe from BBC transmitting facilities.

Like other international services, Radio Canada International directs its signal to a number of target areas including Eastern and Western Europe, Africa, the United States and Mexico, Latin America, and the Caribbean Basin. In addition, Radio Canada International is placing a growing emphasis on the Pacific Rim Nations. The program "Canadian Journal" is a weekly business-oriented program produced in Montreal and transmitted via satellite for rebroadcast on Hong Kong Commercial Radio. Other programs in the Japanese language are fed via satellite to Tokyo's Radio Tanpa and then rebroadcast throughout Japan. Pre-recorded programs are also available for rebroadcast by stations in other countries.

THE BBC WORLD SERVICE

The BBC receives respect and attention from a global audience, much of it gained through the BBC World Service broadcasts. Using long-, medium-, and shortwave transmitters, the BBC World Service reaches most of the areas of the globe. Overseas relay stations are located in such places as Germany, Cyprus, Hong Kong, Singapore, Canada, the United States, Brazil, and elsewhere. "This Is London" uses thirty-nine different languages to broadcast news, information, cultural, and entertainment programming. The World Service transmits values "of a society governed by laws voted democratically, yet willing to listen to dissidents, both within its own frontiers and outside them. It mirrors a national community retooling itself economically and ideologically for the 21st century."[2] The BBC radio and television regional services transmit to Northern Ireland, Scotland, Wales, and the English regions served by television.

THE NETHERLANDS: RADIO NEDERLAND

Operating from the Netherlands is Radio Nederland, an extensive short-wave broadcasting system that beams Dutch programming in nine languages through main transmitter sites and relay stations (Figure 14–1).

Early Amateur Contacts

Radio Nederland's roots date back to the beginning of Netherland broadcasting in 1927 when amateur radio operators established shortwave contact with the island of Java and broadcasting to the Dutch East Indies began. Efforts were improved through the development of the Philips Physics Laboratory in Eindhoven, which came up with an improved transmitting system. Out of this technology the Philips-Holland-Dutch East Indies Broadcasting Organization was formed.

Nazi Occupation and Radio Orange

In 1940, during the Nazi occupation when the Dutch government moved to London, Radio Orange, named after the Royal House of Orange, was established to broadcast "messages of freedom" from London back to the Netherlands.[3] In 1946, just after the end of World War II, Radio Nederland was founded as the official international service. It was charged with "The Production and Preparation of Radio Programmes, Intended for Reception Outside the European Territory of the Realm."[4] The first program, in Bahasa, Indonesia, aired the same year.

Operations

Radio Nederland is headquartered in Hilversum and broadcasts original programming in addition to operating a distribution system that produces and distributes programs on tape, disc, and cassette. RNTV (Radio Nederland Television) produces documentaries that are sold to international markets. A training center for radio and

FIGURE 14–1 An antenna system of Radio Nederland, which broadcasts Dutch programming to Europe and overseas. Radio Nederland's history dates back to 1927 when amateur radio operators established contact with the island of Java and broadcasting to the Dutch East Indies began. In addition to broadcasts, Radio Nederland programming is distributed internationally through prerecorded tapes and discs.

television journalists from the third world is also located at Hilversum. Radio Nederland is funded by license fees and advertising revenues from domestic services. The training center is funded by the Dutch government.

SWISS RADIO INTERNATIONAL

Swiss Radio International operates an extensive program of European and overseas broadcasting. Its high-powered transmitters beam news, information, and cultural programs to all the continents of the world. The Swiss, because of their reputation for neutrality in international affairs, have substantial credibility among many nations, whereas other countries may be perceived as having vested interests in the content of their programming. The Swiss service is well accepted, especially in third-world countries and developing nations.

Broadcasts are heard in German, French, English, Spanish, Portuguese, Arabic, Esperanto, and Romansch. Entertainment programs are also part of the worldwide service, and recorded programs are mailed to over 300 stations throughout the world. News and information programs about events in Switzerland and elsewhere are part of the regular schedule as are cultural and documentary programs.

GERMANY

In West Germany, Deutsche Welle (DW) radio broadcasts worldwide on shortwave in more than thirty languages and Deutschlandfunk (DLF) radio broadcasts to northern Europe. DW and DLF, which were established under German law in 1960, are members of ARD and have full voting rights in ARD, the government agency overseeing German broadcasting.

In East Germany, Radio Berlin International is the external service of East Germany with medium-wave and shortwave transmitters providing a European and overseas service with programming in such areas as news, commentaries, listener mailbag shows, sports, and science and technology.

USSR: RADIO MOSCOW

The USSR's external and overseas service is the responsibility of Radio Moscow. Broadcasting in more than sixty languages, Radio Moscow programs focus mostly on life in the USSR, the Soviet view of international issues, and Russian drama and entertainment (Figure 14–2). Programs are distributed free of charge to stations in other parts of the world.

The North American Service

Typical of programs are those of the Radio Moscow North American Service, which is directed to listeners in the United States and Canada.[5] "News on the Hour," from the Soviet perspective, is read by Russian commentators speaking in English. "Home in the USSR" consists of information about local events, domestic issues, and in-depth coverage of news in the Soviet Union. A question-and-answer program titled "Moscow Mailbag" is based on listeners' letters. "Top Priority" is a weekly panel discussion with top-level Soviet foreign-policy experts. "People" is a talk show about Soviet citizens.

The World Service

The Radio Moscow World Service broadcasts in English to all parts of the world. Features on science and engineering; an audio book club, which discusses Russian classical and contemporary literature; special features on Asia and the Pacific; and musical programs with everything from jazz to rock-and-roll highlight the World Service programming.

The Era of Perestroika

A reflection of the increasing cooperation between the Soviet Union and non-Communist countries is found in the program

FIGURE 14–2 Radio Moscow operates from the USSR and provides programs overseas through short-wave and transcription services. The content closely follows the government's political philosophy, which has been more open since the advent of *perestroika*. Radio Moscow and its programming is ultimately controlled by the state under the USSR's Council of Ministers.

"Perestroika," which began in the late 1980s. As described in the programming schedule of the World Service:

> This is a time of change in the Soviet Union. A time symbolized by two words which have become international. They are "perestroika" and "glasnost." This weekly feature will contribute to your better understanding of what these words stand for. Through investigative press reports, on-the-spot coverage, thought-provoking interviews PERESTROIKA will provide the insight you need to see where the Soviet Union is going.[6]

That change, however, did not stop an episode of one of the most popular radio and television programs, "Vzglad," from being cancelled. The program had been called a "lightning rod for conservative criticism," and some Soviets called the cancellation of the program censorship when the last episode of 1989 was pulled from the schedule.[7] The success of glasnost will determine whether Soviet media drifts toward more openness and less control or whether it returns to a tightly controlled propaganda instrument.

RADIO BEIJING

Radio Beijing, the external service of the People's Republic of China, began transmitting in 1947. Today the service broadcasts in more than thirty languages. English-language broadcasts began in 1984 and were originally beamed to Beijing proper to serve tourists as well as diplomatic missions.

The "Task" of Radio Beijing

The "task" of Radio Beijing as stated by the service is to "explain China's positions and policies on domestic and international issues. . .and promote friendship with the people of other countries; to support the just struggles of the peoples all over the world."[8] Each language service produces programs for its target audience and each transmission includes news and commentary on international issues, as well as feature and cultural programs about China.[9] Radio Beijing is divided into three areas: (1) central newsroom and editorial departments, (2) language departments, and (3) administrative departments.

The Home Service

In addition to Radio Beijing, the domestic or "home" service of China broadcasts to Taiwan through the Central People's Broadcasting Station (CPBS). In addition to news and entertainment, such programs as "Letterbox for Relatives and Friends," "Travel Across the Country," "Friends on the Air," "Our Motherland," "Sports World," "Homeland of the Dragon," "At Your Service," and "Folks on the Mainland" can be heard.[10]

VOICE OF FREE CHINA

The international broadcasting service of the Republic of China, situated on the island of Taiwan, is the Voice of Free China.

History

The history of the Voice of Free China dates back to 1928 when the Central Radio Station was founded.[11] Under the auspices of the Central Radio Station Administration the station grew in size and importance and began international broadcasting in 1932.

Under a revised constitution of the Republic of China, which took effect in 1947, the responsibility for regulating the communications industry fell to the Communications Ministry. The Central Radio Station Administration became a private entity and was renamed the Broadcasting Corporation of China (BCC).

The BCC moved with the government of the Republic of China to Taiwan in 1949 and began overseas broadcasts to mainland China under the call sign VOFC, the Voice of Free China.

Mail as a Measurement of Effectiveness

As with many external services, a concerted effort is made to involve listeners in the station's programming, and a measure of success is determined by the amount of listener mail received.[12] Computerized mailing lists to record listeners' names and addresses, mailbag programs where listeners' letters are answered, and listening clubs in a number of foreign countries support this effort. To serve listeners in Asia more effectively, the Voice of Free Asia was founded in 1979 and broadcasts in such languages as English, Mandarin, Thai, and Indonesian.

RADIO JAPAN

Overseas broadcasting in Japan dates back to 1935 when a one-hour transmission was beamed to the West Coast of North America. While overseas broadcasting from Japan was suspended after World War II, it resumed in 1952 as Radio Japan. From two languages, English and Japanese, the service, which is operated by NHK, has grown to a worldwide enterprise broadcasting in more than twenty languages. With a staff of more than 200, Radio Japan airs two categories of programming: General Service and Regional Service.

General and Regional Services

The General Service broadcasts globally in mostly one-hour-and-thirty-minute programs. Both news and general information features on Japan are included in daily programming. The service expands coverage when major news developments or sporting events take place.

The Regional Service broadcasts to specific regions, two examples being Europe and Southeast Asia. Programming is sensitive to the language and culture of each region, the characteristics of society, and the region's ties with Japan.

Radio Japan Programming

Programming on Radio Japan includes news followed by commentary. Other programs present features on life in Japan. NHK's profile of programming includes:

> "Meet the People," which presents interviews with Japanese opinion leaders in many different fields. "One in a Hundred Million" introduces individual Japanese from all walks of life. "Radio Japan Journal" focuses on important current events or topics. "Science Today" describes the latest research and its practical applications in Japan. "Our Heritage" and "Japan Panorama" give listeners a fresh look at life and culture, both traditional and modern, while "Japan Travelogue" presents a different part of our country every week.[13]

Another popular program is "Let's Learn Japanese," which is broadcast in all languages employed by the overseas service. Radio Japan has cooperative arrangements with other countries where Radio Japan transmitters are located. A West African station relays signals to Europe, the Middle East, Africa, and South America. A Radio Canada relay station broadcasts Radio Japan signals to central and eastern North America.

JAPAN'S INTERNATIONAL NEWS RELAYS

Japan's international ties are further evidenced by its efforts to beam its domestic news to international audiences. For example, one of the four major commercial stations in Tokyo, Asahi National Broadcasting Co. Ltd., operates channel 10 but sends programming via satellite to other parts of the world including Honolulu, San Francisco, Los Angeles, and New York, where the evening news from Tokyo's channel 10 can be seen in the United States shortly after it airs in Japan. Another Toyko-based commercial broadcaster, the Nippon Television Network Corporation (NTV), operates channel 4 in Tokyo but owns an American television production company. NTV also has satellite links throughout major world centers for the purpose of exchanging news and other programs between Japan and other countries.

RADIO AUSTRALIA

The international service of the Australian Broadcasting Corporation is Radio Australia.[14] Radio Australia is an especially dominant force in the Pacific and Asian region and includes a China, French, Papua-New Guinea, Japan, Vietnam, Thai, and Indonesian service. Especially close cooperation exists between Radio Australia and the People's Republic of China. Program exchanges exist with Chinese radio stations, and agreements have been forged between Radio Australia and the Radio Beijing Publishing House for support materials accompanying Radio Australia broadcasts. Radio Australia personnel also assisted China's Radio Shanghai to set up its English-Language Service.

Radio Australia's Japanese Service airs programs that promote Australia tourism and business opportunities for Japanese who want to invest in Australia. The Vietnam Service, facing reception problems, maintains programming directed at Vietnam. Events of interest to Thailand, such as a Thai Studies Conference, are covered by the Thai Service. Close cooperation between Australia and Indonesia has resulted in staff from the Indonesian Service traveling to Indonesia for coverage of Indonesian politics and cultural events.

VOICE OF AMERICA

Under the auspices of the United States International Communication Agency (USICA), the Voice of America (VOA) broadcasts over 900 hours of programming per week in English and approximately forty other languages. Shortwave and medium-wave transmitters reach an estimated 67 million listeners and tens of millions more listening in China. News and news analyses are the basic programming formats, supplemented with information on American society and American popular-music programming, notably jazz. The domestic and overseas transmitters have a combined power in excess of 20.5 million watts. Radio programs from VOA are made available to local stations in many countries.

THE USICA FILM AND TELEVISION SERVICE

In addition to overseeing the Voice of America, USICA acquires and produces videotape programs and films. These are shown by USICA posts to audiences overseas and are sometimes distributed through foreign television stations and commercial theaters. USICA also provides foreign television stations with news clips of events in the United States.

RADIO FREE EUROPE AND RADIO LIBERTY

Radio Free Europe and Radio Liberty (RFERL) are funded by the U.S. Congress and broadcast from Germany to Eastern

Europe. Radio Liberty targets the Soviet Union with a dozen different languages. Products of the Cold War era, the two services in recent years have concentrated on educating listeners to democracy, something countries of Eastern Europe are learning first-hand after the fall of the Communist majority in many East European nations. One of the service's aims is the teaching of Western journalistic ethics and free-press reporting procedures to the newly democratized nations. News on both Radio Free Europe and Radio Liberty reflects international events as well as events occurring in the countries where RFERL broadcasts are targeted.

CNN INTERNATIONAL

Ted Turner's Cable News Network (CNN) operates a 24-hour international satellite news service, CNN International. Beamed from the Atlanta headquarters of CNN, the news service draws from CNN's international bureaus in such places as Beijing, Cairo, Tokyo, Nairobi, and other world centers. Business, weather, sports, and breaking stories are part of CNN International's regular programming. Stories from other foreign news sources also appear on CNN International.

SATELLITE AND CABLE CHANNELS

Many households, especially in Europe, have access to an increasing number of satellite and cable services. Among the channels the Direct Broadcast Satellite Service (DBS) carries are Sky Channel with family entertainment, Sky News with 24-hour news, and Eurosport with international sports programming. Cable subscribers also receive Sky Television services. TV5 Europe (Figure 14–3) broadcasts via satellite a wide variety of French Language programming throughout Europe and through satellite relay systems to many other parts of the world.

BRITISH SATELLITE BROADCASTING

British Satellite Broadcasting operates the DBS system for the Independent Broadcasting Authority.[15]

DBS Service

The DBS service is funded by subscription fees and advertising. It began with the support of a number of investors and was later listed on the stock exchange. Among a planned service, subject to IBA approval, is a channel devoted to Britain's youth culture with concerts, videos, and rock music part of the programming fare. Other channels include a movie channel, a family entertainment channel with daytime magazine programs, and a sports channel.

Datavision

A specialized satellite service of British Satellite Broadcasting called *Datavision* serves the telecommunication needs of business and professional users. Datavision provides a satellite link for companies to communicate with outlying offices and manufacturing sites, to provide data and textual information, and to serve educational and training needs.

THE MARCUCCI GROUP AND SUPER CHANNEL

It is doubtful if grandfather Marcucci envisioned in 1930 when he ran his bread-baking business in Chicago that his family

FIGURE 14–3 Numerous satellite channels operate in different parts of the world. The logo of France's TV5 Europe is familiar in many countries outside France where TV5, an international French-language satellite channel, is available. TV5 is seen in Canada, for example, where such programs as a "news review" of the week in Canada originate. TV5 stresses French culture in its entertainment programming.

would one day control one of the most popular satellite and cable programming channels in Europe. The road from the bread-baking business to television began first in Italy, where the business was moved in the 1950s. The family later bought a pharmaceutical firm, which proved profitable enough to launch other businesses.

The company bought parcels of land on mountaintops where eventually television towers were erected and established an Italian television network, which began in 1976. The company moved into outer interests including the popular Italian-based music channel, Videomusic. A majority interest in the ailing Super Channel was bought by the Marcucci family, which secured the agreement of creditors that the new owners would only have to pay off one-fourth of the debts of the financially troubled channel. To refuse would have meant selling off the channel and even less return to creditors.

Today the Super Channel reaches more than 16 million cable homes in sixteen countries with a potential viewing audience of 40 million Europeans. A multilingual news service with world news on the hour and co-productions of live broadcasts are part of Super Channel's program offerings. Behind much of the business side of Super Channel is Marialina Marcucci, part of the large and extended family that traces its roots to, among other places, the Chicago bread-baking business of the 1930s.[16]

THE CHILDREN'S CHANNEL

Children in many northern European countries can receive a cable channel devoted to children's programming (Figure 14–4). In the United Kingdom, for example, the Children's Channel attracts 90 percent of children's weekday morning viewing against broadcast TV's and AM radio's 10 percent.

FIGURE 14–4 Worzel Gummidge, the mischievous scarecrow come to life on The Children's Channel. A popular satellite channel, The Children's Channel broadcasts primarily in English, but with increasing attention to other languages. Subtitles are used in some non-English-speaking countries and an accompanying teletext service carries program schedules.

Supported by both advertising and subscription fees, the Children's Channel transmits primarily in English but efforts to accommodate other languages, such as Dutch, are being made. In Holland and other countries, subtitles are used. A teletext service integrated with the regular programming carries program schedules, birthdays, news, and other information.

During weekdays, such programs as "Huva," a monkey who strives to improve, "The Magic Corner," which offers programs for parents and children, and "Stories Without Words," which are visual tales for youngsters, can be seen. Weekends include programs on such subjects as games, drama, wildlife, and science. Programs are selected from those produced by the Children's Channel staff and other sources from throughout the world. Strong emphasis on participation results in thousands of letters from youngsters being sent to the Children's Channel. The letters in turn become the source of many ideas for programming decisions.

INTERACTIVE AND PREMIUM CHANNELS

Cable subscribers, in Europe, for example, have access to premium and interactive channels that provide alternatives to standard broadcasting. TeleClub is a commercial Swiss pay-TV premium channel service.[17] First introduced in Zurich, Switzerland, on a continuing experimental basis, TeleClub has now expanded to other areas of Europe and receives its income from subscriber fees. It is prohibited by law from selling commercials in its programming.[18] Concentrating on movies, the service provides approximately seven motion pictures daily. About half of the programs are first-run presentations on television.

Interactive television is available through channels such as the Cable Jukebox, which programs music videos. Viewers can select different video tracks with a library of songs growing at about a hundred per month. The service promotes itself with the statement: "The widest range of music is on offer 24 hours a day without the distraction of video jockeys or advertisements." Headquartered in London, the service is delivered through different interactive technologies. For example, some viewers can access the system through terminals on their TV sets, others can access different music video channels by telephoning their selections to a central number where the choice is verified and instructions given on when the track will be played. The five channels are organized into "pop," "disco," "guitar breaks," "soul," and "golden oldies."

THE FRENCH MINITEL SYSTEM

In France, an extensive telephone database permits consumers to connect their telephone to a video display terminal with a small computer. Operating much like a small personal computer with a modem, the French Minitel system permits electronic access to thousands of services from telephone directories, to stock prices, to stores which accept orders for merchandise sent to the store via Minitel. Operated by France Telecom, the French telephone company, Minitel will become especially important in Europe as trade barriers are removed in the European community and as the technology of world telephone systems becomes standardized.

EASTERN EUROPE: CHANGES IN BROADCASTING AND TELECOMMUNICATION

The movements toward more democratic forms of government in Eastern Europe may open up new avenues for private commercial broadcasting and telecommunication. Already, changes in such countries as the United Kingdom have not gone unnoticed among European nations, especially with the elimination of international trade barriers in Europe in 1992. Countries having strong commercial media, especially satellite broadcasting systems with the ability to reach large populations, may also enjoy economic strength. These nations will have the ability to use broadcasting for commercial advertising, thereby providing manufacturers and distributors of goods and services a medium to reach potential consumers. Moreover, a commercial broadcasting system will permit test marketing of new products to determine their appeal in the larger European marketplace.

Broadcasting as an Agent for Economic Change

The mass media are powerful agents of economic change and free market enterprise. While broadcasting has been an instrument of propaganda and political stability in countries under more authoritarian rule, it has also been a two-edged sword for such rule. For example, once the movement toward democracy began in many Eastern European nations, television in the hands of reformers became a unifying force for those wanting to institute democratic change. Now broadcasting may take on an equally important role, becoming an instrument of economic change that will be necessary if political reform is to succeed.

Broadcasting can reach the masses very quickly. While issues of cultural intrusion and commercialism are sensitive topics to both critics and policymakers, commercial media nevertheless can bring to a mass public an awareness of goods and services, therefore creating a demand in the marketplace.

Overcoming Obstacles

Other obstacles must still be overcome, however. A solid currency and enough wealth to consume goods and services are important. The Soviet Union, for example, which is an agricultural and industrial force, suffered for years because the Russian ruble was not tied to other currencies, thus making an exchange of goods or services with Western nations difficult. Some broadcast stations, such as Radio Bridge in Hungary, operate with Swedish and Canadian capital, which overcomes some of the currency exchange problems. Expertise will also be necessary. Experienced commercial broadcasters, people with the executive talent necessary to operate successful stations, are not readily available in Eastern Europe. Moreover, stations carrying news programming will gradually begin to operate more like their free counterparts in other parts of the world, whereas in the past the news was controlled by the government. But as talent develops, and as political and economic changes occur, and as trade barriers are lifted, the face of broadcasting in Eastern Europe will also change.

Commercial Roots in Eastern European Broadcasting

While commercial broadcasting has a long way to go to even begin to approach the developed systems elsewhere in the world, some toeholds do exist, and others are growing.[19] In Hungary, for example, commercial radio and television stations exist. The stations were operating even while Soviet troops were still stationed in Hungary after the nation overthrew Soviet rule. By the early 1990s, applications for commercial stations in Hungary outstripped the number of frequencies. Many applicants wanted to put signals on the air that were strong enough to reach Austria, and therefore sell advertising in Austria.[20] In the Soviet Union, the increase in American consumer goods has resulted in some commercials for American products such as Pepsi-Cola on the Soviet's Channel One.[21]

EMERGING TELECOMMUNICATION SERVICES IN EASTERN EUROPE

Economic change, if it is to succeed, will also demand a modern system of telephone and telecommunication services. Many East European countries, however, have telephone systems operated as state monopolies.

Telecommunication Services for Economic Growth

Lack of a modern system of communications in Eastern Europe makes it difficult for foreigners who wish to invest there. Add to that the currency-exchange problems and the telecommunication marketplace becomes even more problematic. This has not discouraged some companies, however, from becoming involved in Eastern Europe's telecommunication marketplace. Despite the monopolies, the nations of Eastern Europe know they must have an efficient telecommunications system to compete economically in a global marketplace. Something as simple as good telephone service is important. High-speed data capabilities to engage in the exchange of pricing and stock information, and to control inventories in remote regions are also important. The result is a more open mind toward foreign investment, but most importantly technological expertise offered by the West, the Japanese, and other countries.

The Role of the U.S. Regional Telephone Companies

Poised in Europe, and in Eastern Europe if the economic and political climate presents itself, are American telephone companies such as the "Baby Bells."[22] They bring a number of assets to the region, not the least of which is their billion-dollar cash flows. Some of the first investments include the approval for U.S. West, Inc. to lay fiber-optic cable across the Soviet Union and to enter the cellular phone business in Hungary. NYNEX also has ties in Hungary and Poland. Especially attractive to these American regional telephone companies is the opportunity to move into such businesses as cable television. Regulations at home restrict U.S. telephone companies from entering the American cable market, but overseas markets are still open. Moreover, the regional companies see entry into Europe as getting in on the ground floor of a telecommunication industry that is sure to have high growth and profits. Wiring with fiber optics also places them in a position to offer voice, data, and video services in the future.

The Baby Bells are also using their movement into Europe as means of applying pressure on American policymakers. The regional telephone companies want more freedom to offer services in the United States. By claiming Europe will pass the United States in the quality of its telecommunication system, and by moving dollars overseas, the Baby Bells are sending a message to Congress, the FCC, and the courts that they want deregulation in the United States.

SUMMARY

In addition to domestic services, many countries operate external and overseas broadcasting services as well as satellite channels.

Radio Canada International began in 1945 and targets its broadcasts to such areas as Eastern and Western Europe, Africa, the United States, Mexico, and the Caribbean Basin. Growing emphasis is being placed on the Pacific Rim countries.

The BBC World Service is one of the most extensive with international transmissions and special broadcasts to the northern regions of the United Kingdom.

In Europe, Swiss Radio International enjoys substantial credibility because of Switzerland's neutrality. Radio Nederland broadcasts from the Netherlands. In West Germany, Deutsche Welle (DW) broadcasts worldwide and Deutschlandfunk (DLF) beams signals to northern Europe. In East Germany, Radio Berlin International operates an external service with medium-wave and shortwave transmissions.

In the Soviet Union, Radio Moscow broadcasts in more than sixty languages and operates both a North American Service directed to the United States and Canada and a World Service with special features for target areas such as Asia and the Pacific. The era of perestroika has brought new programs stressing the theme of cooperation between the Soviet Union and more democratic nations.

Both the People's Republic of China and the Republic of China on Taiwan operate external services, with some programs directed at each other. Radio Beijing serves the People's Republic; the Voice of Free China, with an exceptionally well-developed external support system of magazines and other materials, serves the Republic of China.

Japan operates an external service, Radio Japan, with targeted regional broadcasts to such areas as Europe and Southeast Asia. In addition, Japan operates a satellite news relay service bringing news from Tokyo to America for Japanese-speaking citizens living in major United States metropolitan areas.

Radio Australia also serves Indonesia, Asia, and Japan with special broadcasts.

The United States is engaged in both government-supported and private external services. The Voice of America, the USICA Film and Television Service, Radio Free Europe, and Radio Liberty are examples of government-financed broadcasting services. A developing international presence is being felt by Ted Turner's CNN, which broadcasts CNN International to countries overseas.

Satellite and cable channels serve much of Europe. British Satellite Broadcasting operates a DBS service with a specialized satellite service for business. Super Channel and the Children's Channel also provide satellite programming. Premium and interactive channels are in service. Two examples are the Swiss TeleClub and Cable Jukebox.

Events in Eastern Europe are resulting in a number of changes in this region's broadcasting and telecommunication systems. Increased commercialization of broadcasting and the involvement of American regional telephone companies are two examples of changes taking place as Eastern Europe becomes more democratic and trade barriers begin to disappear starting in 1992.

OPPORTUNITIES FOR FURTHER LEARNING

ADAMS, W. C., ed. *Television Coverage of International Affairs*. Norwood, NJ: Ablex, 1982.

ALEXANDRE, L. *The Voice of America: From Détente to the Reagan Doctrine*. Norwood, NJ: Ablex, 1982.

ALISKY, M. *Latin American Media: Guidance and Censorship*. Ames: Iowa State University Press, 1981.

BARR, T., ed. *Challenges and Change: Australia's Information Society*. New York: Oxford Unviersity Press, 1989.

FEJES, F. *Imperialism, Media and the Good Neighbor: New Deal Foreign Policy and United States Shortwave Broadcasting to Latin America*. Norwood, NJ: Ablex, 1986.

HARRISON, P., AND PALMER, R. *News Out of Africa: Biafra to Band Aid*. London: Hilary Shipman, 1986.

HOWELL, W. J., JR. *World Broadcasting in the Age of the Satellite*. Norwood, NJ: Ablex, 1986.

LARSON, J. F. *Television's Window on the World: International Affairs Coverage on the U.S. Networks*. Norwood, NJ: Ablex, 1985.

LEMAHIEU, D. L. *A Culture for Democracy: Mass Communication and the Cultivated Mind in Britain Between the Wars*. New York: Oxford University Press, 1988.

MATTELART, A., AND SCHMUCLER, H. *Communication and Information Technologies*. Norwood, NJ: Ablex, 1985.

MICKIEWICZ, E. *Split Signals: Television and Politics in the Soviet Union*, New York: Oxford University Press, 1988.

NORDENSTRENG, K. *The Mass Media Declaration of UNESCO*. Norwood, NJ: Ablex, 1984.

STEVENSON, R. L., AND SHAW, D. L. *Foreign News and the New World Information Order*. Ames: Iowa State University Press, 1984.

TIFFEN, R. *News and Power: The Role of the Media in Australian Politics*. New York: Oxford University Press, 1989.

15 EARLY ATTEMPTS AT GOVERNMENT CONTROL

By the early twentieth century, the British Marconi Company had built a well-developed worldwide corporate empire. Ships were comfortable with wireless and had demonstrated its effectiveness in numerous cruises. Experimental radio stations were popping up everywhere, and ham-radio operators were toying with a hobby that would soon became a major social influence. As we learned earlier in the text, the 1920s saw some of the historical giants of broadcasting take to the air—WHA at the University of Wisconsin, KDKA in Pittsburgh, WBZ in Springfield/Boston, Chicago's WGN, and WWJ in Detroit. The public was being entertained with Big Ten football, live orchestras, election-night fervor, and presidential speeches, but its enjoyment of radio was being inhibited by higher-powered stations, interference, and rampant competition. There simply was not enough room on the electromagnetic spectrum for everyone to jump on without someone being pushed off.

The government first became concerned about radio's impact when Marconi started prohibiting ships and shore stations from communicating with each other unless they were equipped with Marconi equipment. This may seem incredible by today's standards, but Marconi managed to get away with it for some time. Germany was especially affected by Marconi's antics, since it housed the competing Slaby-Arco system.

The Germans finally took the initiative and called a conference in Berlin in 1903, at which a protocol agreement was reached on international cooperation in wireless communication. Three years later Berlin hosted the first International Radiotelegraph Convention, at which an agreement was signed by twenty-seven nations. In the United States the stage was now set for domestic legislation that would embody the spirit of the Berlin agreement and foster safety and cooperation among American shipping interests. It is with this 1910 legislation that we will begin a study of the development of government involvement in broadcasting.

THE WIRELESS SHIP ACT OF 1910

In 1910 there were few visions of commercial broadcasting stations as we know them today. Transatlantic experiments were less than a decade old, and Congress was thinking only about the safety applications of the new medium, especially to ships at sea. Some ships, although by no means all, had installed wireless apparatus. It was in this atmosphere that the Wireless Ship Act of 1910 was passed.[1] Containing only four paragraphs, it set the stage for maritime communication. Among other provisions, the act made it illegal for a ship carrying more than fifty persons not to be equipped with radio communication. The equipment had to be in good working order and under the direction of a skilled operator. The range of the radio had to be at least a hundred miles, day or night. Exempted from the provisions were steamers traveling between ports less than 200 miles apart.

The act also specified that the "master of the vessel" see that the apparatus could communicate with both shore stations and other ships. Violations meant a $5,000 fine, and a vessel could be fined for every infraction of the law and cited in the district court having jurisdiction over the port where the ship arrived or departed. Enforcement of the law was clear: "The Secretary of Commerce and Labor shall make such regulations as may be necessary to secure the proper execution of this Act by collectors of customs and other officers of government."

THE RADIO ACT OF 1912

By 1912, wireless had achieved international recognition and cooperation. Yet the United States had been lax in participating in agreements with other nations to control wireless, partially because in the United States wireless was not under total government control as it was in some other countries. That all changed on an April night in 1912 when an iceberg took the American liner *Titanic* to the bottom of the North Atlantic with heavy loss of life.

The *Titanic* Impact

The following days and months were filled with news of the sinking and of the role of wireless in the tragedy. Reports focused on everything from how shipboard wireless operators might have prevented the sinking to the brilliant performance of the medium in relaying news of survivors. Ironically, four months before the tragedy the provisions of the 1906 Berlin treaty had been taken out of congressional mothballs for discussion in Senate committees. Those discussions, hastened by the sinking of the *Titanic*, led to the passage in August 1912 of a second Radio Act.

The Radio Act of 1912 was much more comprehensive than the 1910 legislation. It defined authority between federal and state governments and established call letters for government stations. The law did not claim to regulate intrastate communication: "Nothing in this Act shall be construed to apply to the transmission and exchange of radiograms or signals between points situated in the same State."

Wavelengths and Licenses

Along with providing clauses for revocation of licenses and fines for violators, the act established the assignment of frequencies: the license of each station would "state the wavelength or the wavelengths authorized for use by the station for the prevention of interference and the hours for which the station is licensed to work." In addition to these specified wavelengths, however, stations could still use "other sending wavelengths." Licenses were to be granted by the secretary of commerce and labor "upon application thereof." The president of the United States was given the power to control stations during wartime but had to compensate the station's owners when doing so. The act also established the famous SOS distress signal and allowed it to be broadcast with a maximum of interference and radiation. It was required to reach at least a hundred miles. Moreover, it was the first time an act of Congress had defined radio communications: "any system of electrical communication by telegraphy or telephony without the aid of any wire connecting the points from and at which the radiograms, signals, or other communications are sent or received."

Other provisions of the act included secrecy-of-message restrictions designed to protect government stations' signals, rules for ship-to-shore communication, and a ban on any station refusing to receive messages from stations not equipped with apparatus manufactured by a certain company.[2]

Even though it was passed specifically in reaction to the sinking of the *Titanic*, the 1912 Radio Act was a valiant effort to control wireless communication. But few legislators could have foreseen the huge popularity of wireless, and even if they had, legislative processes could not have begun to keep up with the new technology. It was not long before the regulatory framework began to crumble.

THE NATIONAL RADIO CONFERENCES: THE 1912 LAW IN TROUBLE

By 1917, both the United States and radio were involved in World War I. For the U.S. Navy, this meant hurried construction of wireless towers aboard warships. Having taken over the country's radio stations, the government stopped all radio developments except those designed for wartime service. But the end of the war was like the uncapping of a bottle. All the pent-up enthusiasm for the new technology of radio was released, and experimenters flocked eagerly to their equipment. Although the Radio Act of 1912 survived the war, it was headed for rough sailing in a radio industry exploding with popularity and technology. By the 1920s chaos reigned. In 1922 alone, receiving-set sales climbed 1,200 percent. The airwaves were flooded with everything from marine military operations to thousands of amateur-radio experimenters. Added to this was the advent of commercial radio and its powerful stations booming onto the air.

The First Conference: Grappling with Interference

On February 27, 1922, groups of government officials, amateur-radio operators, and commercial-radio representatives met for the First National Radio Conference in Washington, D.C.[3] The conference was addressed by representatives of all the opposing factions. Amateur-radio operators were afraid their privileges were going to be curtailed under the influence of such large commercial firms as General Electric and Westinghouse. The large commercial firms feared their privileges were going to be delegated to the military. After the rhetoric subsided the conference split into three committees: amateur, technical, and legislative. Since interference was still the biggest problem, it was not surprising that the technical committee's recommendations received the most attention. Based on that report, legislation was introduced in Congress in 1923, but it never emerged from a Senate committee.

The Second Conference: Recommendations for Zoning and Allocations

The second radio conference began on March 20, 1923. This one reaffirmed the problems of interference and recommended discretion on frequency allocations. Taking into account the commercial interests of the new medium, the conference suggested that allowing more stations on the air would only aggravate the young industry's already shaky financial condition. By today's standards of competition among almost 8,000 stations, that proposal seems inappropriate. Realizing that different geographical areas had different problems, the second conference suggested splitting up the country into zones, each of which would tackle its own problems locally. As he had done after the first conference, Congressman Wallace White of Maine introduced legislation, but again it did not emerge from committee.

The Third Conference: Monopoly and Government Intervention

Deafening interference still characterized the airwaves when the Third National Radio Conference convened on October 6, 1924. Two major developments captured the attention of the delegates. Network broadcasting had become a reality. AT&T's wire system and Westinghouse's shortwave system were proving interstation connection was not only possible but also potentially successful. Almost simultaneously, David Sarnoff announced that RCA was going to experiment with the concept of superpower stations crisscrossing the country. It is little wonder that the third conference recommended resolutions opposing monopoly and even encouraged government intervention. Nevertheless, the conference supported the development of network broadcasting. But although it agreed to let the superpower experiments proceed, it warned that they "should only be permitted under strict government scrutiny."[4] At the request of Secretary of Commerce Herbert Hoover, Congressman White refrained from introducing legislation. A third defeat would have been bad politically, and so the decision was made to wait until still another conference was called.

The Fourth Conference: Prelude to the Radio Act of 1927

Convening on November 11, 1925, the Fourth National Radio Conference ended with proposals that became the foundation of the Radio Act of 1927. This conference suggested a system of station classifications and admonished Congress to pass some workable broadcasting legislation. The delegates recommended preventing monopoly, installing five-year terms for licenses, requiring stations to operate in the public interest, providing for licenses to be revoked, and giving the secretary of commerce the power to enforce regulations. They also wanted to guard against government censorship of programming, to provide for due process of law, to give the president control of stations in wartime, and to prevent broadcasting from being considered a public utility. But the good intentions were too late.

JUDICIAL SETBACKS TO THE RADIO ACT OF 1912

Despite the radio conferences' valiant efforts to make the Radio Act of 1912 workable, two lawsuits and an opinion from the United States attorney general made it clear that the law was in serious trouble. Highlighting the problem in 1923 was *Hoover* v. *Intercity Radio Co., Inc.* Intercity had been engaged in telegraph communication between New York and other points under a license issued by the secretary of commerce and labor. Upon its expiration, Intercity applied for a renewal but was denied because there was no space available on the spectrum for a frequency assignment that would not interfere with government and private stations.

The Court Rules on *Intercity*

The issue went to court, and the judges ruled that the secretary had overstepped his bounds in refusing to renew Intercity's license. Cited as justification was a statement by the chairman of the Committee on Commerce when the bill was passed that "it is compulsory with the Secretary of Commerce and Labor that upon application, these licenses shall be issued." The court ruling meant that the secretary of commerce and labor, although having the power to place restrictions on licenses and to prevent interference, could not refuse to issue a license as a means of reducing that interference. The court stated that "in the present case, the duty of naming a wavelength is mandatory upon the Secretary. The only discretionary act is in selecting a wavelength within the limitations prescribed in the statute, which, in his judgment, will result in the least possible interference." The court went on to define the relationship between the restrictions and the license: "The issuing of a license is not dependent upon the fixing of a wavelength. It is a restriction entering into the license. The wavelength named by the Secretary merely measures the extent of the privilege granted to the licensee."[5]

For Secretary of Commerce Hoover the ruling was extremely frustrating. Broadcasting had progressed beyond the experimental and military stages. Hoover was faced with regulating a limited resource, and the court was telling him that he had to give a part of the resource to everyone who wanted it. The 1912 act had given the secretary broad responsibilities, but not the power to implement them.

The *Zenith* Case

This was only the first of Hoover's setbacks. Three years later came the case of *United States* v. *Zenith Radio Corporation et al.* Zenith had received a license authorizing it to operate on a "wavelength of 332.4 meters on Thursday night from 10 to 12 P.M. when the use of this period is not desired by the General Electric Company's Denver station." Zenith clashed with the secretary when it operated at other times and on another, unauthorized frequency. Yet the court ruled in favor of Zenith. The legal catch was a section of the 1912 law reading: "In addition to the normal sending wavelength, all stations may use other sending wavelengths."[6]

The Chicago Federation of Labor Request

The crowning blow came when Acting Secretary of Commerce Stephen Davis denied a request from the Chicago Federation of Labor for a license.[7] Before the application even reached Washington, Davis wrote the federation telling it that all the wavelengths were in use, and that even if the federation constructed a station there would be no license forthcoming. Davis put the blame on the Fourth National Radio Conference, but it did not belong there since the conference did not have the power to dictate policy.

Opinion of the U.S. Attorney General

Some politicians became concerned over the deterioration of the radio industry, and the stations continued to interfere with each other. Finally, the Office of the Secretary of Commerce sought an opinion from the attorney general. In a letter dated June 4, 1926, the secretary asked the attorney general for a definition of the secretary's power. The questions in the letter, as restated by the attorney general, were these:

> 1. Does the 1912 Act require broadcasting stations to obtain licenses, and is the operation of such a station without a license an offense under that Act?
>
> 2. Has the Secretary of Commerce authority under the 1912 Act to assign wavelengths and times of operation and limit the power of stations?
>
> 3. Has a station, whose license stipulates a wavelength for its use, the right to use any other wavelength, and if it does operate on a different wavelength, is it in violation of the law and does it become subject to the penalties of the Act?
>
> 4. If a station, whose license stipulates a period during which only the station may operate and limits its power, transmits at different times, or with excessive power, is it in violation of the Act and does it become subject to the penalties of the Act?
>
> 5. Has the Secretary of Commerce power to fix the duration of the licenses which he issues or should they be indeterminate, continuing in effect until revoked or until Congress otherwise provides?[8]

The attorney general's answers made it clear that the problems were going to get worse, not better. The answer to the first question was affirmative. The act definitely provided for stations to be licensed, and stations operating without a license were clearly in violation. To the second question the attorney general replied that the secretary of commerce had the right to assign a wavelength to each station under one provision of the act, but for the most part the stations could use whatever other frequency they desired, whenever they wanted. The attorney general also stated that with the exception of two minor provisions, the secretary had no power to designate hours of operation. Also lost was the contention that a station's license limited its power. The act stated that stations should use "the minimum amount of energy necessary to carry out any communication desired." The attorney general said, "It does not appear that the Secretary is given power to determine in advance what this minimum amount shall be for every case; and I therefore conclude that you have no authority to insert such a determination as a part of any license."

The answer to the third question was obvious: stations could use any other wavelength they desired. The act and the courts had confirmed that. That answer in turn settled question four. Since the secretary could not limit a station only to the power and operating time stipulated in its license, stations were free to use other wavelengths, power outputs, and times. Finally, the attorney general replied to the fifth question that he could "find no authority in the Act for the issuance of licenses of limited duration."

Clearly a law that only a decade earlier had seemed firmly in control of the new medium was now almost worthless. Four months later, on December 7, 1926, President Calvin Coolidge called on Congress for legislation that would remedy the chaos that threatened to destroy radio broadcasting.[9] The next day he signed a joint resolution of Congress placing a freeze on broadcasting until more specific legislation could be passed.

THE RADIO ACT OF 1927

Congress had been working on the Radio Act of 1927 before President Coolidge's message. The act passed both houses of Congress and received the president's signature on February 23, 1927. The Radio Act of 1927 was administered by the secretary of commerce and provided for the formation of a Federal Radio Commission (FRC) that would oversee broadcasting. The act was intended to remain in force for only a year, but was subsequently extended until 1934. With court decisions to guide it, Congress did an admirable job of plugging the holes left by the 1912 law.

Organization of the Federal Radio Commission

The formation of the FRC was the most important provision of the 1927 act. It was to be "composed of five commissioners appointed by the President, by and with the advice and consent of the Senate, and one of whom the President shall designate as chairman."[10] The law specified that each commissioner be a citizen of the United States and receive compensation of $10,000 for the first year of service. The commissioner system, as well as many other provisions of the 1927 legislation, became part of the Communications Act of 1934.

Other provisions in the 1927 act divided the United States into zones represented by individual commissioners. No more than one commissioner could be appointed from any one zone. One zone covered New England and the upper tip of the Middle Atlantic states and included the District of Columbia, Puerto Rico, and the Virgin Islands. The second zone included the upper Middle Atlantic states west to Michigan and Kentucky. The third zone covered the South, and the fourth and fifth zones the Great Plains and the West, respectively.

General Provisions

The act provided for the licensing of stations, but only for a specified time, and gave the government considerable control over the electromagnetic spectrum. The act also set out to define states' rights in regard to communication. Keep in mind that federal regulation of intrastate commerce, for which wireless was used, was not popular. So it was not surprising that the Radio Act of 1927 tried to avoid direct control of intrastate communication while retaining control of communication crossing state borders. The act stated that the law's jurisdiction would extend "within any State when the effects of such use extend beyond the borders of said State." The most quoted provision was the statement in Section 4 that stations should operate "as public convenience, interest, or necessity requires."

Section 4 also prescribed "the nature of the service to be rendered by each class of licensed station and each station within any class." Control over frequency, power, and times of operation was covered by the act, which gave the FRC the power to "assign bands or frequencies or wavelengths to the various classes of stations, and assign frequencies or wavelengths for each individual station and determine the power which each station shall use and the time during which it may operate." Coverage areas for stations were to be fixed by the FRC, and the commission was to have power over "chain" or network broadcasting. Stations were also required to keep operating logs.

Administrative Power, Qualifications, and Call Letters

In addition to regulating the industry, the 1927 act gave the commission "the authority to hold hearings, summon witnesses, administer oaths, compel the production of books, documents, and papers and to make such investigations as may be necessary in the performance of its duties." The secretary of commerce was empowered "to prescribe the qualifications of station operators, to classify them according to the duties to be performed, to fix the forms of such licenses, and to issue them to such persons as he finds qualified." He was also given the authority to issue call letters to all stations and to "publish" the call letters. But before issuing a license, the government made certain that the prospective licensee gave up all rights "to the use of any particular frequency or wavelength." Once granted, station licenses were limited to three years.

To close the wavelength loophole of the 1912 legislation, the 1927 law stated that "the station license shall not vest in the licensee any right to operate the station nor any right in the use of the frequencies or wavelength designated in the license beyond the term thereof nor in any other manner than authorized therein." The act also discouraged monopolies and prohibited the transfer of licenses without the commission's approval. Furthermore, it empowered the commission to revoke the licenses of stations that "issue[d] false statements or fail[ed] to operate substantially as set forth in the license."

Political Broadcasting

The wording of the famous Section 315 of the Communications Act of 1934 came from the 1927 legislation: "If any licensee shall permit any person who is a legally qualified candidate for any public office to use a broadcasting station, he shall afford equal opportunities to all other such candidates for that office." Commercial broadcasting for its part gained instant recognition and regulation with the requirement that paid commercials were to be announced as having been paid or furnished by the sponsor.

Putting a station on the air was governed by another important provision of the act. Specifically, the act stated that "no license shall be issued under the authority of this Act for the operation of any station, the construction of which is begun or is continued after this Act takes effect, unless a permit for its construction has been granted by the licensing authority upon written application thereof." The law acknowledged that construction permits for stations would specify "the earliest and latest dates between which the actual operation of such station is expected to begin, and shall provide that said permit will be automatically forfeited if the station is not ready for operation within the time specified."

Guarding Against Censorship

An anticensorship provision, later to become incorporated into Section 326 of the Communications Act of 1934, was also included. Ironically, that provision was immediately followed by the statement that "no person within the jurisdiction of the United States shall utter any obscene, indecent, or profane language by means of radio communication."

We can see immediately the conflicts that could develop not only between these two provisions but also in the "convenience, interest, and necessity" clause. It was not long before the broadcasters and the government were indeed arguing. Yet keep in mind that the 1927 law was the very foundation of contemporary regulation of broadcasting. It was simple and straightforward, and the courts gave it strong support.

From 1927 to 1934 the Radio Act of 1927 withstood challenges from all sides. It achieved the ability to regulate effectively the expanding medium of "wireless," which now blanketed the nation with entertainment and news programming envisioned by few of the 1910 pioneer regulators. It is little surprise that the 1927 law was liberally quoted in the Communications Act of 1934, the law governing contemporary broadcasting.

THE COMMUNICATIONS ACT OF 1934

The 1934 communications act removed broadcasting from the supervision of the U.S. Department of Commerce and gave it separate status under an independent agency of government. It had become clear that broadcasting needed a new and more comprehensive regulatory agency. The FRC was still limited in scope, having to share responsibilities with the commerce department. Although the department had at one time been an appropriate home, the public demand for radio was now overshadowing radio's maritime and amateur uses. Guarding the public's convenience, interest, and necessity was no small task.

After examining a number of proposals to coordinate regulation, President Franklin D. Roosevelt sent to Congress on February 26, 1934, a proposal for a separate agency known as the Federal Communications Commission (FCC). Roosevelt told Congress that the FCC should be invested with the authority "now lying in the Federal Radio Commission and with such authority over communications as now lies with the Interstate Commerce Commission—the services affected to be all of those which rely on wires, cables, or radio as a medium of transmission."[11]

Congress responded within five months by passing the Communications Act of 1934. With it came the Federal Communications Commission, which was eventually to reign over everything from citizen's band radios to satellite communication, and from intrastate to international communication. The scope of the FCC had already been hammered out in court challenges to the 1927 law. In fact, much of the 1927 law was left intact in the act of 1934, including the guiding phrase "public convenience, interest, or necessity," which was retained as a nebulous but very powerful concept.[12] There were a few minor changes in wording. "Wavelength" was changed to "frequency," and whereas the 1927 law was concerned with "wireless communication," the FCC was to govern both wire and wireless communication.

As with most laws, the 1934 legislation has been amended many times.

SUMMARY

This chapter traces the government's role in early broadcasting. The Wireless Ship Act of 1910 was an outgrowth of the international radio conferences held in Berlin in 1903 and 1906. It provided early safeguards for ships at sea, requiring them to be equipped with radio apparatus that could communicate with other ships and with shore stations. Violations meant possible fines and court proceedings. The Radio Act of 1912 expanded the 1910 legislation but could not even begin to deal with radio's explosive growth in the 1920s. Four National Radio Conferences discussed how to bring the new medium under government control in a way that was acceptable to the industry yet permitted the orderly use of the electromagnetic spectrum. The combination of these conferences and two landmark court cases that threatened the legality of the 1912 legislation generated enough support in Congress for passage of the Radio Act of 1927. This act created the Federal Radio Commission, which was renewed on a year-to-year basis while it fought a series of battles to affirm its control over radio. The Communications Act of 1934 established the Federal Communications Commission, an independent government regulatory agency.

OPPORTUNITIES FOR FURTHER LEARNING

LE DUC, D. R. *Beyond Broadcasting: Patterns in Policy and Law.* White Plains, NY: Longman, 1987.

MIDDLETON, K., AND MERSKY, R. M., compilers. *Freedom of Expression: A Collection of Best Writings.* Buffalo, NY: William S. Hein & Co., 1981.

MUELLER, M., *Property Rights in Radio Communication: The Key to the Reform of Telecommunications Regulation.* Washington, DC: Cato Institute, 1982.

POWE, L. A., JR. *American Broadcasting and the First Amendment.* Berkeley: University of California Press, 1987.

Few government agencies have had such a direct effect on the public as the Federal Communications Commission. Nearly everything we watch on television and hear on radio is in some way touched by the FCC's control over broadcasting stations, cable, satellites, even the telephone systems. A descendant of the Federal Radio Commission, the FCC is an independent regulatory agency accountable directly to Congress. In this chapter we will learn about the jurisdiction of the FCC, how it conducts business, its organization, and its enforcement powers. We will also examine other agencies of government that affect broadcasting and telecommunication.

PRIMARY RESPONSIBILITIES

The FCC's thirteen areas of responsibility are:

1. The orderly development and operation of broadcast services and the providing of rapid, efficient nationwide and worldwide telephone and telegraph service at reasonable rates.
2. The promoting of safety of life and property through radio, and the use of radio and television facilities to strengthen national defense.
3. Consultation with other Government agencies and departments on national and international matters involving wire and radio

communications, and with state regulatory commissions on telephone and telegraph matters.

4. Regulation of all broadcast services—commercial and educational AM, FM, and TV. This includes approval of all applications for construction permits and licenses for these services, assignment of frequencies, establishment of operating power, designation of call signs, and inspection and regulation of the use of transmitting equipment.

5. Review of station performance to assure that promises made when a license is issued have been carried out.

6. Evaluation of stations' performance in meeting the requirement that they operate in the public interest, convenience, and necessity.

7. Approval of changes in ownership and major technical alterations.

8. Regulation of cable television. . . .

9. Action on requests for mergers and on applications for construction of facilities and changes in service.

10. The prescribing and reviewing of accounting practices.

11. Issuance of licenses to, and regulation of, all forms of two-way radio, including ship and aviation communications, a wide range of public safety and business services, and amateur and citizen's radio services.

12. Responsibility for domestic administration of the telecommunications provisions of treaties and international agreements. Under the auspices of the State Department, the Commission takes part in international communications conferences.

13. Supervision of the Emergency Broadcast System (EBS), which is designed to alert and instruct the public in matters of national and civil defense.[1]

As we can see, the commission's functions cover much more than just radio and television. Telephone, telegraph, and cable are all within the FCC's jurisdiction, as are applications of communication to public safety, transportation, industry, amateur radio, and citizen's service. The regulation of some of these services is shared with other government agencies, such as local municipalities in the case of cable. A television station in New York City and a CB radio in Wyoming are both within the FCC's domain. This domain stretches beyond the fifty states into Guam, Puerto Rico, and the Virgin Islands.

COMMISSIONERS

At the top of the commission hierarchy are five commissioners, one of whom is a chairperson. Appointed by the president of the United States and confirmed by the Senate, commissioners are prohibited from having a financial interest in any of the industries they regulate. This prohibition applies even to industries that are only partially FCC-regulated businesses, such as parent corporations that may own broadcasting stations in addition to publishing companies. Appointees who fill the unexpired term of a commissioner may or may not be reappointed when that term expires.

FCC OFFICES AND SUPPORT AREAS

Under the commissioners are FCC bureaus and offices that are responsible for the day-to-day operations of the FCC.[2]

Managing Director

The Managing Director serves as the FCC's general manager, overseeing the operations of the agency and carrying out the administrative responsibilities. "The Managing Director provides managerial leadership to, and exercises supervision and direction over, the FCC's bureaus and staff offices in management and administrative matters."[3]

Along with serving as the FCC's Director of National Security and Emergency Preparedness, the Managing Director also plans and manages such administrative matters as personnel, data processing, and budgets.

Office of Plans and Policy

Operating as the Office that advises the FCC on major economic and technical matters is the Office of Plans and Policy (OPP). The OPP coordinates the policy and research agenda for the FCC and recommends budgets that are approved by Congress and managed by the FCC's Managing Director. The OPP also handles contract research for the FCC.

Office of Public Affairs

Working with the public and the news media is the Office of Public Affairs. Consumer assistance also falls under the Office of Public Affairs. If you need information about the FCC and its operations, you would contact the Office of Public Affairs. Beyond radio and television stations, many small businesses use radio communication, which is regulated by the FCC. Many of these smaller enterprises use the Office of Public Affairs as a vehicle for communicating with the FCC and having questions answered about where to go in the FCC to obtain specialized information.

Office of Legislative Affairs

Implementing the FCC's legislative programs and relations with Congress is the Office of Legislative Affairs. The Office "is responsible for informing the Congress of the FCC's regulatory decisions, responding to congressional inquiries, and providing or responding to proposed changes in existing law,"[4] which has an affect on the FCC's operations and policies.

Office of Administrative Law Judges

Hearings before the FCC are conducted under Administrative Law Judges and can be appealed to the full commission or into a federal appelate court.

The Review Board

The Review Board reviews decisions issued by the Administrative Law Judges.

Office of the General Counsel

The General Counsel is the FCC's attorney and represents the FCC in the courts. The General Counsel also advises the commission on legal matters and assists in interpreting and implementing policy. It drafts FCC decisions where the courts may have become involved in an issue, and it reviews the decisions of the Review Board and Administrative Law Judges.

Office of Engineering and Technology

The Office of Engineering and Technology provides the FCC with scientific and technical support. It offers technical advice to the FCC staff and handles such tasks as testing new equipment to determine whether it meets FCC standards and coordinating frequency assignments, which are shared by both private and government services. The Office of Engineering and Technology also represents the FCC in international meetings concerned with allocation of frequencies on the electromagnetic spectrum.

FCC BUREAUS

The decisions made by the FCC offices are implemented by the FCC bureaus, which perform the day-to-day administration of the thousands of broadcast stations and licensees. There are four bureaus: the Private Radio, the Field Operations Bureau, the Mass Media Bureau, and the Common Carrier Bureau. The bureaus concerned most directly with broadcasting are the Mass Media and Field Operations bureaus.

The Mass Media Bureau is responsible for regulating AM, FM, and television stations and enforcing the cable TV rules as well as private radio and microwave facilities.

The Field Operations Bureau maintains a number of field offices in the larger cities across the United States. Special investigative teams are assigned to make on-location inspections of stations. The field offices are also placed where the public can get information about the FCC and the communications industry. In addition, this bureau is responsible for administering FCC license examinations.

The Common Carrier Bureau oversees telephone and telegraph, and the Private Radio Bureau supervises aviation and marine communication.

ENFORCEMENT POWER

The Communications Act of 1934 specified that violators of its provisions would be penalized, and the commission has at its disposal a number of enforcement measures. Depending on the type of violation, the commission may take one of the following actions: a simple letter, a cease-and-desist order, a forfeiture (fine), a short-term license renewal, a license revocation, or a denial of renewal.

Letters of Admonition

Letters of admonition are used in less serious matters or in cases in which the FCC accepts assurance that the violation will cease. Letters of admonition are used in lieu of imposing a fine or harsher action. Letters can be used to reprimand stations for failures of programming or failure to submit required FCC documents, such as employment reports or exhibits for a license-renewal application. The letters are not always a reprimand. In the case of license renewal, for example, they state that renewal is being withheld pending receipt of the required exhibit, and that after a certain date the license will be forfeited.

Forfeitures

The most common sanction imposed on a station is a *forfeiture*, usually for a technical-rule violation or the more serious offense of fraudulent billing (the latter can also set the stage for a license revocation). The forfeitures vary, not only with the violation but also with the ability of the station to pay. They can rise as high as $20,000 for serious violations by major-market stations.

Short-Term Renewals

Next to renewal denials and revocation, the most severe sanction that can be imposed on a station is a short-term license renewal.[5] The purpose of these renewals is to give the commission an early opportunity to review alleged past deficiencies. Equal employment opportunity (EEO) violations, unfound business practices, or serious rule violations, can all result in short-term license renewals.

Renewal Denials and Revocation

The most serious penalty the FCC can impose on a licensee is to deny it the right to operate by denying a license renewal or outright revocation of a license.

THE FEDERAL TRADE COMMISSION

As noted earlier, in addition to the FCC, other agencies can play a part in the regulation of broadcasting and telecommunication. The Federal Trade Commission was formed in 1914 by the FTC Act designed to guard against "unfair methods of competition in commerce."[6] Closely related to the FTC Act was the Clayton Act, also passed in 1914, which guarded against corporate mergers that would lessen competition. Since 1914 the FTC Act has been amended many times. Some of the most familiar pieces of legislation that have amended it are the 1966 Fair Packaging and Labeling Act and the 1969 Truth in Lending Act, which requires full disclosure of credit terms. The FTC has five commissioners, who are appointed, like those of the FCC, by the president with the advice and consent of the Senate for seven-year staggered terms. No more than three commissioners can be from the same political party. The president designates one of them as chairperson.

Organization

The primary components of the FTC are the commissioners and the various bureaus. The FTC is not unlike the FCC in that it is organized around office and bureaus and functions with enforcement power. Three key bureaus handle most of the tasks that affect both consumers and practitioners of broadcast advertising. The Bureau of Competition is responsible for enforcing the antitrust laws. The Bureau of Economics advises the commission on the economic impact of its decisions. The Bureau of Consumer Protection is charged with investigating trade practices alleged to be unfair to consumers. The Bureau of Consumer Protection is one of the closest allies of the public, helping to guard it against deceptive advertising. Formed in 1971, the bureau brought under one roof all of the various consumer-related activities that had been performed by the FTC.

Processing an FTC Complaint

To better understand the enforcement procedures used by the FTC, let's imagine that you are about to receive a complaint from the FTC alleging that you are airing false and deceptive commercials.[7] The first notice you would probably receive from the FTC would be a letter. You would then have the opportunity to reply to that letter and explain your position.

The FTC at this point may decide that your arguments have merit and simply decide not to pursue the matter further. But if it is not satisfied with your arguments, it may proceed to subpoena all pertinent records, such as the details of any product testing you may have undertaken.

Examining the records takes us to step 3 in the process. If the records clearly show your claims not to be deceptive, then the FTC may consider your case closed.

If, on the other hand, it is not content with your test results and still feels the advertising to be deceptive, enter step 4, the beginning of negotiation. Two developments will normally take place during this phase. First, you may offer a consent order, stating that you will agree to remedy the problem, perhaps by taking your commercials off the air. The FTC then has an opportunity either to accept or to reject your consent agreement. If the commission accepts your agreement, it will be placed in the public record. During that time, other parties can file pro or con comments on the agreement. And if the evidence builds up against you, the FTC can actually withdraw from the consent agreement and begin formal proceedings.[8] Second, if the consent order is approved by both the Commission and the advertiser, that usually ends the matter.

Although you may feel you have been overwhelmed by the power of a federal agency, the FTC would contend that such safeguards are for the benefit of the public. For the commission, enforcement powers are a stern warning to advertisers to see that their advertising meets the standards of truth and accuracy. A broadcasting station hyping a rating, misrepresenting a coverage map, or participating in unfair competition faces not only the wrath of the FCC but an equally arduous battle with the Federal Trade Commission.

THE INTERNATIONAL TELECOMMUNICATION UNION

The *International Telecommunication Union (ITU)* is a United Nations' organization responsible for coordinating the use of telecommunications among nations.[9] It does not have the enforcement powers of the Federal Communications Commission or of the Radio and Television Commission in Canada. Rather, it is a collective body of sovereign states and is only as strong as the willingness of those states to abide by its treaties. In other words, if a country violates an ITU agreement, no field office will revoke licenses or impose forfeitures. ITU's sovereign states view it not so much as an independent agency but as an arena in which to negotiate the uses of telecommunications,[10] and it has been effective in that role.

Early History of the ITU

The history of the International Telecommunication Union (ITU) dates back to 1849, when the impact of the telegraph was dawning on Europe. In that year, Austria and Prussia signed a treaty whereby they joined their telegraph lines. The treaties that were subsequently signed and the technology that was developed prompted twenty European states to meet in Paris in 1865 to approve an agreement titled the International Telegraph Convention. Included in that agreement was a set of telegraph regulations. A series of Telegraph Conferences grew out of the Paris agreement, and at the Vienna Conference in 1868, the International Bureau of Telegraph Administrations was formed. Located in Berne, Switzerland, it became known as the Berne Bureau and was staffed and funded mostly by the Swiss. It was charged with a variety of administrative functions. The 1865 Convention, the periodic conferences, and the Berne Bureau collectively became known as the International Telegraph Union in 1875. The International Telegraph Convention was its charter. By 1885 the union was involved with the telephone as well as the telegraph.

Effects of the Marconi Monopoly

At this time, Marconi was tinkering with the new technology that would soon revolutionize communication. The rapid corporate development of the British Marconi Wireless Company created a worldwide monopoly. As we saw in the last chapter, the German government convened a conference in 1903 to resolve some of the problems resulting from the monopoly, specifically the failure of ships equipped with Marconi apparatus to communicate with ships equipped with apparatus manufactured by other companies. Six of the eight sovereign states in attendance signed an agreement, which, although mostly protocol, became the foundation for international radio regulations. The agreement called for wireless stations to "operate, as far as possible, in such a manner as not to interfere with the working of other stations."[11]

Further international cooperation emerged from the first International Radiotelegraph Conference in 1906. There, twenty-seven nations adopted the Radiotelegraph Convention and specific Radiotelegraph Regulations. Realizing that radio was a rapidly changing technology, the nations also made provisions to meet at periodic administrative conferences. The Berne Bureau, already serving the telegraph and telephone interests, was designated to handle the administrative duties that concerned radio.

The ITU Emerges

Gradually the Radiotelegraph Convention and the periodic conferences together came to be called the Radiotelegraph Union. Except for the fact that they shared the Berne Bureau, the International Telegraph Union and the Radiotelegraph Union operated independently until 1932. In that year the International Telegraph Convention and the Radiotelegraph Convention were combined into a unified agreement called the International Telecommunication Convention. The International Telegraph Union and the Radiotelegraph Union also merged, becoming the International Telecommunication Union. The International Telecommunication Convention was its charter.

Functions

The primary functions of the ITU include:

1. Effective allocations of the radio frequency spectrum and registration of radio frequency assignments;
2. Coordinating efforts to eliminate harmful interference between radio stations of different countries and to improve the use made of the radio frequency spectrum;
3. Fostering collaboration with respect to the establishment of the lowest possible rates;
4. Fostering the creation, development, and improvement of telecommunication equipment and networks in new or developing countries by every means at its disposal, especially its participation in the appropriate programs of the United Nations;
5. Promoting the adoption of measures for ensuring the safety of life through the cooperation of telecommunication services;
6. Undertaking studies, making regulations, adopting resolutions, formulating recommendations and opinions, and collecting and publishing information concerning telecommunications matters benefiting all Members and Associate Members.[12]

World Administrative Radio Conferences

The *World Administrative Radio Conferences (WARC)* that are called by the ITU leave an indelible impression on international radio regulation. These conferences meet periodically to consider either limited or general topics of importance to member nations and the world use of communications. A conference will review the entire international use of the electromagnetic spectrum and establish policy for that use.

Because each WARC country has one vote, the superpowers do not necessarily control policy the way they do in other international negotiations. The conferences examine frequency use in different regions of the world and strive for the greatest latitude of spectrum use without interference.

SUMMARY

In this chapter we discussed the operation of the Federal Communications Commission and allied agencies. We learned that the FCC has thirteen areas of responsibility, among them the orderly development and operation of broadcast services; control over AM, FM, TV, telephone, common-carrier, cable, and satellite communications; new stations and transfer of ownership of those already operating; domestic administration of the telecommunications provisions of treaties and international agreements; and supervision of the Emergency Broadcast System. The FCC is organized around office, bureaus, and support areas. Among the four bureaus, those most responsible for broadcasting are the Mass Media Bureau and the Field Operations Bureau.

No government agency can function effectively as a regulator without enforcement powers. The FCC is no exception. At its disposal are such measures as letters of admonition, cease-and-desist orders, forfeitures, short-term renewals of licenses, renewal denials, and revocations of licenses.

The Federal Trade Commission is organized much like its communication counterpart, the FCC. The executive director is the chief administrator. Administrative Law Judges conduct trials in complaint cases. The three FTC bureaus concerned most directly with broadcasting are the Bureau of Competition, the Bureau of Economics, and the Bureau of Consumer Protection.

The International Telecommunication Union (ITU) establishes and administers agreements among countries on the use of the electromagnetic spectrum. A United Nations agency, the ITU is an outgrowth of the telegraph era of the mid-1800s. It has gradually evolved through a series of telegraph and radiotelegraph conventions into its current role as a coordinator of telecommunication policies and applications throughout the world.

OPPORTUNITIES FOR FURTHER LEARNING

Horwitz, R. B. *The Irony of Regulatory Reform: The Deregulation of American Telecommunications*. New York: Oxford University Press, 1989.

Katsh, M. E. *The Electronic Media and the Transformation of Law*. New York: Oxford University Press, 1989.

Paglin, M. D., ed. *A Legislative History of the Communications Act of 1934*. New York: Oxford University Press, 1990.

Ray, W. B. *FCC: The Ups and Downs of Radio-TV Regulation*. Ames: Iowa State University Press, 1990.

Rothblatt, M. A. *Radiodetermination Satellite Services and Standards*. Norwood, MA: Artech House, 1987.

Snow, M. S., ed. *Marketplace for Telecommunications: Regulation and Deregulation in Industralized Democracies*. New York: Longman, 1986.

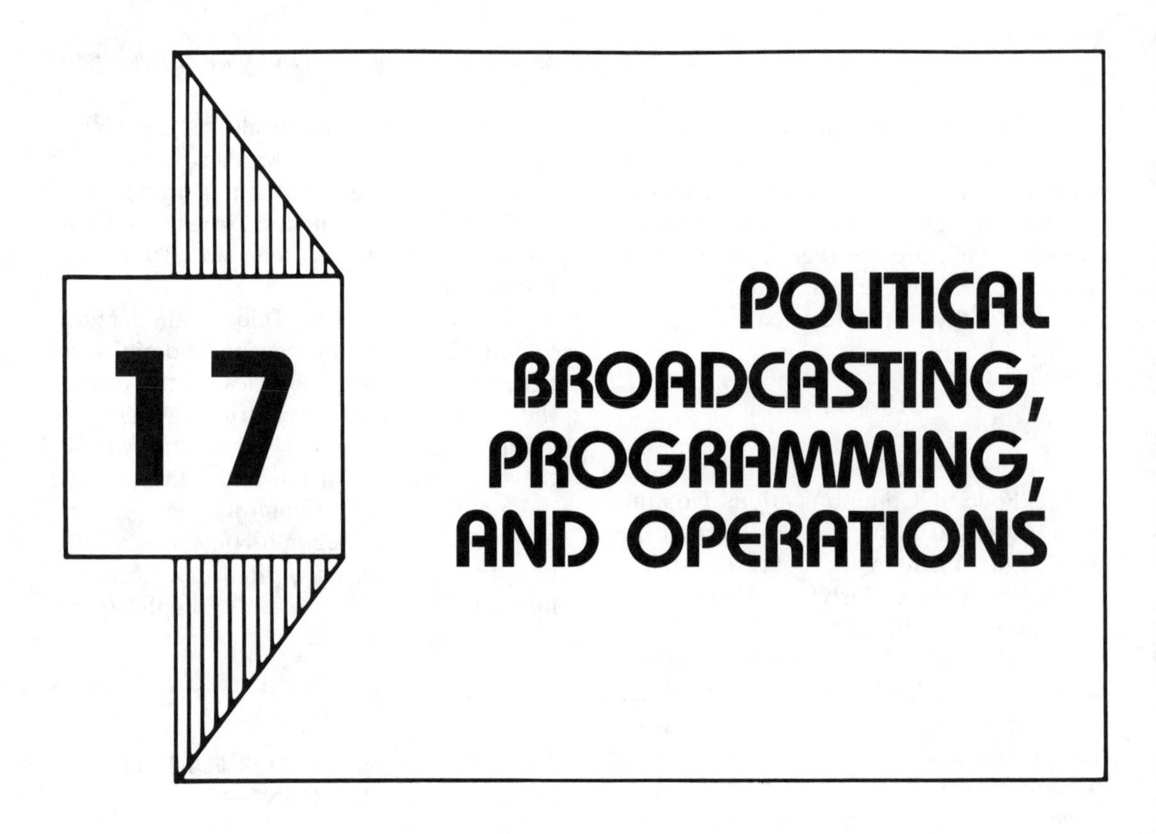

In the last two chapters we have examined the historical foundations of regulatory control of broadcasting and looked at such regulatory agencies as the Federal Communications Commission and the Federal Trade Commission. We now turn to rules and regulations that affect the programming decisions made at radio and television stations and in some cases the content of programming. Our discussion in this chapter focuses on over-the-air broadcasting. In the next chapter we will examine wired systems and such issues as common-carrier and cable regulations.

THE RATIONALE FOR REGULATORY CONTROL

The control of broadcasting is a function of supply and demand. We know that if there is a great demand for a product in short supply, people will write certain rules for obtaining it in order to avoid chaos. Imagine that a group of children all want a piece of candy but there are only half as many pieces of candy as there are children. Who gets the candy? Perhaps the children who have perfect behavior records. Perhaps only those who agree to share their candy with others.

Perhaps those who eat responsibly and do not gobble. Or perhaps only those who can afford to buy the candy. Our example illustrates the need for controls that will both regulate the allocation of the product and maintain order.

Limited Spectrum and Mass Influence

Now transpose our example to the allocation of frequencies on the electromagnetic spectrum. The spectrum has only so much space within which radio and television stations can operate. Consequently, certain rules governing the allocation and operation of stations are necessary. This limited-resource concept is the reason behind much broadcast regulation.

The second important reason is that broadcasting influences a great number of people. The citizens-band radio that sends out a five-watt signal to a passing motorist has little impact on a *mass audience*. If the operator decides to sing songs into the microphone, chances are the FCC will not be overly concerned. On the other hand, if a local television station decides to forego all its regular programming for a steady diet of jokes and traffic reports, it will have a difficult time justifying its privilege to operate. Messages that are broadcast to the public have a considerable effect on it. Thus, to assure that society is protected from abuse, we make certain rules.

At this point you may say, "Fine, we set up certain rules, people follow the rules, and the system functions." Unfortunately, it is not that simple. Everyone from FCC commissioners to citizens' groups to broadcasters argue the legitimacy of the regulatory process. Part of the discussion focuses on the legal philosophy upon which our society operates. America is considered a free country. Harold Nelson and Dwight Teeter have written that "Seventeenth and Eighteenth Century thought in much of Western Europe and America turned to faith in man's reason as the safest basis for government." Lee Loevinger describes the practical application of this philosophy as negative or proscriptive rather than positive or prescriptive. Law in America, for example, forbids behavior that might harm society, but it does not require behavior that society has determined to be beneficial.[1] Nor does it require the best behavior of which one is capable, or even behavior that is socially desirable. At first, it might seem that this attitude would undermine the good of society. Not so, Loevinger assures us, for "when the law prohibits antisocial conduct, it leaves an extremely wide area of personal choice and individual liberty to the citizen."[2]

Control Versus Noncontrol

From the standpoint of broadcasting, we can see the head of regulatory conflict beginning to protrude. Although we must control the allocation of frequencies on the electromagnetic spectrum, to control programming on those frequencies is to go against traditional American legal philosophy.

The arguments concerning control of broadcasting run between two extremes. One point of view suggests a total lack of control; its supporters point out that the First Amendment assures free press and free speech. Some legal scholars even suggest that one freedom embodies the other.[3] The other point of view supports total control of broadcasting. Its advocates base their case on four assumptions: (1) there is a reliable and authoritative basis for determining program quality; (2) the public interest can be determined in one broadcast without reference to all other broadcasts; (3) there are programs that meet the assumed authoritative government standards; and (4) if the government commands it, then quality programs will be produced.[4] The debate over all these arguments has led to a regulatory system that greatly affects what radio and TV stations will program.

POLITICAL BROADCASTING: SECTION 315 OF THE COMMUNICATIONS ACT

Section 315 of the Communications Act of 1934 regulates political broadcasting. It instructs the broadcaster and the candidate for office in how the electronic media are to be used as part of our political system. Along with the *Fairness Doctrine*, to which we will turn next, it affects how we, the consumers of broadcast communication, are informed of our electoral process.

Definitions Guiding the Equal-Time Provision

Section 315's "equal-time" provision states that "if any licensee shall permit any person who is a legally qualified candidate for public office to use a broadcasting station, he shall afford equal opportunities to all other such candidates for that office in the use of such broadcasting station." The Communications Act and the courts have grappled with the definition of what exactly is a legally qualified candidate. While specific cases may have different interpretations, generally an individual must announce for public office as well as meet the criteria to be placed on the ballot to be considered legally qualified.[5]

Hundreds of state and local statutes further clarify political eligibility. Broadcasters are prohibited from deciding themselves who is legally qualified. It makes little difference whether the candidate has a chance of winning. If he or she is qualified under the law and has publicly announced his or her candidacy, then the equal-time provision will apply. That provision also applies to cable-TV systems.

The Anticensorship Provision

As a further safeguard against unfair treatment of political candidates, Section 315 expressly prohibits the broadcaster from censoring the content of any political message: the licensee "shall have no power of censorship over the material broadcast under provisions of this section." Broadcasters were confused by the noncensorship rule, fearing it was only a matter of time until some candidate blatantly libeled an opponent and the station was sued for damages. The event occurred in 1956 in North Dakota, when U.S. senatorial candidate A. C. Townley charged on the air that the North Dakota Farmers' Union was Communist-controlled. The Farmers' Union sued the station and Townley for $100,000. But the North Dakota Supreme Court ruled that the station was not liable and that the suit should have been brought against Townley alone. Undoubtedly the Farmers' Union had thought about that, but since Townley made only $98.50 a month, the prospect for recovering damages was not bright.[6] The union then appealed to the U.S. Supreme Court. Justice Hugo Black, in delivering the opinion of the Court, stated, "Quite possibly, if a station were held responsible for the broadcast of libelous material, all remarks even faintly objectionable would be excluded out of an excess of caution . . . if any censorship were permissible, a station so inclined could intentionally inhibit a candidate's legitimate presentation under the guise of lawful censorship of libelous matter."[7]

Exemptions from the Equal-Time Provision

Exempt from the equal-time provision are appearances by candidates on these types of news programming:

1. bona fide newscast,
2. bona fide news interview,
3. bona fide news documentary (if the appearance of the candidate is incidental to the presentation of the subject or subjects covered by the news documentary), or
4. on-the-spot coverage of bona fide news events (including but not limited to political conventions and activities incidental thereto).

Also included in the list of exemptions are candidate debates and news conferences carried in their entirety. Debates are considered bona fide news events if, based on the broadcaster's good faith effort, they are designed to inform the public and not favor a particular candidate.

Selling Time: The Lowest Unit Charge

Besides granting equal time to candidates, Section 315 of the Communications Act spells out how much candidates are to be charged for the use of broadcast facilities:

> The charges made for the use of any broadcasting station by any person who is a legally qualified candidate for any public office in connection with his campaign for nomination for election, or election, to such office shall not exceed
>
> (1) during the forty-five days preceding the date of a primary or primary runoff election and during the sixty days preceding the date of a general or special election in which such person is a candidate, the lowest unit charge of the station for the same class and amount of time for the same period, and
>
> (2) at any time, the charges made for comparable use of such station by other users thereof.

This is known as the "lowest unit charge" rule. To understand it more clearly, assume that you are the sales manager for a television station. The station's rate card charges an advertiser $1,000 to buy a single one-minute commercial in prime time. An advertiser purchasing two commercials receives a discount and is charged only $850 per commercial. We will assume that the rate card permits an advertiser purchasing twenty-five commercials to receive an even bigger discount: each commercial will cost $500. Along comes John Doe, who is running for municipal judge. Doe wants to buy just one commercial to remind his friends that he is running for office. He wants it to run in prime time. What will you charge him for the cost of his one commercial? You will charge him $500. Even though he is buying only one commercial, the law states that you must charge him the "lowest unit charge." If he wanted to purchase a commercial in a fringe-time period during which the rates are lower, then you would charge him the "lowest unit charge" for that time period.

THE FAIRNESS DOCTRINE: POLICY IN TRANSITION

The *Fairness Doctrine* was first issued in 1949 as an FCC report to broadcasters on handling controversial issues with fairness to all sides.[8] The FCC has periodically examined the Fairness Doctrine. Most recently, in 1987, it said it would no longer enforce the doctrine, but some members of Congress were adamant in keeping the doctrine and are still trying to make it law by an act of Congress. Regardless of whether the Fairness Doctrine is in force at any particular time, broadcasters still have an obligation to adhere to the spirit of the doctrine under their "public interest" responsibilities as defined in the Communications Act. Moreover, the broadcaster who disregards the spirit of the doctrine is operating on dangerous ground and may be asking for an open invitation to a license challenge at renewal time, a challenge that the broadcaster may lose.

Early Decisions on Editorializing and Controversial Issues

Attention to the fairness issue crystallized in 1941 in the so-called Mayflower decision, which involved station WAAB in Boston. The Mayflower Broadcasting Corporation petitioned the FCC to award it the license for WAAB, which was up for renewal. The issue was editorializing. Although the FCC ruled in favor of WAAB, it strongly criticized the station for its practice of "editorializing." The FCC stated that it was "clear that with the limitations in frequencies inherent in the nature of radio, the public interest can never be served by a dedication of any broadcast facility to the support of his own partisan ends."[9]

While the Mayflower decision was stifling editorials, the Code of the National Association of Broadcasters (NAB) was stifling discussion of controversial issues by prohibiting the purchase of commercials airing those issues. Station WHKC in Columbus, Ohio, had adhered to the NAB Code, believing it was operating in the public interest, but found itself in a dispute with a labor union. Claiming the station had refused to sell the union time and had censored the scripts it submitted, the union filed a petition against WHKC's license renewal. The union and the station agreed to a compromise. The agreement broke with the code by prohibiting any further censorship of scripts, and requiring the station to drop its policy of not selling time for controversial issues. The FCC, which approved the compromise, stated that the station must be "sensitive to the problems of public concern in the community and . . . make sufficient time available on a nondiscriminatory basis, for full discussion thereof, without any type of censorship which would undertake to impose the views of the licensee upon the material to be broadcast."[10]

Clearly, the WHKC decision was, in spirit, the reverse of the Mayflower decision.

Further support for airing controversial issues came in 1946, when Robert Harold Scott of Palo Alto, California, filed a petition asking the FCC to revoke the licenses of three California stations. Scott claimed that he wanted time to expound his views on atheism and thereby balance the stations' "direct statements and arguments against atheism as well as . . . indirect arguments, such as church services, prayers, Bible reading and other kinds of religious programs."[11] Scott did not get the stations' licenses revoked, but in its decision the FCC stated,

> The fact that a licensee's duty to make time available for the presentation of opposing views on current controversial issues of public importance may not extend to all possible differences of opinion within the gambit of human contemplation cannot serve as the basis for any rigid policy that time shall be denied for the presentation of views which may have a high degree of unpopularity.[12]

After a convoluted history of policy over editorializing, the FCC stepped in and held hearings on the matter.

Issuing the Doctrine

From the hearings came a statement issued by the FCC on June 1, 1949, under the heading *In the Matter of Editorializing by Broadcast Licensees*. It was to become known as the *Fairness Doctrine*. The statement reasserted the commission's commitment to free expression of controversial issues of public importance, as stated in the WHKC and Scott decisions. It also reversed the Mayflower decision by supporting broadcast editorials. The commission came "to the conclusion that overt licensee editorialization, within reasonable limits and subject to the general requirements of fairness . . . is not contrary to the public interest.[13]

Specifically the Fairness Doctrine, when it is enforced, places two primary obligations on the broadcaster:

> First, a broadcaster must devote a significant amount of time to coverage of "controversial issues of public importance." Second, a broadcaster must ensure that its coverage of any given "controversial issue of public importance" is not grossly out of balance; i.e., that a "reasonable opportunity" is afforded in the station's overall programming for presentation of significant contrasting or opposing viewpoints on each issue.[14]

REGULATING OBSCENE, INDECENT, AND PROFANE MATERIAL

One of the most complex areas of broadcast regulation is obscene and indecent programming. The statutes governing such programming have evolved from both the Radio Act of 1927 and the Communications Act of 1934. The former provided for penalties of up to $5,000 and imprisonment for five years for anyone convicted of violating the act, including its obscenity provisions. The Communications Act changed this to $10,000 and two years in prison, and stated that the violator's license could be suspended for up to two years. In 1937 the penal provisions covering obscenity were amended to include license suspension for those transmitting communications containing profane or obscene "words, language, or meaning." The license suspension was no longer limited to two years, and the word meaning became even more appropriate as television became more popular.[15]

The U.S. Criminal Code

In 1948, Congress took the obscenity provisions from the Communications Act of 1934 and placed them into the United States Criminal Code. Section 1464 of the code states that "whoever utters any obscene, indecent, or profane language by means of radio communication shall be fined not more than $10,000 or imprisoned not more than two years or both."[16] "Radio communication" includes television. Both the Department of Justice and the FCC have the power to enforce Section 1464. Penalties include forfeiture of a license or construction permit and fines of $1,000 for each day the offense occurs, not to exceed a total of $10,000. The Justice Department can also prosecute under Section 1464 and send a licensee to jail.

The Seven Dirty Words Ruling

Among the cases in which the FCC has acted against stations that have broadcast obscene, indecent, or profane material, two stand out. One concerned an Illinois station's "topless" format and the other a New York radio station's broadcast of a monologue by comedian George Carlin.

On the afternoon of October 30, 1973, WBAI-FM in New York warned its listeners that the broadcast to follow included language that might be offensive. What they heard was an excerpt from George Carlin's album "George Carlin: Occupation Foole." Carlin's monologue satirized seven "four-letter" words that could not be used on radio or television because they depicted sexual or excretory organs and activities. A month later the FCC received a complaint from a man who said he had heard the broadcast while driving with his son. It was the only complaint received about the broadcast, which had been aired as part of a discussion on contemporary attitudes toward language.

The FCC issued a declaratory ruling against WBAI-FM, stating that such language "describes, in terms patently offensive as measured by contemporary community standards for the broadcast medium, sexual or excretory activities and organs, at times of the day where there is a reasonable risk that children may be in the audience."[17] The commission also argued that broadcast media should be treated differently from print media in the regulation of indecent material, because broadcast media are intrusive.[18] The commission reiterated that it was not in the business of censorship but that it did have a statutory obligation to enforce those provisions of the criminal code that regulated obscene, indecent, or profane language. The decision was appealed and upheld by the U.S. Supreme Court.[19]

It was now clear that there were at least seven words that would cause broadcasters much trouble if they decided to use them on the air.

Topless Radio

On February 23, 1973, a radio station in Oak Park, Illinois, broadcast a call-in program on oral sex. Female listeners called moderator Morgan Moore with graphic descriptions of their experiences. The format, also employed at other stations, was known as "topless radio." Female listeners were not the only ones to contact the station. The FCC notified it of an apparent liability of $2,000 for violating both the indecency and obscenity clauses of the criminal code.[20] The ruling was upheld on appeal to the circuit court.

Broadening the Indecency Guidelines

With over-the-air television competing with cable and the decline in AM listenership, it was perhaps not surprising that the FCC in 1987 broadened its indecency guidelines beyond the seven dirty words of the WBAI-FM ruling. Specifically, the FCC has encompassed the broad interpretation of the WBAI-FM ruling, which defined indecency as "language or material that depicts or describes, in terms patently offensive as measured by contemporary community standards for the broadcast medium, sexual or excretory activities."[21] The commission is enforcing the interpretation if there is a reasonable risk that children will be present in the audience.[22]

Indecency vs. Obscenity

Still left open to interpretation is the difference between indecency and obscenity. Obscenity is generally determined by the definition applied in the U.S. Supreme Court case of *Miller* v. *California*.[23] The *Miller* case considered material obscene if the average person applying contemporary community standards, would find the work (1) appeals to the prurient interest, (2) describes or depicts, in a patently offensive manner, sexual conduct as defined by state law, and (3) taken as a whole, lacks serious literary, artistic, political, or scientific value.

PRIME-TIME ACCESS

Concern over the dominance of network programming prompted the FCC in 1971 to take measures assuring that alternative programming would be aired during the evening hours. Out of these measures came the prime-time access rule (PTAR), which affects television.

Provisions of PTAR

Stations in the top fifty markets that are either network-affiliated or network-owned cannot air more than three hours of network programming between the prime-time hours of 7 P.M. to 11 P.M. E.S.T and P.S.T. and 6 P.M. to 10 P.M. C.S.T. and M.S.T. Programs that were formerly aired on a network also come under the PTAR. The rule is designed (1) to give independent producers and syndicators a market for their programming and (2) to encourage local stations to develop creative programming. By applying the rule to the top fifty markets, the FCC has successfully covered the nation. Yet the rule has been more successful in providing time for syndicated programming than in stimulating local creativity. The result has been a plethora of quiz and game shows in the 6:00-to-8:00 P.M. time slot.

General Exemptions

PTAR allows a variety of exemptions. In very general terms, stations can broadcast network or off-network documentary, public-affairs, and children's programming.[24] Feature films can also be broadcast, as can fast-breaking news that would be of interest to the viewing audience. In other words, if a network provides its affiliates with coverage of a major news event, such as an assassination or a natural disaster, the local affiliates can carry the program and have it count as prime-time access. If a television station produces an hour of local news (for example, from 6:00 to 7:00 P.M.) immediately before the prime-time access hour, then it can carry network news up to one-half hour into the access period, or until 7:30 P.M.

Sports Exemptions

Sports programming is also exempt. If a sports event is scheduled to end at the beginning of prime-time access but lasts longer, stations are permitted to continue covering it. Major sports events whose coverage requires all of prime time, such as a New Year's Day football game or an Olympic contest, receive the same exemption.

BROADCAST ADVERTISING

Advertising provides the economic lifeblood of the American system of broadcasting; this is not the case with the government-financed systems existing in many other parts of the world. But federal, state, and even municipal regulations can oversee this lifeblood. We will now concern ourselves primarily with state and federal jurisdiction over commercial radio and television.

State and Federal Jurisdiction

Although we tend to think of radio and television as being governed by federal law, as expressed in the Communications Act of

1934, state laws play an important part when advertising is involved. In a landmark case that applied state jurisdiction to broadcast advertising, a court upheld a New Mexico statute that prohibited a New Mexico radio station from accepting advertising from Texas optometrists. The Texas advertising violated a New Mexico law regulating optometric advertising. The U.S. Supreme Court upheld the New Mexico law, rejecting the contention that it interfered with interstate commerce and was thus preempted by federal law.[25] Some states have given protection to broadcasters who in good faith broadcast an advertisement that turns out to be deceptive.

The two principal federal agencies affecting broadcast advertising are the Federal Communications Commission and the Federal Trade Commission. The FCC can call upon its blanket "public interest" clause to move in on an unscrupulous broadcaster involved in a deceptive advertising scheme.

Federal Trade Commission Controls

The most pervasive of the agencies that control advertising is the Federal Trade Commission. Through its Bureau of Consumer Protection, the FTC keeps watch on advertising practices affecting both the broadcasting and print media. The quickest way for an advertiser to get into trouble with the FTC is to violate one of its six "basic ground rules":

1. *Tendency to deceive*. The Commission is empowered to act when representations have only a tendency to mislead or deceive. Proof of actual deception is not essential, although evidence of actual deception is apparently conclusive as to the deceptive quality of the advertisement in question.

2. *Immateriality of knowledge of falsity*. Since the purpose of the FTC Act is consumer protection, the Government does not have to prove knowledge of falsity on the part of the advertiser; the businessman acts at his own peril.

3. *Immateriality of intent*. The intent of the advertiser is also entirely immaterial. An advertiser may have a wholly innocent intent and still violate the law.

4. *General public's understanding of controls*. Since the purpose of the Act is to protect the consumers, and since some consumers are "ignorant, unthinking and credulous," nothing less than "the most literal truthfulness" is tolerated. As the Supreme Court has stated, "laws are made to protect the trusting as well as the suspicious." Thus it is immaterial that an expert reader might be able to decipher the advertisement in question so as to avoid being misled.

5. *Literal truth sometimes insufficient*. Advertisements are not intended to be carefully dissected with a dictionary at hand, but rather are intended to produce an overall impression on the ordinary purchaser. An advertiser cannot present one overall impression and yet protect himself by pointing to a contrary impression which appears in a small and inconspicuous portion of the advertisement. Even though every sentence considered separately is true, the advertisement as a whole may be misleading because the message is composed in such a way as to mislead.

6. *Ambiguous advertisements*. . . . Since the purpose of the FTC Act is the prohibition of advertising which has a tendency and capacity to mislead, an advertisement which can be read to have two meanings is illegal if one of them is false or misleading.[26]

EQUAL-EMPLOYMENT OPPORTUNITY

The federal government's insistence on increasing the proportion of women and minorities in the work force has been translated into action by the Equal Employment Opportunity Commission and the requirement that affirmative-action measures be taken by business and industry throughout the United States. Although the Federal Communications Commission is not directly responsible for enforcing affirmative-action programs, it has taken steps to assure that broadcasting stations do not fall behind in their commitments to affirmative action. An extensive explanation of how a station administers its affirmative-action program is required at the time of license renewal. And when considering a license renewal, the FCC will compare the current affirmative-action program with the one in the previous license renewal. By using the "public interest" clause of the Communications Act, the FCC is able to put some teeth into its requirements. Its power is based on the rationale that a "broadcaster who refuses to hire minority and women employees will face a difficult, if not insurmountable, obstacle to the presentation of programming to meet the problems, needs and interests of minorities and women."[27]

Model Affirmative-Action Plan

To better understand how an affirmative-action plan works, let us assume we are operating a station and want to be in compliance with FCC requirements, not only at license renewal time but also at other times. A model affirmative-action plan would contain the following elements.[28]

Statement of General Policy The first part of our program would consist of a statement committing the station to affirmative action in all areas of station business, which would include not only hiring employees but also promoting, compensating, and terminating them. Take note of the word *terminating*—if we aren't going to discriminate in hiring, then we can't do so in firing. Overall, the program must be a positive effort, assuring equal opportunity without regard to sex, race, national origin, color, or religion.

Responsibility for Implementation Our next responsibility would be to implement our commitment. We would want to appoint someone at the station as our affirmative-action administrator. If we have delegated the responsibility for firing and hiring to another administrator, such as a sales manager or news director, then we will want to make sure that person adheres to our commitment.

Policy Dissemination But it is not enough merely to have an affirmative-action program. We need to publicize it through such means as posters, which tell applicants or employees where to write if they feel they have been discriminated against. The Department of Labor makes available posters containing such warnings. We could also put an affirmative-action statement on the station's employment application.

Recruitment Hiring is usually the easiest task in an affirmative-action program. What takes work is obtaining a pool of applicants from which to choose. We will need to recruit people by advertising our job openings. And in each ad we will want to include a statement identifying our station as an equal-opportunity employer. Potential female applicants can be reached through ads in newsletters such as Matrix, published by Women in Communications, Inc., and *News and Views*, from American Women in Radio and Television. Minorities can be reached through similar publications. Employment agencies and the placement services at local colleges are two additional avenues. Keep in mind that we will need to provide the FCC with a list of the organizations we contacted and the number of applicants received from each one.

Training If our station is small, developing a full-scale minority-training program may be difficult. On the other hand, an internship program initiated with a local college can at least show a good-faith effort within our means.

Availability Survey In order to compare the success of our program with the work force in our local area, we will need a recent availability survey. Such a survey discusses such factors as the percentages of women and minorities in the work force from which we can directly recruit—usually the metropolitan area in which the station is licensed, or in some cases the county in which it is located.

Job Hires We will also want to note the number of women and minority employees hired in the past twelve months. If in our opinion not enough minority applicants are applying for positions, we will want to explain how we are going to improve our recruiting practices in the future.

Promotion A responsible affirmative-action program deals not only with hiring but also with promotion. When openings develop within our organization we should always scan our current personnel to see who might be qualified for the jobs.

Current Employment Survey In addition to our model EEO program, our station must file an annual employment report. To become part of the public file, this report details the number of women and members of minority groups who are employed by the station and notes how many occupy top-management positions. We may want to supplement the employment report with a description of women and minority employees in all job classifications within the station. In 1990, the U.S. Supreme Court affirmed once again the FCC's enforcement of strong affirmative action policies, including affirmative action in the awarding of broadcast licenses.

SEXUAL HARASSMENT

Broadcasters are becoming increasingly aware of the effects of sexual harassment both on individuals and on the overall operation of stations. Recent court rulings consider sexual harassment a form of sexual discrimination under Title VII of the 1964 Civil Rights Act. What exactly constitutes sexual harassment is something that must be determined by the circumstances surrounding each incident. Moreover, because such incidents often occur in private, the testimony of the plaintiff and the defendant without the benefit of other witnesses makes the sexual-harassment area of discrimination law particularly difficult to rule on.

Specifically, the Equal Employment Opportunity Commission considers unlawful sexual harassment to occur:

> (1) When submission to such sexual conduct is "explicitly or implicitly" a condition of an individual's employment;
>
> (2) When submission to or rejection of such sexual conduct becomes the basis of employment decisions "affecting" an employee; or
>
> (3) When such sexual conduct has the purpose or effect of substantially interfering with an individual's job performance or creating an intimidating, hostile or offensive working atmosphere.[29]

Because sexual harassment can be treated as a violation of a station's affirmative-action policy, communications attorneys advise stations to try to make employees aware of what constitutes sexual harassment, what the penalties are for those engaging in sexual harassment, and how to file complaints of sexual harassment.

STARTING A NEW STATION

Even though in many communities frequencies are getting harder and harder to find, enterprising entrepreneurs have not been deterred from seeking out locations for new stations. Let us briefly review the steps that one must go through to start a new station.

Preliminary Steps

The first step in starting a new station is to find an area where a frequency is available. For an AM radio station, the search will involve not only consulting the engineering data of stations already in the market but also having a qualified engineer conduct a *frequency search*. The frequency search entails checking the exact broadcast contours of stations presently serving the area and determining what type of signal will not interfere with those currently operating. Thus, researching possible wattage, contour patterns, and available frequencies must all precede the application process.

Starting an FM radio or TV station is a bit different. An applicant for an FM radio license must select either an available frequency already assigned by the FCC to the area where the applicant wants to operate or a place within a specified radius where no FM frequency has been assigned. TV applicants must request a UHF or VHF channel, assigned either to the community or to a place where there is no channel assignment within fifteen miles of the community.

Once the frequency search has been completed, the next step is a community-needs and ascertainment survey.

From Construction Permit to License

Once the community-needs and ascertainment survey is completed, the applicant applies to the FCC for a *construction permit*. The applicant must also possess the wherewithal to operate the station for at least one year after construction. Notice of the pending application must be made in the local newspaper, and a public-inspection file must be kept in the locality where the station will be built. After the applicant has filed with the FCC, others have the opportunity to comment on the application or, in the case of competing applicants, file against it. If necessary, the FCC will schedule a hearing on the application. Following the hearing, the FCC administrative law judge will issue a decision, which can be appealed.

If everything in the application is found satisfactory and there are no objections, the FCC then issues the construction permit. Construction on the station must begin within sixty days of the date the construction permit is issued. Depending on the type of station, a period of up to eighteen months from that date is given for construction. If the applicant cannot build the station in the time allotted, then he or she must apply for an extension.

After the station is constructed the applicant applies for the license. At this time the applicant can also request authority to conduct program tests. These tests will usually be permitted if nothing has come to the attention of the FCC that would indicate that the operation of the station would be contrary to the public interest. When the license is issued the station can go on the air and begin regular programming.

Although the procedure is somewhat systematic, putting the station on the air is anything but simple. Completing the paperwork, dealing with engineers and communications attorneys, and securing the financing necessary not only to buy land and equipment but also to keep the station running for a year can all be difficult and time-consuming obstacles to overcome. If objections or competing applications become an issue, the court costs involved can discourage an applicant from completing the application process. Still, for those who succeed, the rewards can be substantial, both monetarily and in terms of personal satisfaction.

SUMMARY

Government control of over-the-air programming is a major concern of the broadcasting industry. It is based on the fact that the electromagnetic spectrum over which radio and television waves travel is a limited resource that must have safeguards if it is to be responsibly used. Coupled with this is the tremendous influence of radio and television, an influence that with the aid of satellites can cross international boundaries.

Section 315 of the Communications Act is concerned mainly with political broadcasting and assures that candidates for the same public office will have the same opportunity to gain access to the broadcast media. Key parts of Section 315 include its definitions of equal time, its anti-censorship provisions, and its lowest-unit-charge rule.

The Fairness Doctrine traces its roots to the 1940s, when the FCC prohibited editorializing. The FCC reversed itself in 1949, and since then the doctrine has been revised considerably, mostly through FCC policies and court decisions. Although in 1987 the FCC said it would no longer enforce the Fairness Doctrine, some members of Congress remained committed to making it law by an act of Congress, and broadcasters still have a responsibility to adhere to the spirit of the doctrine under their obligation to operate in the "public interest."

The FCC has found one of its strongest footholds in the control of obscene, indecent, and profane material. Supported by the U.S. Criminal Code, the FCC has levied sanctions against numerous stations for violations in this area. Two of the most famous cases were the frank discussions of sex found in "topless radio" formats and the broadcasting of comedian George Carlin's monologue on words prohibited on radio and television.

The prime-time access rule (PTAR) is designed to encourage alternatives to network programming during certain hours preceding prime-time programming. Although PTAR makes available certain exemptions, it has created a market for syndicated programming and some non-network programs produced by local stations.

Broadcast advertising is controlled mostly by the Federal Trade Commission, though in some areas it falls under both state and federal jurisdiction. Six areas in which violations can quickly draw FTC scrutiny are the tendency to deceive, the immateriality of knowledge of falsity, the immateriality of intent, the general public's understanding of controls, the insufficiency of literal truth, and ambiguous advertisements.

Of all the federal agencies, the FCC has been one of the most vigorous in enforcing affirmative-action programs. A typical affirmative-action program consists of a statement of general policy, responsibility for implementation, policy dissemination, recruitment, training, availability survey, current employment survey, a job-hires summary, promotion, and effectiveness of the affirmative-action plan.

Allied to a station's affirmative-action plan is its policy statement on sexual harassment. Because sexual harassment can be regarded as a violation of a station's affirmative-action policy, communications attorneys advise stations to make employees aware of what constitutes sexual harassment.

Starting a new station is a systematic process. The owner must first obtain a construction permit and then a license to operate.

OPPORTUNITIES FOR FURTHER LEARNING

BAKER, C. E. *Human Liberty and Freedom of Speech*. New York: Oxford University Press, 1989.

BARBER, S. *News Cameras in the Courtroom: A Free Press–Fair Trial Debate*. Norwood, NJ: Ablex, 1987.

BOLLINGER, L. C. *The Tolerant Society*. New York: Oxford University Press, 1988.

CARTER, T. B., FRANKLIN, M. A., AND WRIGHT, J. B. *The First Amendment and the Fifth Estate: Regulation of Electronic Mass Media*, 2nd ed. Westbury, NY: Foundation Press, 1989.

DONAHUE, H. C. *The Battle to Control Broadcast News: Who Owns the First Amendment?* Cambridge, MA: MIT Press, 1989.

MERRILL, J. *The Dialectic in Journalism: Toward a Responsible Use of Press Freedom*. Baton Rouge: Louisiana State University Press, 1989.

MONTGOMERY, K. *Target: Prime Time Advocacy Groups and the Struggle Over Entertainment Television*. New York: Oxford University Press, 1989.

O'DONNELL, L., HAUSMAN, C., AND BENOIT, P. *Radio Station Operations*. Belmont, CA: Wadsworth, 1989.

RUSHER, W. A. *The Coming Battle for the Media: Curbing the Power of the Media Elite*. New York: William Morrow, 1988.

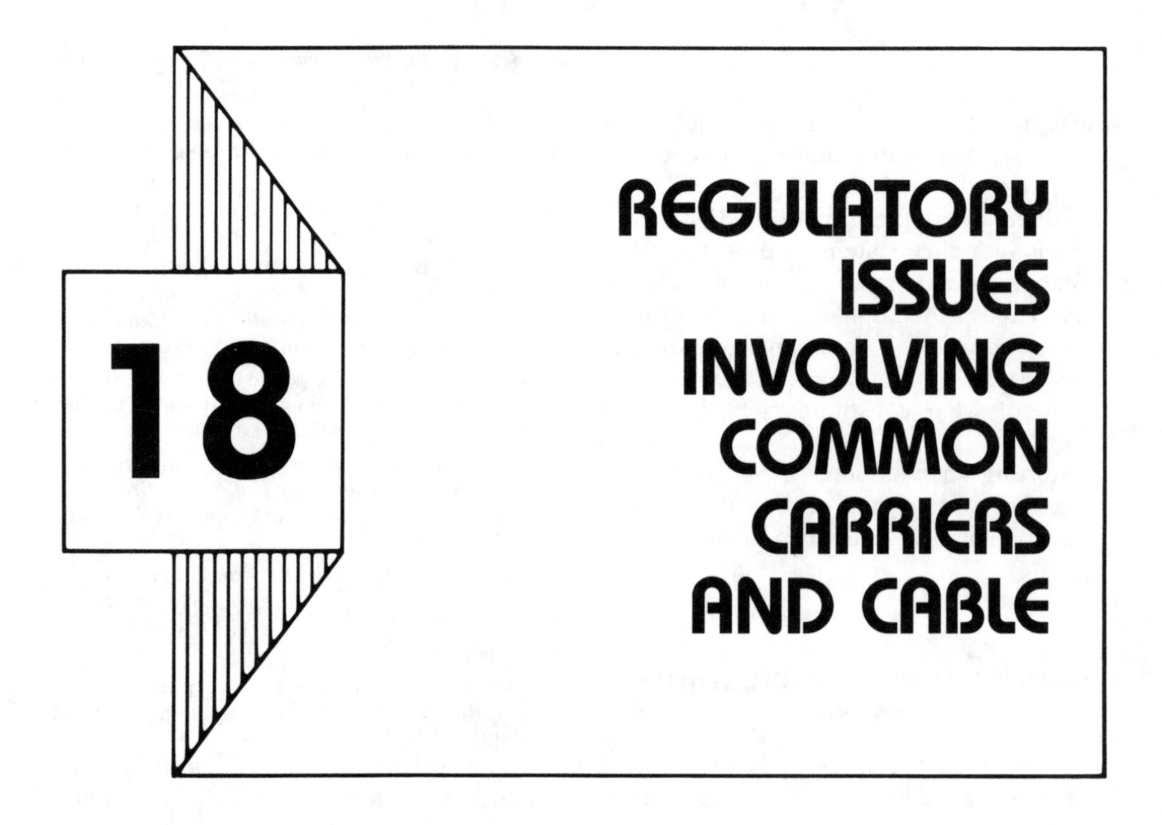

18 REGULATORY ISSUES INVOLVING COMMON CARRIERS AND CABLE

The microwave link between the island of Ocracoke and the mainland and the satellite relaying data from an ocean research vessel both operate on regulated channels of communication. These channels, which are leased to the public, are called *common carriers*. Under the law, a communications common carrier is defined as "one whose services are open to the public for hire for handling interstate and foreign communications by electrical means."[1] An over-the-air radio or television station that sends us the evening news is not a common carrier. The services of the station are not open to the public for hire. On the other hand, the telephone lines that come into our home are common carriers since anyone can "lease" these lines from the telephone company, and in fact we use these leased lines every time we make a telephone call. We pay a monthly fee for this service, and the Federal Communications Commission regulates the telephone company just as it does the radio or television station.

With the advent of information technologies, a knowledge of common carriers has become important to a total understanding of telecommunication regulation. When we *interface* our personal computer with a central data bank through our telephone and call up information on a videotex terminal, we are using a common carrier. When we purchase a machine that automatically answers our telephone calls when we are not at home, we become subject to the regulations that guarantee us the freedom to purchase and connect to our telephone any answering machine we like as long as it meets certain technical standards prescribed by the FCC.

THE COMMON-CARRIER CONCEPT: HISTORICAL BASIS OF CONTROL

Two pieces of legislation regulated interstate communication long before modern applications of telecommunication—the Post Roads Act of 1866 and the Mann-Elkins Act of 1910.

The Post Roads Act of 1866

The system of telephone lines and cables that intertwine across vast stretches of landscape was encouraged by the Post Roads Act of 1866. At the time of its passage, however, the telegraph was the new electrical device being introduced to long-distance communication. The act granted the government the right-of-way over public land, thereby permitting it to install telegraph lines over the most accessible routes. The legal basis for this system was inherited by the telephone companies and even the railroads. Essential to a nationwide system of electrical communication was this capacity to install lines uninhibited by private property.

The Mann-Elkins Act of 1910

The authority of the FCC, and state utility commissions, to set the rates common carriers charge for their services is based in the Mann-Elkins Act of 1910. The act extended certain provisions of the Interstate Commerce Act to cover common carriers using both wire and radio communication.

Today, through a series of amendments to the Communications Act of 1934, common carriers are regulated by the FCC. Internationally they are regulated by treaties administered through the International Telecommunication Union in cooperation with various government bodies such as the United States Department of State.

STATE VERSUS FEDERAL CONTROL OF COMMON CARRIERS

The FCC does not have exclusive control over common carriers. Much control is shared by state governments. To understand this division of regulatory responsibility it is necessary to understand the differences between *fully subject* and *partially subject* carriers.

Partially Subject Common Carriers

These are carriers that are only partially subject to the controls of the FCC and the Communications Act of 1934. Partially subject carriers do not engage in "interstate or foreign communication except through connection with the wire, cable, or radio facilities of nonaffiliated carriers."[2] An intrastate common carrier (one that is entirely within a state's boundaries) is generally subject not to FCC jurisdiction but to the state utility commission.

Fully Subject Common Carriers

These carriers are engaged in interstate and international communication and come under FCC jurisdiction. Before constructing, acquiring, or operating common-carrier facilities for such communication they must receive FCC approval. In addition, they cannot discontinue or curtail service without FCC approval. Charges and practices must be reasonable, and the carriers must file rate schedules with the FCC, which reviews and has final authority over them.

AREAS OF FCC JURISDICTION OVER COMMON CARRIERS

The areas where the FCC has jurisdiction over common carriers can be divided into (1) operations and (2) licensing and facilities.

Operations

Fully subject carriers are subject to FCC control over their basic operating practices. For example, the FCC determines the forms, records, and accounts that fully subject carriers employ. This uniform system of accounts includes establishing and maintaining uniform records for cost accounting, property, pension cost, and depreciation. Much like radio and television stations, common carriers must maintain certain records over time.

The FCC also has the authority to determine the depreciation rates the larger carriers use for equipment and facilities. Monthly and annual reports required of the carriers provide the FCC with operating and financial information. The FCC, moreover, "regulates the interlocking of officers and directors of carriers fully subject to the Act, it being unlawful for any person to hold office in more than one such carrier unless authorized by the FCC."[3]

Licensing and Facilities

Common carriers that use radio waves, including microwave and satellite communication, must obtain a license. The FCC guards against interference of these radio signals just as it does the signals of standard broadcast stations. Licenses are limited to citizens of the United States and denied to corporations in which any officer or director is an alien or of which more than one-fifth of the capital stock is owned by aliens or foreign interests. The FCC is charged with seeing that facilities are adequate but not excessive and that rates are "reasonable and prudent."

REGULATING COMMON-CARRIER INTERCONNECTION DEVICES

With the increased use of videotex devices, more and more companies are becoming involved in marketing equipment that connects to a telephone in the home. The opening up of this market, once primarily the domain of the telephone companies, has been made possible by key court and FCC decisions. In *Hush-a-Phone* v. *United States* the courts overruled the FCC, claiming that the public should be protected against "unwarranted interference with the telephone subscriber's right to use his telephone in ways which are privately beneficial without being publicly detrimental."[4] In another case the FCC ruled that prohibiting the use of interconnection devices that do not adversely affect the telephone system was unreasonable and unlawful.[5] In litigation over General Telephone's acquisition of the Hawaiian Telephone Company a court held that a telephone company could not limit purchase of equipment from the company's subsidiary.[6] Such action was an unreasonable restraint of trade.

THE BASIS FOR REGULATING CABLE

Originally the FCC exercised considerable control over cable systems.[7] Its authority over cable began as early as 1962, and rules for regulating cable were approved by the FCC in 1965. FCC control was based on cable's use of microwave systems: since the FCC had control over microwave, it gradually acquired control over cable systems as well. In 1966 the FCC passed rules controlling cable systems that did not employ microwave links.

United States v. *Southwestern Cable Co.*

Knowing that a court case would soon test its jurisdiction over cable, the FCC decided to prepare for the inevitable when it issued a decision limiting the signals that a San Diego cable system could import from Los Angeles. The test case came in *United States* v. *Southwestern Cable Co.*, in which the Supreme Court upheld the FCC's right to regulate cable as part of its mandate under the Communications Act to regulate "interstate commerce by wire or radio."[8] By 1968 the FCC had started developing comprehensive regulations for cable, which it finally issued in 1972.[9]

Gradually, however, local government began to exercise control over local cable systems. It became clear that this new medium was much better controlled at the local level than by the federal government. Regulatory conflicts between agencies of federal and local control (and in some cases of state control) began to develop, and today the FCC has delegated much of its authority to local communities.

CABLE'S LOCAL REGULATORY FRAMEWORKS

The concepts that regulate cable are based at the local level. Unlike over-the-air broadcasting, cable can be regulated by its local community, which has the authority to place certain service and operational requirements upon it, to levy fees, and to determine community-access channels. The types of local control vary considerably.[10] The first is an *administrative office*, where the local government establishes a regulatory agency much like the FCC. It might be found in the mayor's office or in the city planner's office. A second type is the *advisory committee*, which can be appointed by the mayor or the city council to "advise" city government on cable regulation. Closely related to the advisory committee is an *advisory committee* with *administrative office*, which "combines an appointed advisory committee with a full-time salaried executive office." Sparkes points out that the executive usually works independently of the advisory committee, which advises the city council on policy matters. A fourth organization calls for the creation of an *independent regulatory commission*, which administrates and participates in rule-making. A fifth plan provides for an *elected board*, which answers to the electorate on cable regulations rather than to another elected body.

RECOMMENDED FRANCHISE STANDARDS FOR CABLE SYSTEMS

The *franchise* is the contractual agreement between the local governmental unit and the cable company. Although the FCC has kept a regulatory distance between itself and the local authorities who govern cable systems, it has recommended standards that communities can follow in dealing with local cable systems. The FCC suggests that any cable franchise contain the following provisions:

1. The franchising authority should approve a franchisee's qualifications only after a full public proceeding affording due process;
2. Neither the initial franchise period nor the renewal period should exceed 15 years, and any renewal should be granted only after a public proceeding affording due process;
3. The franchise should accomplish significant construction within one year after registering with the Commission and make service available to a substantial portion of the franchise area each following year, as determined by the franchising authority;
4. A franchise policy requiring less than complete wiring of the franchise area should be adopted only after a full public proceeding, preceded by specific notice of such policy; and
5. The franchise should specify that the franchisee and franchisor have adopted local procedures for investigating and resolving complaints.[11] The FCC also recommends that local franchisees adopt a local complaint procedure, identify a local person who will handle complaints, and specify how complaints can be reported and resolved. The FCC recommends further that the franchisee identify, by title, the office or person who is responsible for the continuing administration of the franchise and the implementation of complaint procedures.[12]

STATE REGULATION OF CABLE

State government also plays a major role in controlling cable.[13] However, state control is not widespread and varies in degree. State laws can be classified into three categories.

Preempt Statutes

These are the strongest laws, and they take precedent over local regulations. If subject to preempt statutes, cable will fall under the jurisdiction of the public utility commission or public service commission in some states. Preempt statutes give considerable clout to a state commission, permitting it to issue and enforce a separate set of state cable regulations. These rules can govern everything from the day-to-day operation of the cable system to collecting fees on gross revenue to demanding financial collateral before allowing construction.

Appellate Statutes

Here, local municipalities retain some control over franchising, but the state has the power to review local agreements and be the final arbiter of disputes. Everything works fine until the state and a municipality disagree. Then the municipality stands a less than even chance against the state.

Advisory Statutes

These are more popular with cable systems and municipalities, for they have neither the clout nor the enforcement power of a state commission. Some serve as general guidelines for local government.

Arguments for State Control

Proponents of state control argue the need for consistency among cable systems within a state. Such arguments gain support when two municipalities cannot resolve their jurisdictional differences over a cable system or when significantly different fee structures provoke public outcry. Control of cable can also be a political plum for legislators because it means control of a communication system, and communication influences public opinion. As cable commissions can have a significant effect on cable growth within a state, appointment to the commission can be a sweet political reward for a member of the party in power.

Arguments Against State Control

Opponents of state control are equally vociferous, asserting that it presents an unnecessary duplication of law. States are sometimes caught between local and federal control, and meeting the requirements of one can violate the requirements of the other. Opponents claim that state control throws local interests into a political arena with representatives who are looking out for their own interests, not for those of the local community. The Big Brother argument also pops up: when a state becomes involved in direct programming it will be oriented more toward propaganda than toward the public interest.

Despite the existence of state statutes, local municipalities seem to continue to have fairly firm control over local cable systems. Moreover, with the tremendous diversity among the systems and the communities they serve, governance at the municipal level appears to have significant advantage over state control.

COPYRIGHT AND CABLE: THE COMPULSORY LICENSE

Under the new copyright law, cable systems must obtain a compulsory license; this permits cable systems to carry over-the-air signals pursuant to FCC rules. This license should not be confused with the contracts or other agreements instituted by local or state governments with cable systems, which govern the actual operations and fee schedules of the systems.

Structure of Compulsory Licensing

The copyright law views cable systems as commercial entities involved in the "performance" of copyrighted works, and as such they must pay copyright fees under the compulsory-licensing system. For example, under the law and compulsory licensing, broadcasters receive copyright protection from infringement by cable systems.

The Statement of Account

Every six months the cable system must send the Copyright Office a "Statement of Account Form," depending on the amount of the system's "gross receipts." In the very simplest terms, think of "receipts" as income the cable system earns for the accounting period. Each "Statement of Account" must also be accompanied by the payment of a royalty fee covering retransmissions during the preceding six months.[14]

Primary- and Secondary-Transmission Services

An important distinction is made in the compulsory license copyright liability between primary- and secondary-transmission services.[15] Primary-transmission services "include broadcasts by radio and television stations to the public that are retransmitted by cable systems to their subscribers."[16] Secondary-transmission service is the basic service of retransmitting television and radio broadcasts to subscribers. "The statute requires all U.S. cable systems, regardless of how many subscribers they have or whether they are carrying any distant signals, to pay some copyright royalties. However, instead of obliging cable systems to bargain individually for each copyrighted program they retransmit, the law offers them the opportunity to obtain a compulsory license for secondary transmissions."[17] The secondary-transmission service does not include "transmission originated by a cable system (including local origination cablecasting, pay-cable, background music services, and originations on leased or access channels)."[18]

Restrictions Contained in the Compulsory License

Although many benefits are granted by the compulsory license, such as not having to negotiate individual copyright licenses for retransmission of TV and radio broadcasts, there are also certain things the license does not permit, among them:

Originations. A cable system's compulsory license extends only to secondary transmission (retransmissions). It does not permit the system to make any originations of copyrighted material without a negotiated license covering that material.

Nonsimultaneous Retransmissions. In general, to be subject to compulsory licensing under the copyright law, a cable retransmission must be simultaneous with the broadcast being carried. As a rule, taping or other recording of the program is not permitted. Taping for delayed retransmission is permissible only for some (not all) cable systems located outside the 48 contiguous states; and, even in these exceptional cases, there are further limitations and conditions that the cable system must meet.

FCC Violations. The broadcast signals that a cable system can carry under a compulsory license are limited to those that it is permitted to carry under FCC rules, regulations, and authorizations. If signal carriage is in violation of FCC requirements, the cable system may be subject under the Copyright Act to a separate action for copyright infringement for each unauthorized retransmission.

Foreign Signals. In general, the copyright law does not permit a cable system to retransmit signals of foreign television and radio stations under a compulsory license. The only exceptions have to do with the signals of certain Mexican and Canadian stations. Unless foreign signals fall within these exceptions, their carriage would not be authorized under a compulsory license, even if permissible under FCC rules.

Program Alteration or Commercial Substitution. Cable systems are not permitted to alter the content of retransmitted programs, or to change, delete, or substitute commercials or station announcements in or adjacent to programs being carried. There is only one exception: under certain circumstances, substitutions involving "commercial advertising market research" may be permitted.[19]

Forfeiture of the Compulsory License

Somewhat like a station license, a cable system's compulsory license can be revoked. For example, failure to file the required "Initial Notice of Identity" or "Notice of Change" can result in loss of the compulsory license.[20] Other violations can include failing to file the "Statements of Account or royalty fees; taping for delayed transmission; carrying signals in violation of FCC requirements; carrying certain foreign stations; and altering programs or substituting commercials."[21]

If a cable system goes as far as to disregard the copyright laws and not obtain a license, it can be sued by a copyright owner. In the case of willful infringement the owner can attempt to collect actual damages and profits, or statutory damages. Moreover, civil and criminal penalties as well as injunctions can be served on the cable system.[22]

SUMMARY

Increased emphasis on information technologies has made common carriers, cable, and copyright important areas in the study of telecommunication regulation.

The basis for the development of common carriers can be found in two early pieces of legislation, the Post Roads Act of 1866 and the Mann-Elkins Act of 1910. The Post Roads Act established right-of-way over public lands, and the Mann-Elkins Act authorized the establishment of rate structures for common carriers that are involved in interstate communication.

Common carriers are subject to state and federal control. Carriers engaged only in intrastate communication are regulated by state public utility commissions and are classified as partially subject carriers. Fully subject carriers are "fully" covered by the Communications Act of 1934 and are regulated by the federal government through the FCC. Fully subject carriers are involved in interstate and international communication.

Both the operation and the facilities and licensing of common carriers are regulated by government.

Court decisions have assured manufacturers that devices that connect to telephone equipment in the home can be produced, marketed, and sold in a competitive marketplace.

Ever since *United States* v. *Southwestern Cable Co.* the FCC has gradually shifted control of cable systems to local communities. Registration requirements exist for cable systems. Local regulatory frameworks for cable vary.

State regulation of cable takes place within three frameworks: preempt statutes, appellate statutes, and advisory statutes.

Cable systems must meet copyright statutes by obtaining a compulsory license. The license can be revoked if the terms of the license are not met and the payment of copyright royalty fees are not made.

OPPORTUNITIES FOR FURTHER LEARNING

BAUGHCUM, A., AND FAULHABER, G.R., eds. *Telecommunication Access and Public Policy*. Norwood, NJ: Ablex, 1984.

BRANSCOMB, A.W. *Toward a Law of Global Communications Networks*. New York: Longman, 1986.

LAWRENCE, J. S., AND TIMBERG, B., eds. *Fair Use and Free Inquiry: Copyright Law and the New Media*. Norwood, NJ: Ablex, 1980.

SCHILLER, D. *Telematics and Government*. Norwood, NJ: Ablex, 1982.

VIAN, K. *Communications Technologies and Political Control*. Norwood, NJ: Ablex, 1983.

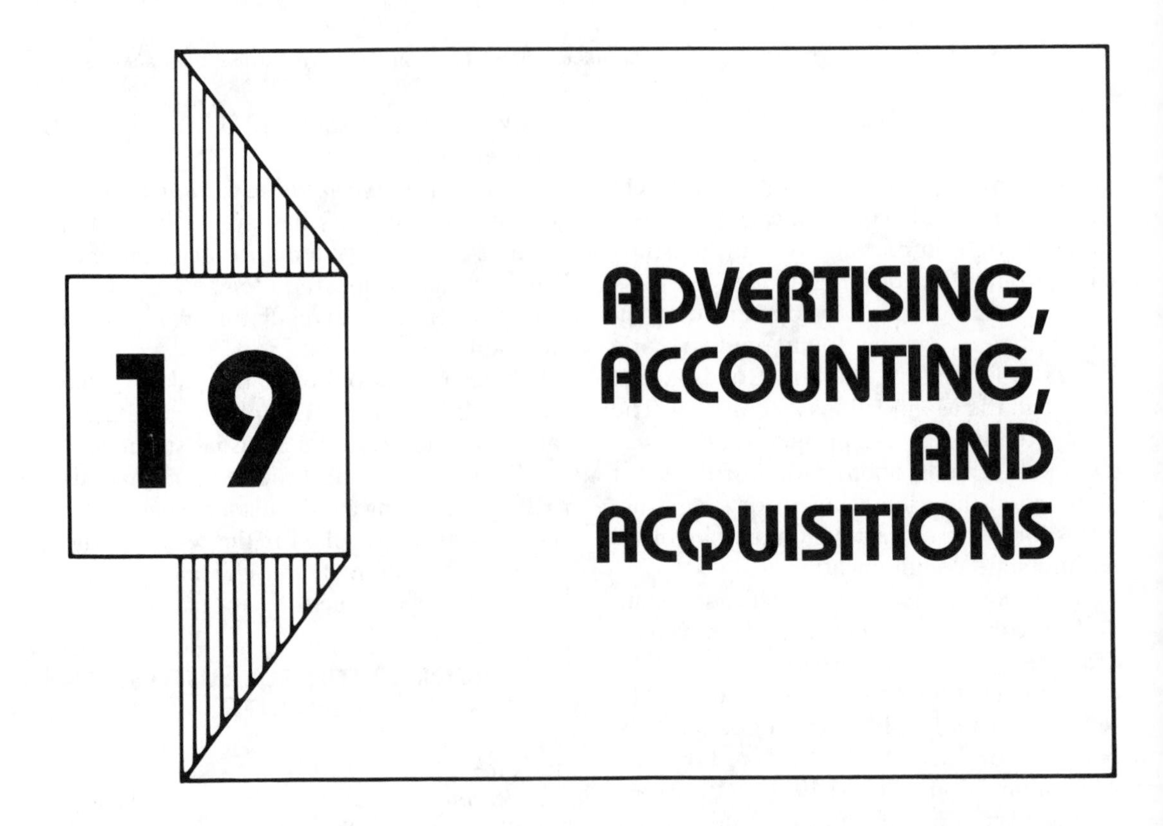

H. P. Davis was a vice-president at Westinghouse in 1928 when in a lecture delivered to the Harvard University Graduate School of Business Administration he said,

> In seeking a revenue returning service, the thought occurred to broadcast a news service regularly from ship-to-shore stations to the ships. This thought was followed up, but nothing was accomplished because of the negative reaction obtained from those organizations whom we desired to furnish with this news material service. However, the thought of accomplishing something which would realize the service referred to still persisted in our minds.[1]

Davis's persisting thoughts turned out to be the foundation of American commercial broadcasting. The purpose of this chapter is to examine the different aspects of that foundation: building station revenues, managing station finances, buying and selling stations, starting new stations, and promoting the station.

BUILDING STATION REVENUE

In the United States, advertisers pay commercial stations to broadcast their ads during a certain time to the listening or viewing public. That elusive commodity, *time*, is the product that stations offer. Some times are better and more expensive than others, and a great amount of time is cheaper than a small amount.

Understanding Cost per Thousand

Central to rate cards, building station income, and operating competitive markets is the term *cost per thousand* (abbreviated CPM), which *is the cost of reaching 1,000 people with one commercial announcement.* Let us assume that your station's rate card lists the cost of a one-minute commercial as $100. That same commercial on the competing station costs $150. Now let us assume that the Ace Garden Supply Company wants to advertise topsoil to people twenty-five to forty-nine years of age. You examine your station's latest ratings and find that between 9:00 and 10:00 A.M. you are reaching about 12,000 viewers in the twenty-five-to-forty-nine age bracket. But your competition reaches 14,000 viewers and has convinced the Ace people that they should buy advertising on your competing station. You compare the CPM of your station with that of the competition.

First you figure your station's CPM by dividing the cost of your one-minute commercial by the thousands of people reached by that commercial. You use the equation

Cost divided by thousands of viewers = CPM

Before reading further, substitute the figures from our example above and figure the CPM. The correct answer is $8.33 ($100/12). If you did not get $8.33, it was probably because you divided the $100 by 12,000. That would tell you how much it cost to reach just one person. But remember, CPM is the cost of reaching 1,000 persons, and since there are 12 of those (12,000) we divide the $100 by 12.

Now you need to figure the CPM for your competition. To do this, divide $150 (the cost of the competition's one-minute commercial) by 14 (the thousands of persons viewing). The answer is $10.71. Clearly the most economical way to reach the viewers is on your station. It is now up to you to convince Ace Garden Supply of that.

ESTABLISHING RATES FOR COMMERCIAL TIME

The rates a station charges for its time are listed on two types of rate cards—the local one for businesses in the local community and the national one for advertisers who buy large amounts of time on many different stations.

Local Rates

Figure 19–1 presents one station's local rates for two time classifications, AA and A. Class AA time runs from 6:00 A.M. to 10:00 A.M. and from 3:00 P.M. to 7:00 P.M., Monday through Saturday. These times—the most expensive on the local rate card—represent *drive time*, those heavy radio-listening hours when radio captures not only the home audience but also the people driving to and from work. All other times are class A times, and they are less expensive than drive time because they traditionally do not attract as large an audience.

National Rates: Reps and Ad Agencies

Now compare the rates listed in Figure 19–2 with those in Figure 19–1. You will notice a difference of approximately 17.65 percent. The higher rates in Figure 19–2 are from the same station's national rate card. To eliminate all the paperwork and negotiations involved in placing advertising with every individual station, advertisers pur-

ANNOUNCEMENTS

(AA) 6:00 to 10:00 a.m., Mon./Sat.
3:00 to 7:00 p.m., Mon./Sat.

	60 Sec.	30 Sec.
1	19.00	15.20
52	16.00	12.80
156	13.00	10.40
312	11.00	8.80
624	9.50	7.60
1040	8.50	6.80

(A) All other days and times

	60 Sec.	30 Sec.
1	16.00	12.80
52	13.00	10.40
156	11.00	8.80
312	9.00	7.20
624	7.50	6.00
1040	6.50	5.20

ANNOUNCEMENTS

(AA) 6:00 to 10:00 a.m., Mon./Sat.
3:00 to 7:00 p.m., Mon./Sat.

	60 Sec.	30 Sec.
1	22.40	17.90
52	18.90	15.10
156	15.30	12.20
312	13.00	10.40
624	11.20	9.00
1040	10.00	8.00

(A) All other days and times

	60 Sec.	30 Sec.
1	18.80	15.10
52	15.30	12.20
156	13.00	10.40
312	10.60	8.50
624	8.80	7.00
1040	7.70	6.10

FIGURE 19–1 (left) and **FIGURE 19–2** (right). The Local Rate Card on the left is the one used to sell advertising to local businesses. Local rates are often lower than those listed on a National Rate Card, shown on the right. The National Rate Card incorporates additional commissions paid to advertising agencies and rep firms; therefore, the rates are higher. These two rate cards are examples. Rates differ considerably among different stations and are often heavily negotiated to make a sale.

chase their advertising time through either station representatives, called *reps*, or advertising agencies. Station representatives, as their name implies, represent the station to large advertisers. They also represent more than one station and contract to buy time on many different stations in order to reach the audience the advertiser requests. Advertising agencies, on the other hand, represent the advertiser, but they buy time in the same way that reps do.

A company using a rep or agency typically pays a rate 17.65 percent higher than the rate on the local rate card. Of this increase, the rep or agency may take a commission of 15 percent. The remainder goes to the station. This 2.65 percent increase is considered compensation to the station for the long-distance account, promotion of itself to and through the rep and the ad agencies, the extra bookkeeping, and related costs. The advertiser, in turn, saves the time and cost of placing each individual advertising order, a cost that on large buys would run much more than the 15 percent commission paid to the rep or agency.

Negotiating Rates

Despite the rate cards and agency and rep commissions, stations in highly competitive markets wheel and deal to entice advertisers to buy time on their station. This usually happens when an advertising agency tries to buy the most time for the least money and pits stations against each other to see which one can offer the best price. Sometimes it is successful. There is considerable price slashing: discounts of 35 percent below the rate card are not unusual, and some stations discount 50 percent.

Time for announcements or commercials can be purchased in various lengths, although ten-, twenty-, thirty-, and sixty-second commercials are common. The shorter time periods are common in television, where rates are higher than in radio. This allows smaller businesses to buy television time. Forcing them to buy longer commercials might price them out of the market.

Commercial time is less expensive when bought in quantity. For example, on the local rate card one sixty-second announcement costs $19.00 in AA time. If you purchase 52 announcements, the cost drops to $16.00 per announcement. If you purchase 1,040 sixty-second announcements, the cost drops to $8.50. Similarly graduated discounts are available for thirty-second announcements, and for all announcements made in A time as well.

Revising Rates

Stations periodically revise their rate cards just as supermarkets revise prices of meat and eggs. Successful stations revise theirs upwards in response to inflation and to increases in their audience. The more viewers or listeners a station has, the more people an advertiser can reach and the more the station can charge for its commercial time. Weaker stations may need to revise their rates downward to reflect a loss of listenership or viewership. Another factor in rate changes is the station's market. If there are no other broadcast outlets in the market, the station's rate may be higher than if there were other stations offering competing rates. It is the old rule of supply and demand. The cost of local newspaper advertising can also influence the rate charged by the broadcast media. All media compete for those advertising dollars.

Other time buys are available on most stations. These include time for remote broadcasts, such as live coverage of a store opening, and larger time blocks in which to air entire programs.

STATION REVENUE

Time, that valuable commodity a station sells, can be packaged in many different ways, all leading to sources of station revenue.

Selling Time

The most common source of station revenue is selling time based on the rate of the local and national rate cards. The all-important sales department, which in many stations has separate people assigned to local and national sales, is where the station revenue is generated. Whether a local retailer or a national advertiser, the commodity for sale is the same—time. The only difference is how much time will be purchased, for what rate, and through what means. In addition to the straight sale of time based on rate cards, time is sold in other ways as well.

Trade-out Arrangements

Trade-out arrangements are another way in which stations receive income. Trade-outs are when *the station provides commercial time in exchange for goods or services offered by the advertiser*. These goods and services—anything from appliances to world cruises—may be used by the station at its discretion. Many stations give them away as prizes; others award them to their top account executives.

Most prizes awarded on TV game shows are supplied by manufacturers who pay a small fee for on-air announcements in return for their products' publicity. National television exposure of this kind is a relatively inexpensive way of obtaining advertising time, compared with the usual national TV rate. For the game show, the prizes are for all practical purposes free merchandise, and beyond the salary of an announcer there is almost no cost involved in announcing the products on the air. Incorporated into the programming, the announcements add the elements of excitement and dream fulfillment to the description of the "fabulous prize" the contestant has a chance to win.

Many companies engage in trade-out advertising. Windjammer "Barefoot" Cruises Ltd. trades on a dollar-for-dollar basis, which means that whatever the cruise costs is the amount the station will make available in advertising time. The cruise can be an incentive to the station's account executives to meet sales quotas or can be used as part of a station promotion. For the station, the cost of airing the commercials is minimal since the announcements are usually scheduled in unsold time that would otherwise go unused.

Although trade-out advertising is common, many stations try to avoid it, especially when the advertiser can be persuaded to pay cash instead. Cash looks better than merchandise or services when the station prepares its annual financial statement. Furthermore, when stations are sold, trade-out advertising is usually not considered part of the station's annual income, since the merchandise supplied by the advertiser usually cannot be used to pay the station's bills.

Co-op Advertising

Co-op advertising is an arrangement by which a *local retailer splits part of the cost of advertising with the company whose products are advertised in a commercial*. Let us assume a radio announcement of a sale of Westinghouse appliances at the Ace Appliance Store costs $19.00. In a 10/90 co-op advertising arrangement between Westinghouse and the store, Westinghouse would pay 10 percent of the cost of the announcement, or $1.90, and the store would pay $17.10. The radio station receives the rate card price for the announcement, Ace airs the commercial at a 10 percent discount, and Westinghouse receives local advertising exposure.

Not all companies use co-op advertising, and the amount paid by the manufacturer and the local store varies considerably. Newspapers continue to receive the majority of co-op advertising dollars, about 75 percent, since co-op advertising began with newspapers before broadcast advertising was widespread.[2] Some companies employ full-time co-op managers whose responsibility is to keep track of co-op advertising by retail outlets and to show retailers how to coordinate their advertising with the manufacturer's co-op program.

Barter Arrangements

Stations sometimes receive programming in return for providing advertising time. Such arrangements, called *barter*, are widespread in syndicated programming. With barter, *a station receives a free program, or pays a reduced fee for the program, and in return airs commercials that are already in the program when it is received by the station*. The station may also be able to place additional commercials in the program within available time slots.

While profitable for syndicators, barter programs are sometimes a mixed blessing for stations. Barter programs, while filling programming time, use up available time for commercials. The commercial that runs in the bartered program and is supplied by the syndicator is actually a commercial that in effect may air at a reduced rate when based on the station's rate card. Thus, the station accepting barter programming must consider not only how barter will affect station income but also its programming.[3]

Combination Rates

Combination rates, also called *combos*, are when *commercials on a jointly owned AM and FM station are sold for the same rate*. For example, a local retailer can purchase commercials on WAAA-AM and WAAA-FM for one rate and in effect buys two stations instead of one, although not for the price of one. A number of issues surround whether a station offers combination rates.[4] For example, if the formats of the AM and FM are different and the two stations reach different audiences, each may lose its identity when promoted together. On the other hand, if one station is weaker, combining it with its stronger counterpart increases its image and identity in the marketplace. The two stations combined get a larger share of the ratings as well as advertising dollars than if they were promoted and sold separately. This is especially true when both the AM and FM station have the same call letters.

Time Brokerage

Selling time through *time brokerage* is when *a station sells a block of time to another party who in turn programs the time block and also sells commercials in it*. For example, if you operate a radio station and find it difficult to sell commercial time during a particular time period you might decide to sell that entire time block, say two hours, to another party. That party in turn will program the time slot and sell advertising in it. Any number of negotiable rate structures can exist to make it profitable for both the station and the other party. What is necessary, however, is for the station to retain control over the programming and have the right to refuse programming that would not be deemed in the public interest. Selling the time block to another party does not relieve the station of the programming responsibility of operating in the public interest.

THE ROLE OF PROMOTION IN BUILDING STATION REVENUES

For both new stations and those with long records of operation, station promotion is an important part of the broadcasting business. In major markets, promotion budgets can run into the millions of dollars.[5] Successful station promotion involves much more than a few announcements puffing up a program or personality: it is a well-planned and systematically executed campaign. More and more stations are realizing the importance of station promotion. Many have rested on the laurels of their programming without learning from the example of other businesses that competition requires promotion to convince the public that it should patronize your business or listen to your station. In fact, broadcast promotional campaigns are becoming a necessity in many markets.

Promoting Assets

The most important commodity for any station is its listenership or viewership. Thus, how the station fares in the ratings is the central theme of many promotional campaigns. Although only one station can be first overall, many stations can find their own niche in the ratings data. A certain station, for example, may be first among women, another first among eighteen- to thirty-four-year-olds, another first among men thirty-four to forty-nine, another first during morning drive time, and still another first in late evening news viewership. Each placement in the audience ratings can provide opportunities to promote the station's accomplishments.

Other promotional campaigns can focus on new call letters and format, network, or programming changes. If a station changes hands and subsequently call letters, calling attention to the new identity is important. Similarly, a change in a radio format often results in a new audience. Telling that audience about the new sound and informing advertisers about that new audience are equally important.

Everyone understands awards, and some of the most familiar belong to broadcast journalism. Although publicizing a journalism award promotes only one department—the news department—it can boost other areas of the station as well. Such promotion is especially critical to television, in which the local news can be the single most important determinant of the audience's perception of the station. More than one television station has had its local news pull it to the top in the ratings while its network programming was running a poor third in other markets and nationwide.

Planning Successful Promotion

To be successful, a promotional campaign must have certain qualities. One of the most important is simplicity. Although contests and giveaways can be effective audience builders, successful long-term radio promotions, for example, have a central theme. Such slogans as "Musicradio," "Candlelight and Gold," and "Happy Radio" can be woven throughout the station's programming. If ratings are being publicized, a single concise statement that the station is first is much better than many sentences about the station's share of the audience.

Simplicity also belongs in station logos. Effective designs are uncluttered; they look good in black and white, which is essential to newspaper advertising; they are pleasing on letterheads; and they can be easily recognized and understood (Figure 19–3).

An audio identity is as basic as a visual identity, especially in radio. Most radio stations have a particular "sound" in addition to their musical format. These sounds are usually reproduced in a *jingle package*—a collection of sounds designed around a four-to-eight-note sequence fitting the call letters. Variations in the jingle package are then incorporated into news introductions, bulletins, musical bridges between records, and backgrounds for commercial and public service announcements.

As important as simplicity is consistency. Too many stations make the mistake of constantly changing their promotional theme. Many mediocre promotional efforts have been successful simply because they have continued unchanged long enough for the audience to accept them as household words. To schedule even the very best promotional effort for six months and then disappear is to defeat the purpose of a promotional campaign. Contests and giveaways will change from season to season, but there needs to be a consistent theme that represents the station over time, be it the logo, the jingle package, or a combination of both. Done effectively, a promotional campaign can elevate listener

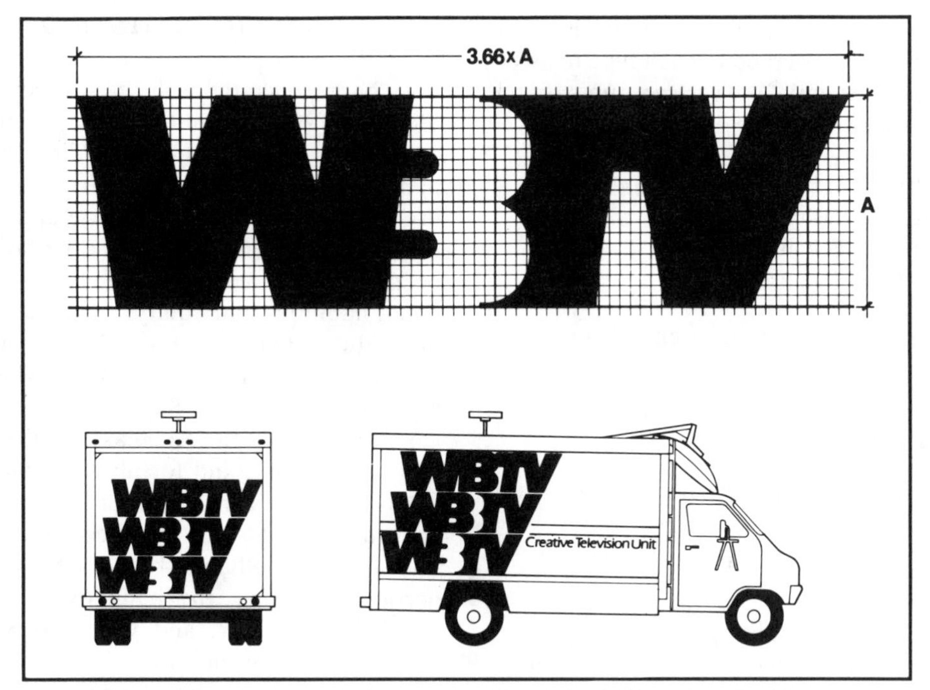

FIGURE 19–3 Promotional campaigns involving billboards, television, and newspaper advertising, as well as specially staged promotional events, are part of the process of marketing a station. Careful attention must be paid to the reproduction of the station logo, which helps to establish and maintain the station's visual identity in the marketplace. The station's call letters superimposed over the graph paper are a guide for artists and designers who must reproduce the logo with exact dimensional requirements. Under the call letters are the specifications on how the logo is to be reproduced on the station's news vehicles.

awareness of the station, and fostering awareness of the station is the first step in building an audience.

FINANCIAL ACCOUNTING: THE CHART OF ACCOUNTS

Once earned, a station's advertising revenue is translated into a series of numbers that managers, bookkeepers, accountants, investors, and bankers spend many hours studying in order to answer some complex questions. How can we improve daytime sales? How can we cut expenses? How much money will we need to borrow? How much money should we invest? These and countless other queries plague station executives as they mull over their charts of accounts and financial statements.

The basic ledger for recording all station finances, both income and disbursements (expenses), is the *chart of accounts*.[6] Although charts of accounts vary, many in broadcasting use a system of numbers whose first two digits (commonly starting with 10 and going up) represent various account classifications. Assume the numbers

100 to 199 represent assets. A specific two-digit prefix is assigned to each type of asset. Here, for example, are cash accounts, which are designated by the 10 prefix:

101 Cash in banks—regular
102 Petty cash
103 Cash in banks—payroll
104 Cash in banks—other

The prefix 14 might represent advances and prepayments:

141 Prepaid insurance
142 Prepaid rent
143 Prepaid taxes and payments of estimated taxes
144 Expense advance to officers and employees

Three-digit accounting systems provide about as much flexibility as the average broadcaster needs, but additional digits can be added wherever appropriate. For instance, a group owner might want a different prefix for each station in the group. WAAA might have the prefix 5. Thus, if we wanted to know the petty cash at WAAA, we would look under account number 5–102, 5 representing the station and 102 petty cash. WBBB might be assigned the number 6, and so on. Some computer programs use additional digits to facilitate more complex data analyses of a station's financial status.

Using the chart of accounts, management can periodically check the station's financial structure as well as pinpoint and plot its activity over time. Such information is critical if management is to ensure the station's profit structure. A station's information and accounting systems are equally important to the broadcaster selling the station, the banker lending the station money, or the accountant preparing its tax returns.

FINANCIAL STATEMENTS

The chart of accounts is the basic accounting tool of the station, but the money recorded on it is translated into many different financial statements. The balance sheet (Figure 19–4) is a stop-action picture of the station on any given day, usually the last day of the year. Notice that assets ($90,000) equal liabilities and equity. Balance sheets must always balance. Why? Because all assets have a claim on ownership, which is equal to those assets. Suppose you drive a car worth $5,000 (a $5,000 asset). The title is in your name, and for all practical purposes it is your car. But what about claims of ownership on your car? We shall assume you still owe the bank $1,000 on the car. In accounting terms, the bank owns $1,000 worth of your car, and you own $4,000 worth. The claims on ownership of your car are $1,000 (the bank's claim) plus $4,000 (your claim), which total the exact value of the car—$5,000.

A comparative balance sheet compares assets and liabilities between two points in time. An income statement shows the amount of income remaining after the debits are subtracted from the credits—in other words, how much money is left after all the bills are paid.

A *comparative income statement* compares income between two points in time. Figure 19–5 is a comparative income statement comparing two years of broadcast operation. Notice that the revenues are listed first, then the separate expenses (indented), and then the total expense. We can see that in year 8 the station's income before taxes was $11,500, on which a federal income tax of $2,530 was paid, leaving an income of $8,970.

If you were the treasurer of a broadcasting station, you would find the cash-flow

WBPE, Inc.
BALANCE SHEET
January 31, Year 1

Cash	4 000
Accounts receivable	2 000
Depreciable assets	50 610
Land	10 000
Intangibles	23 390
ASSETS	90 000
Accounts payable	5 000
Notes payable—sellers	38 000
Notes payable—stockholder	50 000
Liabilities	93 000
Contributed capital	1 000
Loss since inception	(4 000)
Stockholders' Equity (Deficit)	(3 000)
LIABILITIES AND EQUITY	90 000

FIGURE 19–4 The balance sheet is much like a stop-action picture of the station's finances on any given day, usually the last day of the year.

statement especially important. This statement would tell you the amount of money the station needs in order to keep running smoothly. It would also tell you if you needed to borrow money to pay the bills or to invest excess income. A cash-flow statement is looked upon as a statement of the station's health and can help management find where changes in operations may have to be made.

Figure 19–6 is a cash-flow statement comparing one year of operation with the

FIGURE 19–5 The comparative income statement compares income between two points in time, such as this example comparing two years of station operations.

WBPE, INC.
COMPARATIVE INCOME STATEMENT
Year Ending December 31

	Year 8	Year 7
Revenues	168 900	162 200
Technical expense	12 300	12 100
Program expense	48 200	47 600
Selling expense	29 400	28 100
General & administrative	67 500	63 900
Expense	157 400	151 700
Income before tax	11 500	10 500
Federal Income Tax	2 530	
INCOME	8 970	8 190

WBPE, INC.
CASH FLOW STATEMENTS
Years Ending December 31

	Projected Year 9	Year 8
CASH BALANCE AT JANUARY 1	8 505	9 113
Collections from advertisers	190 000	173 460
Cash Available	198 505	182 573
Technical disbursements	15 000	8 229
Program disbursements	50 000	45 861
Selling disbursements	52 000	39 400
Administrative disbursements	74 000	70 825
Agency commission payments	8 000	6 100
Repayment of stockholders loan		886
Dividend payments	200	200
Federal income tax payments	9 100	2 567
Purchase of equipment	20 000	
Total Disbursements	228 300	174 068
CASH BALANCE AT DECEMBER 31, YEAR 8		8 505
PROJECTED CASH AT DECEMBER 31, YEAR 9—overdraft	(29 795)	
DESIRED MINIMUM CASH BALANCE	5 000	
REQUIRED ADDITIONAL FUNDS	34 795	

FIGURE 19–6 This cash flow statement compares one year's operating statement with the next year's projections. Cash flow statements are important in determining the financial health of the station and how it can operate in the future.

projected cash flow for the coming year. Notice year 9. If projections hold true, year 9 will see the station spend more money than it makes, resulting in a cash overdraft of $29,795. As treasurer, you will need to inform management of the necessity to borrow at least $34,795 to make ends meet and still have the desired $5,000 minimum cash balance. If management does not want to borrow the money, it must change that year's projected operating procedures. It might decide to make do with the equipment on hand and save the $20,000 budgeted for purchase of equipment. But as treasurer you point out that never in the station's history has that been done, that the equipment is old, and that the station's transmitter cannot last even through this year. Can you suggest cutting other disbursements? What recommendations can you make, and what will be the consequences? For the individuals who run a broadcasting station these are the questions that financial statements both pose and help answer.

BUYING AND SELLING BROADCAST PROPERTIES

Financial statements are also especially important when a station changes hands. Buying and selling broadcast properties, including CATV, is big business, involving everyone from small-town entrepreneurs to corporate conglomerates. Buying a station requires knowledge of the broadcasting business and considerable money, not only for the purchase itself but for operating the station until it builds an income. The sale of a station must be approved by the FCC, which scrutinizes the buyer's and stockholders' character, their financial worth, their other media investments, their history in managing other enterprises, and more. A record of bankruptcy can quickly close the door to owning a broadcasting property.

The Broadcast Broker

At the heart of over 70 percent of broadcast property sales is the broadcast broker—the real-estate person of the broadcasting business. Most are professionals with years of experience in station transactions. Many people wanting to sell or purchase a broadcast property begin by contacting the broker.[7] On the basis of personal contacts, referrals, advertisements, and in some cases direct solicitation, the broker knows what properties are for sale at what price, and who the buyers are. Because the sale of a broadcasting property can run into millions of dollars, good brokers are highly paid, but most industry professionals would agree that they are well worth their commission.

Commissions

Although commissions vary among brokers, typical fees for a station sale are 5 percent of the first million dollars, 4 percent of the second million, 3 percent of the third million, and 2 percent of the balance.[8] Brokers also perform property appraisals, typically $1000+ a day.

THE TRANSACTION CHECKLIST

Still, the primary business of the broker is handling the sale of broadcast properties. For the buyer, there are many factors to consider before the final transaction.[9] The prospective buyer of a broadcast property should examine them carefully.

Timing

Although there may be a need to dispose of a broadcast property for reasons that are not financial, taking into account the general economic climate is important whenever possible. Selling a property during an economic recession or when the prices of stations are depressed may not be in the seller's best interest.

Financial Checks

The prospective buyer should review the gross-sales record of the station for at least the previous three years. The term *gross sales* refers to all income made by the station but *not* with expenses deducted. At the same time it is important to see how much of the income is from *trade-outs*, whether any promotions have been used to inflate the station income, and what share of the total market revenue the station enjoys.

Profit and cash flow should also be considered including such items as the net profit before taxes, any interest on debts the station pays, and the salaries of officers and directors. Because a buyer might change the compensation levels of such people, this information is important in planning future operating expenses.

Equipment owned by the station needs to be examined. If it is old and outdated the expense of replacement may need to be figured into the selling price. How much is owned and how much is being rented or leased is also important.

Such items as leases, licenses, and contracts must also be examined. For example, contracts with syndicators may run far beyond the date when the station changes hands. News service fees such as those for wire or video service must be figured in as would contracts with rep firms, networks, or advertising agencies.

Legal Checks

The station's attorneys should prepare a list with the status of any unsettled lawsuits or pending legal actions that may be brought against the station. These include any matters pending before the FCC. When possible, legal matters should be settled before the new owners assume control. It cleans the slate and eliminates what can become a sore point in the negotiation of a selling price.

Operation Checks

The salary and position of all employees should be studied as well as any bonus or incentive plans and the commission structure for sales representatives. Vacation policies and the status of the affirmative-action plan also need to be reviewed.

Ratings will tell the prospective buyer the size and composition of the audience. Analyzing the programming of any other stations in the market is necessary to determine if schedules and formats need to be changed and at what cost. A check will also need to be made of the advertising rates of local newspapers or other major publications that compete for revenues. Within the community, such information as past and anticipated population growth, retail sales, major sources of employment, and consumer spending power will provide indicators of the present and future economic climate.

If a buyer finds positive information after reviewing the items in the checklist, then the buy is probably a good one, provided the price is right.

SUMMARY

At the basis of building station revenue is the rate card. Rate cards tell advertisers the cost of commercials. Combined with ratings data this information makes it possible to compare the cost-per-thousand of reaching different target audiences. Local rate cards list the rates charged to local advertisers. The national rate card, which lists rates higher than the local rate card, is used for national advertisers placing orders through station representatives or advertising agencies. Despite the rate card, the cost of commercial time is negotiable, and hard bargaining for the best rate, especially by advertising agencies, is a common practice. Rates are continually revised to reflect both inflation and the position of the station in the market. A station with a good showing in the ratings can charge more than a station with a weak showing because the weak station reaches a smaller audience. In addition to the straight sale of time based on rate cards, time is sold in other ways as well. Trade-out advertising, co-op advertising, barter, combination rates, and time brokerage are additional sources of station revenue.

Promoting a station is becoming essential as more stations sign on the air and competition increases. An effective promotional campaign contains two major qualities—simplicity and consistency.

In the managing of station finances, the chart of accounts is the numerical and descriptive list used to classify income and expenditures. Financial statements usually include the balance sheet, comparative balance statement, income statement, comparative income statement, and cash-flow statement.

People who invest in broadcasting buy stations or start new ones. The broadcast broker is the real-estate person of the broadcasting business and is responsible for arranging many of the transactions. A prospective buyer should examine key areas of a station before purchasing it. Weighing the economic climate is important as is checking the financial, legal, and operational health of the station. If these checks provide positive information for the buyer and the price is right, then the station may be a good buy.

OPPORTUNITIES FOR FURTHER LEARNING

BLUMLER, J. G., AND NOSSITER, T. J., eds. *Broadcasting Finance in Transition: A Comparative Handbook*. New York: Oxford University Press, 1990.

BORTZ, P., WYCHE, M., AND TRAUTMAN, J. *Great Expectations: A Television Manager's Guide to the Future*. Washington, DC: National Association of Broadcasters. 1986.

EICOFF, A. *On Broadcast Direct Marketing*. Lincolnwood, IL: National Textbook, 1988.

LAVINE, J. M., AND WACKMAN, D. B. *Managing Media Organizations: Effective Leadership of the Media*. New York: Longman, 1988.

ORINGEL, R. S., AND BUSKE, S. M. *The Access Manager's Handbook: A Guide for Managing Community Television*. Stoneham, MA: Focal Press, 1987.

PICARD, R. G. *Media Economics*. Newbury Park, CA: Sage, 1989.

SCHULBERG, B. *Radio Advertising: The Authoritative Handbook*. Lincolnwood, IL: NTC Business Books, 1989.

SHERMAN, B. L. *Telecommunications Management: The Broadcast & Cable Industries*. New York: McGraw-Hill, 1987.

ZEIGLER, S. K. *Broadcast Advertising*, 3rd ed. Ames: Iowa State University Press, 1990.

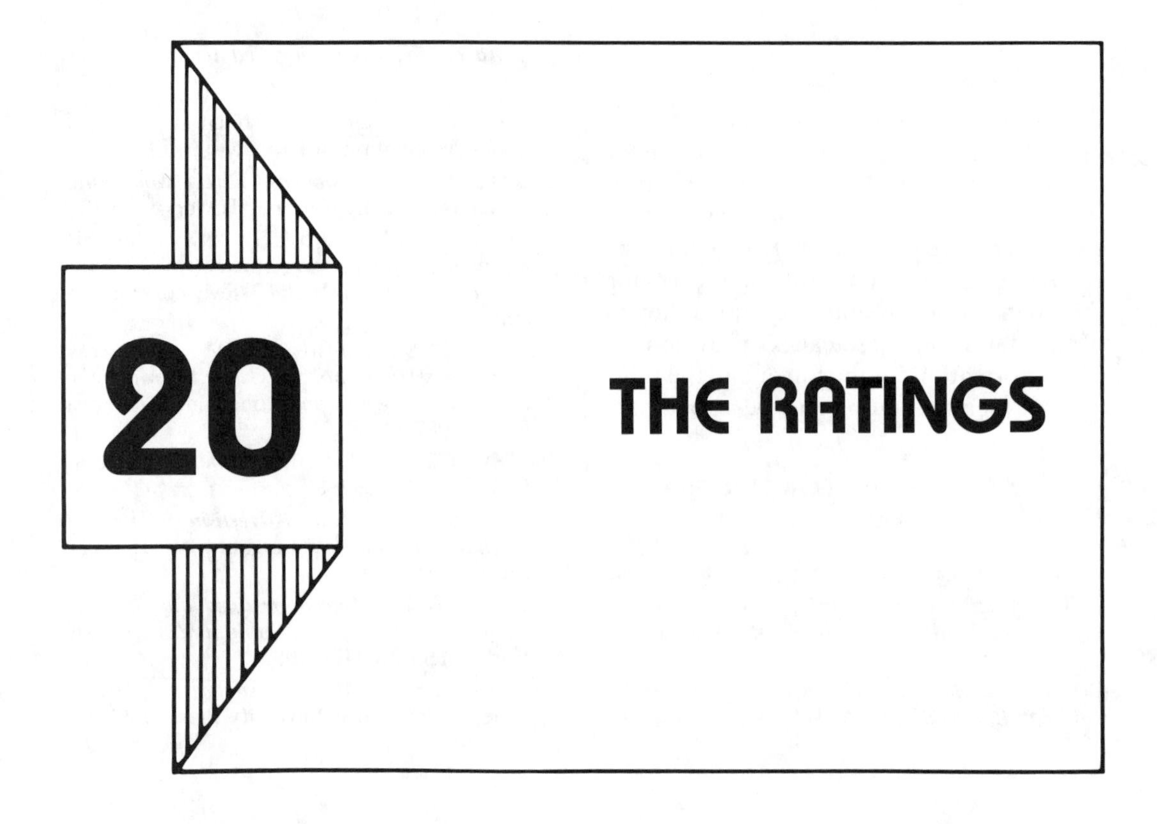

A network pollster gathers data on election day. A radio program director examines data about tape and disc sales in the recording industry while preparing the weekly playlist. Management listens to its employees' opinions. These are examples of the feedback necessary for the successful maintenance and operation of a broadcasting system. Computers analyze viewing-habit data from meters attached to home television sets, from viewers' diaries, from an assortment of answers to questions by interviewers, and from viewing questionnaires returned through the mails. This chapter deals with feedback, specifically, the feedback provided by broadcast ratings.

THE BACKGROUND OF BROADCAST RATINGS

The earliest attempts at the systematic collection of ratings data occurred in radio's experimental era of the 1920s when stations contacted listeners or asked them to reply by mail or phone if they were listening to a particular station. As commercial broadcasting began to develop, advertisers and stations wanted more reliable information on listening habits to set advertising rates and to know who was listening to what commercials. By the 1930s the Association of National Advertisers supported what was known as the Cooperative Analysis of

Broadcasting, one of the first true "rating" services.[1] In the same decade CBS spearheaded audience research under the direction of Frank Stanton, who later headed the network. Using devices developed at the Massachusetts Institute of Technology, Arthur C. Nielsen, Sr., the founder of the well-known Nielsen television-rating service, began measuring radio audiences in 1936.[2] Ten years later in 1946 researchers Harry Field of the University of Denver and Paul F. Lazarsfeld of Columbia University published *The People Look at Radio*, one of the classic studies of audience research.[3]

Since that time, broadcast ratings have been criticized as being inaccurate, unreliable, and arbitrary. Ratings have been accused of canceling quality network programs, determining network policy, and influencing everything we see and hear on the broadcast media. These are all misconceptions. Broadcast ratings are services that tell how many people are viewing or listening to what, when they do it, and how often. Two of the most familiar are the A. C. Nielsen Television Index and Arbitron, which are separate companies and do not have any relationship to the networks. However, stations and networks do use the data from rating surveys to make decisions on what programs to cancel or keep on the air. As a result, the rating services get blamed for canceling programs when it is the stations and networks that make the decisions.

Despite sustaining serious criticism, the ratings have for the most part proved to be very reliable, producing some of the most accurate and sophisticated audience research data available. But they are by no means perfect. Along with the major ratings services, which are controlled and professionally responsible, are minor ratings services that do use some of the methodologies that legitimize criticism of broadcast ratings. It is important from both a consumer and a professional standpoint to learn how ratings work, what function they have, and how reliable they actually are.

THE FUNCTION OF BROADCAST RATINGS

Media industries subscribe to broadcast rating services to determine the size and composition of the viewing and listening audiences. In a sense, these industries are buying detailed feedback on which future decisions are based. For instance, if you were operating a commercial television station in a major city, you would want to know how many people watched your station and how many watched other stations. You would also be interested in detailed demographic information on those viewers, such as their age, sex, education, and income. This information is essential to your advertisers. Similarly, you would want to know when these audiences tune to your station. You would also want to know whether a particular program commanded a larger share of the audience than some other programs. What share of the audience did your newscast capture, compared with those of competing television stations? Your advertisers need all of this information in order to purchase air time—your station's air time—wisely.

Comparing Audiences and Costs

Advertisers also want to compare the cost of attracting viewers. By combining the information found in ratings with the price schedule on your station's rate card, an advertiser might discover that the cost of reaching a thousand people (the cost-per-thousand, or CPM) over your television station is less than the cost-per-thousand of another television station. If it can be proven that your TV station reaches more potential customers than your competition does, the advertiser will probably realize that buying commercials on your station is a wise investment. Rating services provide the necessary proof.

With this proof in hand, you may find that certain programs need to be rearranged or even canceled because of limited viewership. It simply is not profitable to air them, at least not in their current time slot. Notice the word *you*. As a media executive it is you and not the rating service who makes the decision to cancel or reschedule a program.

Data vs. Decisions

Rating services provide station management with information on how many viewers or listeners a station has, compared with other stations in the same community or, on a national scale, other networks. If a station is not attracting the number of listeners or viewers that management feels it should, and if management believes a change in personnel or programming will improve the ratings, then the result may be a change in staff or programs. But keep in mind the many aspects of ratings. Station management may be perfectly comfortable with relatively few viewers if the station has a high income and considerable buying power. Public broadcasting stations, knowing their ratings' position among commercial stations, direct quality programming to an audience with a higher than average educational level. Such a station may be completely satisfied with less than first place.

The entertainment industry is also interested in ratings because ratings specify the kinds of programs that will be hits. If the public demands situation comedies, then producing science fiction for prime-time television might not be a wise decision. Similarly, producers of television series are interested in what share of the audience their program has in cities across the country or around the world.

Public Confidence

Although broadcast management and the entertainment industry generally believe that the larger rating services are accurate and reliable, the public does not. A lack of understanding of the methodologies used creates much of this skepticism. Personal preference is also powerful, and all the mathematical formulas in the world will not convince a devoted viewer that a favorite television program has been canceled because too few people watched. "That's impossible; all my friends watch that program every week!" This enthusiastic conclusion may be mistaken. Among your friends, perhaps everyone does watch the program. But perhaps because your friends watch the program and talk about it in the dormitory their friends do not want to be left out, and so they also watch. Neither "standard" is an accurate indication of how the rest of the viewing audience feels about the program. How, then, is a broadcast rating determined?

JUDGING ACCURACY: THE SAMPLING PROCESS

A typical ratings skeptic will claim there is absolutely no way a small group of people selected to tell what TV programs they watch or what radio stations they listen to can possibly determine the viewing or listening habits of thousands or millions of other people. To some extent they are correct, but only if the group of people polled is extremely small.

Defining Sampling

At the heart of broadcast ratings is a process called *sampling*. Sampling means examining a small portion of some larger portion to see what the larger portion is like. A chef in a large restaurant tastes a tiny teaspoon of soup from a five-gallon kettle to determine if the soup is ready for serving. A doctor can examine a small blood sample from your arm to ascertain the characteristics of the rest of the blood in your body.

If we want to determine the number of people listening to radio in Elmsville, instead of calling everyone in Elmsville which has a population of 10,000, we'll call a sample of, say, 500 people. Based on what we learn about the listening habits of those 500 people we will then estimate the listening habits of all of Elmsville.

Each person in our sample would represent the listening habits of 20 people in Elmsville since 10,000 (the population of Elmsville) divided by 500 (the size of our sample) equals 20.

Random Sampling

The essence of the sampling process is called *random sampling*—sampling such that *each unit of the larger population has an equal chance of being selected*. If the population of Elmsville is being randomly sampled, then each person in Elmsville has an equal chance of being selected. In our example, we selected 500 people out of a population of 10,000. Those 500 people became our random sample.

Sampling Error

You may say, "Okay, I'll get on the phone, randomly select ten people from my hometown telephone directory, ask them what radio station they're listening to, and find out what the rest of the town is listening to." If you do this and your prediction turns out to be true, you would merely be lucky. Because the random sample you selected was so small, its *sampling error* was too large for you to make an accurate prediction. Sampling error is determined by the size of the sample. The larger the sample, the smaller the sampling error.

Centuries ago mathematicians proved that a truly random sample is all one needs to tell the characteristics of a larger population. Moreover, once a certain number of people are chosen for the random sample, increasing that number will not significantly change the outcome.

You may be surprised to find out how small that random sample needs to be. For example, a truly random sample of 600 persons is sufficient to make a prediction about the entire city of New York or London with only a ±4 percent sampling error.[4] That means that if 75 percent of the 600 people you sampled were listening to a certain radio station, then you could predict that somewhere between 71 percent (75 percent − 4 percent) and 79 percent (75 percent + 4 percent) of the entire population were listening to the same station.

Increasing your sample size to 1,000 would decrease the sampling error only to ±3.1 percent. Moreover, increasing the size of the population to include an entire nation would not significantly change that error.

Our random sample of 500 people in Elmsville would have a sampling error of slightly more than ±4 percent.

Reducing Sampling Error

Even with the rules of sampling, some rating services try to reduce sampling error even more. For example, in measuring television viewing, only one housing unit in a specific geographic area is selected as determined by government census data. For example, television usage in a small geographic area tends to be similar due to such factors as cable penetration, signal reception, and *demographic* characteristics. To avoid loading the results of a survey, care is taken to make sure the sample does not cluster in one specific geographic area and miss other households that might better represent the larger population.[5] Such a procedure may seem to deviate from our definition of random sampling, but in reality it further refines the sampling process to avoid an unintentional bias that might not accurately reflect the total population.

DATA COLLECTION

Some of the most critical steps in determining a broadcast rating take place in the data-gathering process. After the random sample has been selected, the rating service must next find people who are willing to provide the information it is trying to collect.

Gaining Cooperation

For a number of reasons, people may be hesitant to cooperate with a rating service. Perhaps they are apprehensive about the stranger at the door or on the telephone who is requesting their help. Other prospective candidates may not be at home or may have moved. To overcome these obstacles, the major rating services employ field representatives who are highly trained in everything from persistence to interpersonal relations.

Interviews

Rating services use three different methods of gathering information: interviews, diaries, and meters. Interviews are either personal or by telephone. Different studies have come to different conclusions as to which is more effective.[6] Some rating services use both. Regardless of whether a rating service uses personal or telephone interviews, it must consider certain variables in the interview process. For example, differences in the way questions are asked can affect how a person responds. Rating services, therefore, conduct sophisticated training sessions to make sure their interviewers are asking the same questions in the same way.

Diaries

Besides personal and telephone interviews, rating services frequently utilize the diary method (Figure 20–1) of collecting data. Here, the viewer or listener keeps a record of the programs and stations he or she tunes in to during a given week. This diary is then mailed back to the company, which tabulates the results of all diaries submitted. In some cases a small monetary incentive is included with the diary.

Meters

The third method of data collection is the meter. The rating service enlists the cooperation of a household in installing on the television set a small, inconspicuous monitoring device connected through a telephone system to a central computer. The computer automatically monitors each meter and records the readings. The monitoring not only reveals what channel is on at any time of the day but also tells if more than one set is in use in the household.

Pre-programmed into the meter is demographic information such as age and sex. A remote-control device with a series of numbers for each member of the household permits viewers to enter their identification number when they begin viewing. The re-

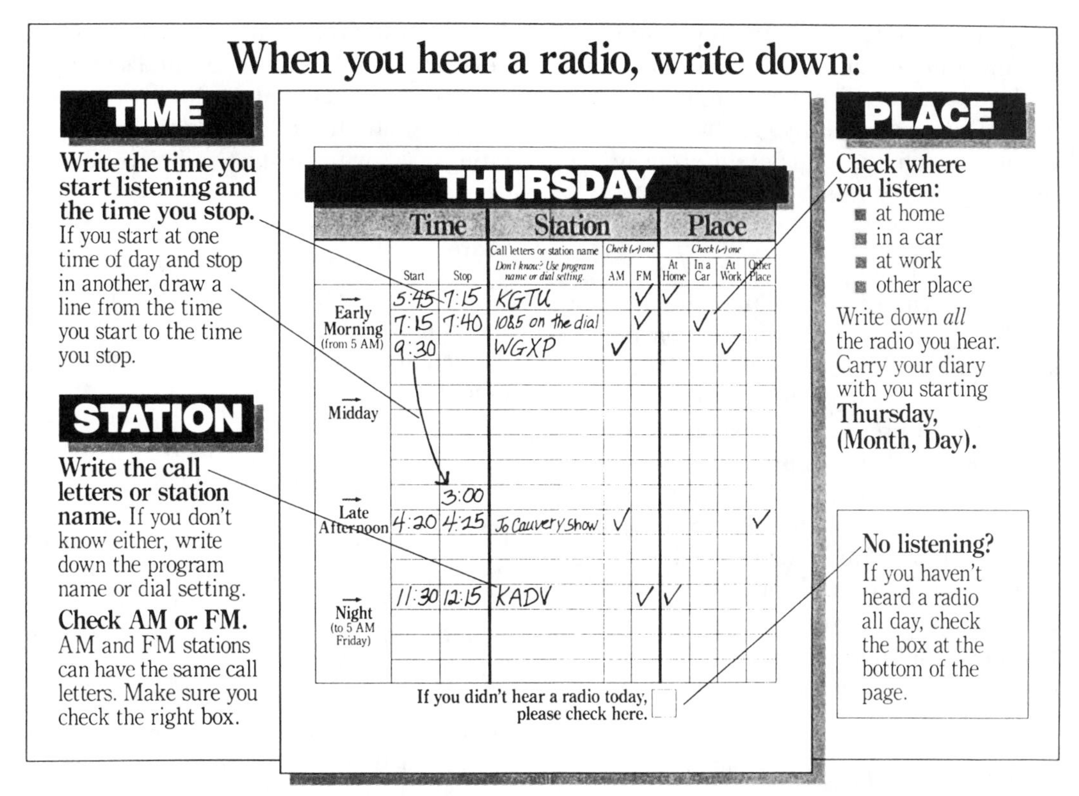

When you hear a radio, write down:

TIME

Write the time you start listening and the time you stop. If you start at one time of day and stop in another, draw a line from the time you start to the time you stop.

STATION

Write the call letters or station name. If you don't know either, write down the program name or dial setting.

Check AM or FM. AM and FM stations can have the same call letters. Make sure you check the right box.

THURSDAY

	Time		Station			Place			
	Start	Stop	Call letters or station name *Don't know? Use program name or dial setting.*	AM	FM	At Home	In a Car	At Work	Other Place
Early Morning (from 5 AM)	5:45	7:15	KGTU		✓	✓			
	7:15	7:40	108.5 on the dial		✓		✓		
	9:30		WGXP	✓				✓	
Midday									
		3:00							
Late Afternoon	4:20	4:25	Jo Cauvery Show	✓					✓
Night (to 5 AM Friday)	11:30	12:15	KADV		✓	✓			

If you didn't hear a radio today, please check here. ☐

PLACE

Check where you listen:
- at home
- in a car
- at work
- other place

Write down *all* the radio you hear. Carry your diary with you starting **Thursday, (Month, Day).**

No listening? If you haven't heard a radio all day, check the box at the bottom of the page.

FIGURE 20–1 Diaries are one method by which rating services collect information about the media audience, and diaries are especially useful for estimating radio listenership. This sample Arbitron diary explains how the listener completes the entries, which are then combined with data from other listeners' diaries. This information is then used to *estimate* the total listenership for a specific market area.

mote-control device also contains numbers that can record when guests are watching as well as the guests' demographic characteristics. Meters are capable of monitoring viewing through any device attached to the TV set, including VCRs, cable converters, and satellite dishes

INTERPRETING THE RATINGS

When the data from whatever collection method is employed reach the rating-service headquarters, banks of computers process the information and provide printouts in the form of ready-to-publish sheets that are bound in booklet form and made available to the local stations and networks. The computer data are also sold to advertising agencies and station sales representatives as an aid in making time buys. If a media buyer for an ad agency in New York wants to know on which stations to buy commercial time in order to reach the female audience eighteen to thirty-four years of age in Seattle, that information is available.

To understand how to use this information we are going to learn how to interpret ratings, learn some of the formulas used in these interpretations, and understand the meaning of the terms commonly used in reporting ratings data. This material is not difficult, and you do not need to know much math. But read carefully. The terms we will encounter sometimes have different meanings to different rating services.[7]

Rating

The first term we will tackle is *rating*. You may say, "But we've already been talking about ratings." That is correct. The term has become a cliché for all the processes employed in predicting viewing and listening habits. But that is a more general definition of the term. More precisely, a rating is the percentage of the people in a given population who are tuned to a radio or television station during a given time period. For example, the formula for determining a rating for a radio station is the population divided into the number of people who are listening.[8]

Station's Listeners divided by Population = rating(%)

To understand what the term "rating" means, assume that the town of Elmsville has a population of 10,000 and supports three radio stations. Using a random sample of the Elmsville population, we estimate 1,000 persons are listening to WAAA radio. Dividing 1,000 by 10,000, we find that WAAA radio has a rating of 10 percent, usually expressed as just "10."

1,000 divided by 10,000 = 10%

In the case of television, the term *television households*, which are *households with at least one television set*, is used in place of population. When a *television set is in use* in the household it is referred to as a *household using television* (HUT) and is used in place of a station's listeners in the formula. For our examples, however, and so you are not jumping back and forth between radio and television to understand the concept of ratings, we will use just radio to help understand the terms.

Share

Share is also expressed as a percentage—the number of a station's listeners divided by the number of all listeners during a given time period:

Station's Listeners divided by Total Listeners = Share

In determining share we do not use the entire population of Elmsville, only those in Elmsville listening to radio during a certain period. Suppose we find through sampling that between 3:00 P.M. and 7:00 P.M., half the population, or 5,000 Elmsville residents, had their radios on. Thus, although we found that WAAA had a rating of 10 percent, its share is higher. We determine this as follows:

1,000 divided by 5,000 = 20%

Another way to understand share is to consider it as part of the listening pie.[9] The entire pie represents all of the listeners to the Elmsville radio stations. A slice of the pie, 20 percent of the total pie, belongs to WAAA, which has 20 percent of the listeners. The other two stations have the remaining 80 percent of the pie.

Average Quarter-Hour Persons

Average quarter-hour persons is an *estimate of the number of persons listening to a station during any quarter-hour in a specified time period.*

Many ratings are based on four-hour time blocks, such as 6:00 A.M. to 10:00 A.M. or 3:00 P.M. to 7:00 P.M. For example, assume that for WAAA these time blocks are important because of the large number of listeners. To define average quarter-hour persons, we'll use the time block of 3:00 P.M. to 7:00 P.M. Between 3:00 P.M. and 7:00 P.M. there are 16 quarter-hours (four hours consisting of four quarter-hours each). Let us assume that our random sample of Elmsville found only twenty of the persons we sampled were listening to station WAAA, and that they *only* listened from 3:00 P.M. until 4:00 P.M. Perhaps a storm knocked the station off the air at 4:00 P.M., but, nevertheless, they listened only during four quarter-hours, the four quarter-hours between 3:00 P.M. and 4:00 P.M.

We also know that each person in our random sample of Elmsville represents twenty residents of Elmsville, since there are 10,000 persons in Elmsville and our sample is 500 (10,000 divided by 500 = 20). Because our sample showed that twenty people from our random sample were listening to WAAA, we can estimate that between 3:00 P.M. and 4:00 P.M., 400 were listening to WAAA (the twenty persons from the sample multiplied by twenty, the number of people represented by each person in the sample).

To find WAAA's average quarter-hour persons between 3:00 P.M. and 7:00 P.M., we take 400 and divide it by 16 (the number of quarter-hours between 3:00 P.M. and 7:00 P.M.). The answer is 25. Notice that our survey discovered that people were listening to WAAA only during the 3:00 P.M. to 4:00 P.M. time block, but since we are figuring average quarter-hour persons between 3:00 P.M. and 7:00 P.M., we projected those figures beyond 4:00 P.M. to 7:00 P.M., which is four hours with a total of 16 quarter-hours. We did this because when the ratings for Elmsville are published, all of the stations will compete in that same evening time block. The published ratings might also note that WAAA was off the air after 4:00 P.M. because of the storm.

Cume Persons

In determining the average quarter-hour persons we also determined the number of *cume persons* (*cumulative audience*), which was 400. *Cume persons* are defined as *the number of different persons listening to a station at least once during a given time period*. (By "once" we typically mean at least five minutes.) Cume persons, also called "reach," refers only to a specific time period and only to people listening at least once during that period. We never count a person more than once when we are figuring cume. For example, if we sampled people who were listening to WAAA between 3:00 P.M. and 4:00 P.M., and who turned their radio back on and listened again between 6:00 P.M. and 7:00 P.M., we would not count those people a second time during that time block.

Cume Rating

If we divide cume persons by the population of Elmsville we arrive at what is called "cume rating." *Cume rating* is *the cume persons expressed as a percent of the total population*. For example, to find the cume rating for WAAA between 3:00 P.M. and 7:00 P.M., we would divide 400 by 10,000. The number 400 represents cume persons, and the population of Elmsville is 10,000. Other calculations provide still additional information of importance to stations and sponsors who need to know who listens to radio and watches television. Other information can be drawn from survey data and offer many more interpretations than we can discuss here. Such information is a prerequisite for intelligent management decision making.

Subdivisions of Data

An actual rating entails many more subdivisions of data than those we have encountered in this discussion. You might compare women between the ages of twenty-five and forty-nine who listen between 3:00 P.M. and 7:00 P.M., or men between twenty-five and sixty-four who listen during that same period. A portion of an actual rating from an Arbitron survey (or *ARB*) of radio listening in Dayton, Ohio, shows how a listener survey is displayed for broadcast clients. The survey represents radio listening during the 3:00-P.M.-to-7:00 P.M. time block on Sunday. Notice how women are broken down into categories of twenty-five to forty-nine and twenty-five to sixty-four years of age. Other age-category breakdowns for both men and women are included in the complete survey.

Not all of the stations listed are actually in the Dayton area. Some are in outlying areas; others, such as WKRQ and WSAI, are in Cincinnati, southwest of Dayton. The survey focused on radio-listening habits in the Dayton area, and because some residents of that area listen to radio stations broadcasting from other communities, those stations are listed in the survey data.

Survey Area

A rating report lists categories for both the *total survey area (TSA)* and the *metro survey area (MSA)*—the geographic areas from which the random samples were drawn. The TSA and MSA are determined by a combination of data, including the signal contours of the stations, audience listening habits, and government census data. The TSA is larger than the MSA, encompassing outlying areas. The MSA corresponds more closely to the metropolitan district of a city. Look closely at the white area and the area containing horizontal lines in Figure 20–2. The area in white is the TSA. The area of the map in which horizontal lines appear is the MSA. The diagonal lines in the map represent the *Area of Dominant Influence (ADI)*, which is also referred to as *Designated Market Area (DMA)*. ADI and DMA apply primarily to television-viewing estimates. They are exclusively defined market areas (no two ADIs or DMAs overlap).

APPLYING RATINGS TO MANAGEMENT DECISIONS

Ratings by themselves are not valuable. Their importance lies in how broadcast management uses them to make decisions.

If you were a television station manager and the ratings showed that your station was second in the local news ratings, you would have a number of options. First, you could do nothing, leaving everything as it is. Your station might be profitable, and you would see little need to change. Another option would be to replace your news department with different personnel. A third option would be to replace the news with another type of programming you feel would be more profitable. Whatever your decision, it could not be made without a careful analysis of the ratings.

You might find that although your station was second in the category of all adults eighteen years and older, you had some other strength to offset this. You discover that among adults between eighteen and thirty-four, your station was first. The other station was reaching mainly adults forty-

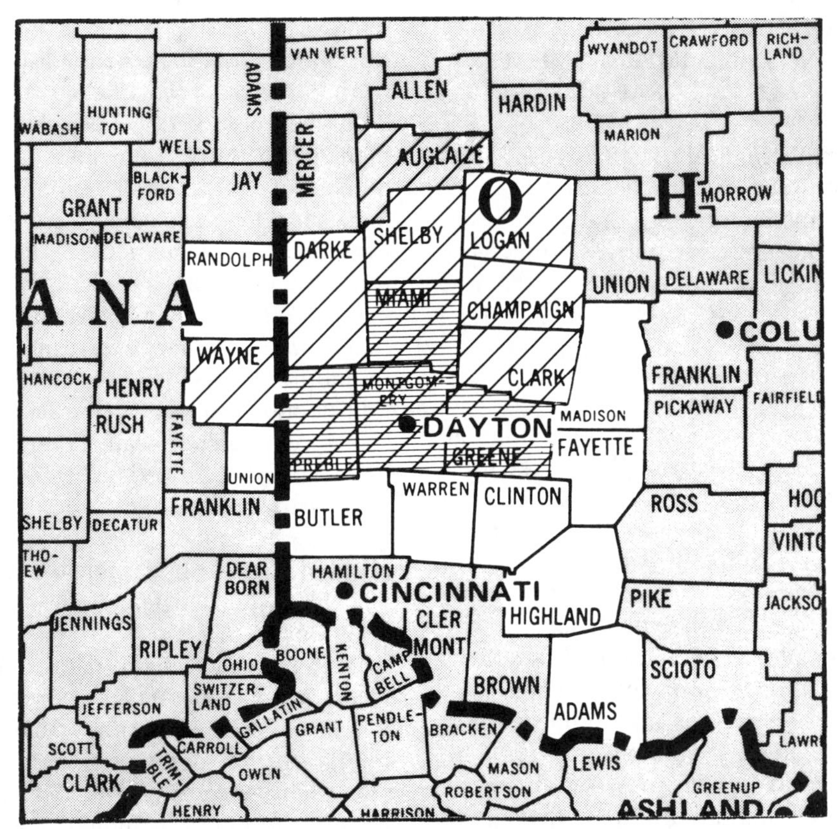

FIGURE 20–2 The area in white represents the Total Service Area (TSA), the area covered by horizontal lines represents the Metro Service Area (MSA), and the area covered by diagonal lines (which includes the MSA) represents the Area of Dominant Influence (ADI).

nine and older. You are very satisfied with your station's showing among the younger group. It is a relatively young audience that still is forming buying habits and making such major purchases as homes, home appliances, and automobiles.

Looking more closely at the ratings, you discover that for the time slot immediately preceding the local news, the other station is far ahead of yours. What does this mean? Perhaps the news department is not to blame for the second-place showing as much as the program that precedes the news is. Some viewers simply would rather stay in their comfy chairs than get up to change channels. So changing the program leading into the news may be all that is needed for your station to capture first place.

As a consumer of broadcast communication, you will immediately recognize that what you see on your TV screen is the result of management's statistical scrutiny. A legitimate criticism of this decision making is that it creates bland programming that will please the general public. That is true. Commercial broadcasting is a business, and nowhere does this become more evident than in the use of broadcast ratings.

CRITICISM OF BROADCAST RATINGS AND IMPROVEMENT OF ACCURACY

Despite mathematical formulas and efforts toward accuracy, the broadcast ratings continue to come under considerable criticism. Some of the criticism derives from ignorance; some is well-founded.

Sampling Error

One criticism of ratings focuses on the actual sample. Statistics tell us that although there may be a given sampling error for a random sample of an entire population, that sampling error increases as the population shrinks. For example, if you have an error of ±4 percent for the total sample, that sampling error grows as you divide the sample into smaller units, such as those of sex or age. Although your total sample may have been 600 persons, it may include 100 teenagers. Thus the sampling error for teenagers is based on 100, not 600. Criticism occurs when stations promote themselves using small samples that have high margins of error. Although advertising agencies and people who work in the media are familiar with sampling error, the small retailer or the public is sometimes fooled into thinking a station has a strong position in the marketplace when in fact it may not.

Data Collection

Data-collection procedures come under criticism, regardless of which method of data collection is used. Diaries, for example, are criticized because people don't fill them out until long after they have watched a program or listened to a particular radio station. When measuring television, this delay can result in only reporting the most popular shows or those that have the most impact on the viewer.[10] In radio, the delay can cause errors in reporting listening away from home or at work, even though diary users may be asked to keep the diary with them when away from home.[11] The large number of choices in television viewing also makes it difficult to complete diary entries. Moreover, for some people, the push buttons on a radio dial and the remote-control keypad of a TV set can be changed so easily that filling out a diary every time you switch to another station is unrealistic.

Telephone surveys are criticized because some are based on recall. Trying to remember what radio station you listened to twelve hours earlier can be difficult. Different interviewer techniques, the way questions are phrased, and not being able to reach people who have unlisted telephone numbers are additional criticisms.

Meters, while reliable data-collection devices, have come under scrutiny because young children sometimes have difficulty operating the meter's remote-control device, thereby causing young children to be underrepresented in audience estimates.[12]

Minority Audiences

Because it is necessary for a station to reach a certain percentage of the total audience before it can even be listed in the ratings books, those that reach small, specialized audiences can suffer. When a station's ratings do not show any listeners, an advertiser's reluctance to spend dollars on that station is understandable. Some rating services use a certain percentage of listenership or viewership, below which a station is not reported. The problem can be especially troublesome where minority audiences are reached by stations with specialized formats that may not be reported. In large markets, the audience for these stations may run into the tens of thousands.[13]

Some criticism has made a difference. One rating service withdrew a market report after receiving complaints that the sample was not adequate to report the Spanish-speaking audience. The rating service attributed the problem to field-staff members who helped viewers fill out diaries. Efforts to correct these alleged deficiencies do exist, such as having diaries printed in Spanish.

IMPROVING BROADCAST RATINGS

Because of the developments in cable television, people's changing lifestyles, and the sophisticated needs of clients, efforts continue toward improving the accuracy of broadcast ratings. Some of these efforts are credited to the services themselves such as the use of more sophisticated metering devices, which can even track VCR use.

Tracking VCR Use

Increased use of VCRs has resulted in sophisticated metering systems that can track VCR recording and playback. Such information is critical in making accurate estimates of viewership for specific programs. In addition, the manufacturer's code placed on videotapes can be read by these meters and that can also help verify VCR use, even of rental tapes. Information on rentals is valuable to advertisers who want to place commercials on rental tapes.

Tracking Cable

Along with VCR use, cable has been a particular problem for the ratings. Not only are channels from distant markets often preempted for local programming, as with over-the-air broadcasting, but some cable operators also supply different stations at different times on the same channel. The major rating services continually attempt to distinguish the viewing habits of cable and noncable subscribers in their reports. Some services have major departments whose sole responsibility is to verify and update cable programming schedules so that the audience estimates for such programming are as accurate as possible.

The Future: Infrared Scanners

In an attempt to develop a truly passive and unobtrusive metering system, researchers are working on infrared scanners. These devices would typically be placed on top of a TV set and silently scan the room to determine when someone was actually watching television as opposed to just having the TV set on and being out of the room. Sophisticated technology may make it possible for the scanner also to distinguish between different people and between people and pets who may also be in the vicinity of the TV set.

Improved Reporting of Demographics

Sampling techniques combined with other information permit rating services to give their clients more detailed *demographic* information. For example, demographic breakdowns such as income, education, occupation, and additional characteristics are available, even within Zip Codes and more specialized geographic areas such as the blocks used by the U.S. Census Bureau. Such specialized target-audience measurements may make specialized rate cards more commonplace. For example, if you own a neighborhood store in a high-income residential area, you may be able to purchase commercials based on your client's viewing habits. A television account executive may show you a ratings book indicating that a certain weekend sports program has a large viewing audience in your customer's Zip-Code area. You thus could receive a special advertising rate based on that program's "broadcast circulation." It would be cheaper than the rate for reaching the entire viewing audience but perhaps more than you might pay for reaching a lower-income audience.

MONITORING QUALITY: THE ELECTRONIC MEDIA RATING COUNCIL (EMRC)

Concern over the accuracy of the ratings and their influence on national programming prompted congressional hearings on these issues in 1963. The publicity generated by the hearings focused on the industry's need for a systematic, self-regulatory body that would assure confidence in broadcast ratings.

Formation of the EMRC

In 1963 the *National Association of Broadcasters (NAB)*, with the blessings of the American Association of Advertising Agencies and the Association of National Advertisers, joined with ABC, CBS, NBC, and Mutual to form the *Broadcast Rating Council (BRC)*. In 1982 the name of the organization was changed to the Electronic Media Rating Council. The council's main duty is to give the industry and the rating services credibility, assuring advertisers that radio and television are indeed reaching the audiences they say they are reaching.

Audits and Operation

Ever since its inception the council has policed the rating services, granting or revoking the seal of approval it gives them and their surveys. The council coordinates a major audit of rating-service practices, paid for by the services themselves. These audits cover all phases of the rating process—the development of sample design, the gathering of data, data processing, and the published report. The audits are unannounced and cover various market areas selected by the auditors. In return for the council's accreditation, the accredited services agree to (1) provide information to the council, (2) operate under substantial compliance with council criteria, and (3) conduct their services as they represent them to their subscribers and to the council.

Although they continue to receive criticism, the rating services are still the only source of reliable, economical broadcast audience data. And rating is done quickly. A network executive can walk into the office on Tuesday morning and find Monday night's ratings on the desk. A local broadcaster can obtain data on local listeners without having to conduct the ratings or to hire, train, and supervise someone else to do it. Continual efforts are being made by broadcasters, networks, and rating services alike to improve the quality of this important area of feedback for the industry.

SUMMARY

Early attempts at ratings occurred in the 1920s and surveyed radio listening habits. Later, the Cooperative Analysis of Broadcasting was the first true rating service and was supported by the Association of National Advertisers. CBS spearheaded audience research in the 1930s and Arthur C. Nielsen began measuring radio listenership in 1936. *The People Look at Radio*, a major national study of attitudes toward radio, was published in 1946.

Media industries subscribe to ratings services to determine the size and composition of the viewing and listening audience. This information is a form of feedback to the industries to assist in making management decisions that eventually affect the consumer. Costs of reaching an audience are usually figured in cost-per-thousand (CPM), of which ratings provide the data on the size and composition of the audience.

The public still feels that ratings are susceptible to error. Although data on audience composition are estimates, data nevertheless are quite accurate when good sampling procedures are used.

Sampling, and specifically random sampling where everyone has an equal chance of being selected, is at the heart of the ratings. From sampling we are able to determine a station's rating and share.

Three primary methods of data collection are used in broadcast ratings: interviews, diaries, and meters. Meters are receiving widespread use in monitoring TV viewership. Meters can be pre-programmed with demographic information and activated by a remote-control device with a keypad on which the person watching the program can log in his or her identification number.

Despite improvements in data collection and reporting, ratings are still criticized in such areas as sampling error, data-collection techniques, and the reporting of specialized and minority audiences. Both subscribers to rating services and the rating services themselves work continually to improve the quality of broadcast ratings. Some of these efforts are centered in the work of the Electronic Media Rating Council (EMRC), which audits member rating services and accredits those that measure up to EMRC standards.

OPPORTUNITIES FOR FURTHER LEARNING

Beville, H. M., Jr. *Audience Ratings: Radio, Television, Cable*, rev. ed. Hillsdale, NJ: Erlbaum, 1988.

Goldman, R. J., and La Rose, R. *Assessment of Audience Feedback Systems for Research and Programming*. Washington, DC: Corporation for Public Broadcasting, 1981.

Myrick, H., and Keegan, C. *Boston (WGBH) Field Testing of a Qualitative Television Rating System for Public Broadcasting*. Washington, DC: Corporation for Public Broadcasting, 1981.

Rating services such as Nielsen and Arbitron publish booklets explaining their services and methodologies.

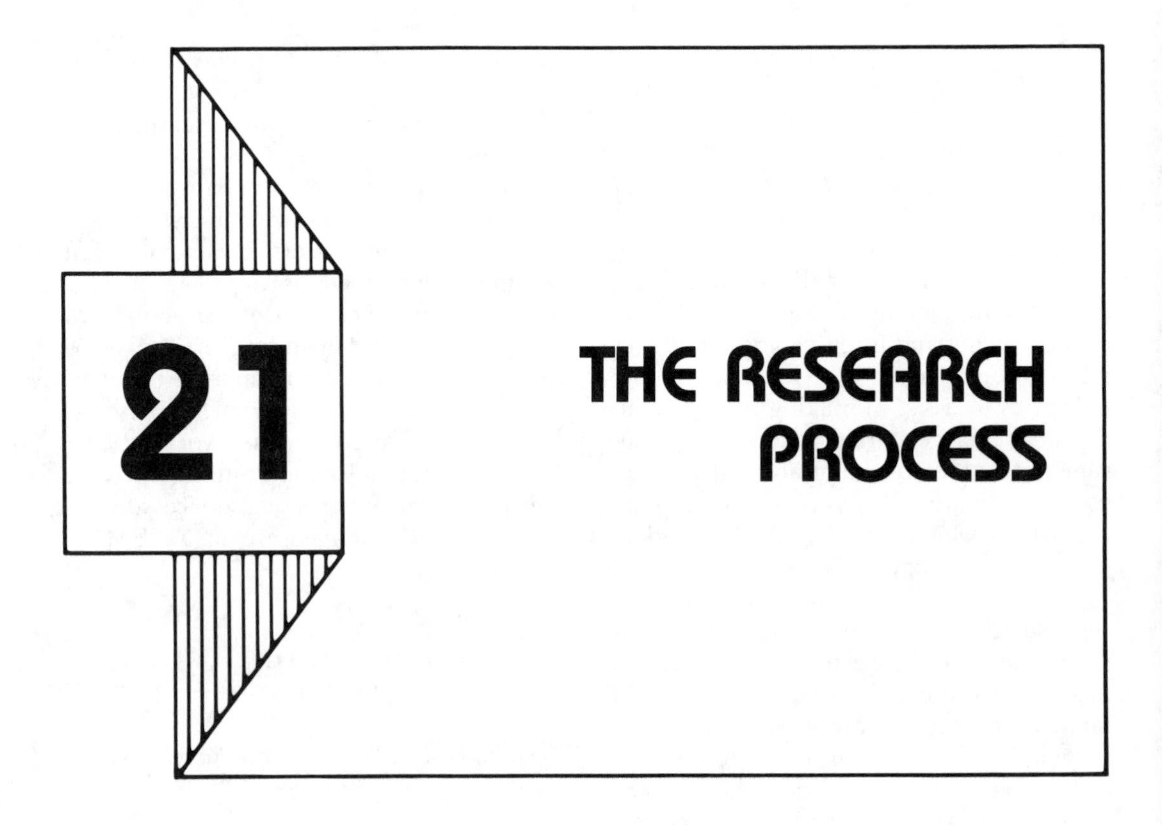

21 THE RESEARCH PROCESS

To many people the word *research* conjures up images of bloodshot eyes poring over unsolved problems, stuffy laboratories with bubbling flasks, incomprehensible calculations, and all-night study sessions for impossible exams. Even related words can cause the research jitters—words such as *theory, numbers, experiments, statistics*, and *computers*.

The jitters would subside quickly if we stopped long enough to realize that we use research every day. We conduct research, we make decisions based on research, and we are affected by research. Turning to the radio–TV section of the newspaper is a form of research. You survey the available programs and decide which ones to watch or not to watch. When you decided to enroll in college you may have examined college catalogues, looked at college guides, written letters to different college admissions counselors, and even had on-campus interviews. In each case you were performing research.

The professional researcher in a major television network does the same sort of thing every day. He or she may look through back issues of annual reports so as to establish the image of the company, much the way you examined college catalogues. The researcher might send letters to affiliate stations asking local managers to write back and describe network service. Just as you wrote to the admission counselors and studied their responses, the researcher will examine the managers' responses. If the researcher needs additional information, he or she may interview the manager over the telephone or even visit the station and talk to the manager in person.

Certainly broadcasting has much more complex and sophisticated research projects than these. However, with some training and the opportunity to conduct research projects of your own, you will be able to direct the same type of research undertaken by many local broadcasting stations. So the next time the word *research* pops up, don't be afraid of it. By learning more about how research takes place in the field of telecommunication, you can open up a whole new frontier of exciting knowledge and experience.

TYPES OF RESEARCH

We first should become acquainted with the different types of research—historical, descriptive, experimental, and developmental.

Historical Research

As the name implies, *historical research* focuses on the past. For example, you might trace the history of one of the radio stations in your community. Perhaps a local broadcaster is nationally famous as an industry pioneer. A study of his or her professional career would be a historical study. Historical studies are not necessarily studies of something ancient. A research study of the first five years of WAAA radio would be considered historical even if WAAA signed on the air in 1986.

Descriptive Research

Descriptive research describes a current condition. The most common type of descriptive research is the *survey*. A survey of local listenership and a survey of television sets in use are examples of descriptive research. Descriptive research is one of the most common types of research within the broadcasting industry.

Qualitative Research

Qualitative research has its roots in the humanities. It is concerned with describing the meaning found in the messages of our culture such as radio and television programming.[1] Research tools such as statistics, used in experimental research, are not usually found in qualitative research. Also referred to as *cultural studies*, qualitative research examines the context within which *mass communication* occurs and the way such communication has been used by people to help organize their lives. For example, if we analyze the production variables, the scripts, and the motives of writers and directors who produce a series of police shows for television, and then make our own assumptions about what meaning these programs have for actual working police officers who watch the shows on television, we are engaging in qualitative research. Qualitative research is primarily the domain of the university researcher, the scholar, the person examing the broader context of media messages in our culture.

Experimental Research

Experimental research is sophisticated research that entails a "controlled" experiment. For example, we might compare children's aggressive behavior before and after they watched TV programs containing violence. We would select two groups of children. One group would watch a TV program containing violence while another watched the same program but with the violence deleted. We would then compare the behavior of the two groups. Two terms used frequently in experimental research are *independent variable* and *dependent variable*. An independent variable is the factor that is manipulated in the research study. In our example, it is the episodes of violence in the television program. The dependent variable refers to the phenomenon of change, which in our example is the children's behavior.

Developmental Research

Developmental research is conducted mostly in the field of instructional television. Continually perfecting an ITV program on teaching tennis is an example of developmental research. The researchers first establish certain objectives. Then they develop an instructional radio or television program, producing, testing, revising, and retesting it until it meets the stated objectives.

THE SCOPE OF TELECOMMUNICATION RESEARCH

Telecommunication research reaches nearly every facet of the industry. Aside from government research studies and those conducted by such research corporations as A. C. Nielsen, telecommunication research extends from colleges and universities to local stations.

Research in Colleges and Universities

Some of the most sophisticated research on every aspect of broadcasting is done in an academic setting. Many colleges have major research centers that concentrate on radio and television. Among these centers are the Communication Research Center, Broadcast Research Center, Division of Communication Research, Center for Media Research, and Institute for Communication Research. These centers range from depositories for current studies and scholarly research papers to institutes that administer major research contracts examining everything from the social effects of broadcasting to technical problems associated with fiber-optic transmission. Doctoral dissertations provide another opportunity for research, as do research projects undertaken by students and faculty. Many individual studies appear in scholarly journals.

Research in academia is funded by a multitude of sources. Major foundations are active in funding broadcasting research, including the Ford Foundation, the Rockefeller Foundation, the Lilly Endowment, and the John and Mary Markle Foundation. The networks, corporations, and the federal government contract with institutions of higher learning to do the research for projects of special interest to the funding organization.

The Networks and the CPB

Although we are more attuned to the networks' prime-time programming, their research arms are an integral part of their total operation.[2] One of the most famous early network research projects was the one on radio listenership started in the early days of CBS by an Ohio State University professor named Frank Stanton. Stanton, who was hired by the network, went on to head CBS. Even in the broadcast networks' early years, research was clearly defined. Duties assigned to NBC's research department in 1948 were broken down into five areas developed under a Master Plan for Television Research:

1. The audience: its size, characteristics, and viewing habits.
2. TV stations and their coverage.
3. Programs: the contribution of research to their better selection and presentation.
4. Advertising: measuring TV's advertising effectiveness.
5. The social impact of TV: its effect on the family and children, the psychological effects of TV, public attitudes toward TV, and TV as an educational medium.[3]

Noncommercial broadcasters also engage in research. The Public Broadcasting Service (PBS) conducts research in the areas of system information, system finance, and audience evaluation. The network (1) provides information to stations and other PBS departments using federal monies distributed by the Corporation for Public Broadcasting to public stations; (2) coordinates information on program funding and calculates such information as PBS broadcast hours, hours of programming produced by independents, programming acquired from foreign producers, programming sold to foreign countries, and programming hours intended for or dealing with minorities and women; (3) answers requests for information from such places as other PBS departments, Congress, the White House, the public, the press, and book publishers; (4) measures the extent of usable signals transmitted by public television stations and the demographic breakdown of the population in each signal area; (5) provides prompt accounting of each affiliate station's use of PBS programs; and (6) publishes reports on PBS activities.[4]

Local Stations

Local stations engage in applied research, and many stations have in-house research departments. Research at local stations is designed to gather information on such things as the demographic characteristics of the audience; the buying habits of the audience; the use of products and services, such as how many people who use a certain product also listen to or watch the station; how many radio or TV sets are in the market served by the station; the size of the station's audience; public-opinion polls; and the effectiveness and popularity of on-air personalities.[5]

STATION-IMAGE SURVEYS

The effectiveness and popularity of on-air personalities is part of the station's image, and station-image surveys keep track of what the audience thinks of a station's programming and personalities. Station-image surveys, while being combined with ratings data, go beyond reporting just the number of people who listen to or watch a given station.[6]

Researching Programming

Imagine you are managing a TV station and discover that certain programs are not getting a satisfactory share of the viewing audience. But you do not know why. Ratings tell only who is viewing what and when. They do not tell why a person likes or dislikes a certain program. A station-image survey seeks to find out why people do or do not watch or listen to a station. It explores such things as attitudes toward, and opinions about, programming and personalities, and it helps management make decisions. For example, if you discover that a program is being watched by a large share of your audience but find in a station-image survey that the audience actually rates the program very low, then you could have a problem. The audience may be watching that program only because the competition is airing even more dismal programming. If the competition were to change its offerings and insert a popular program instead, you could have your audience swept away overnight.

Researching On-Air Personalities

Assume as a program director you discover that, although you seem to be capturing a large share of the audience for most of the broadcasting day, that share tends to dip during the local news. Why? You have already conducted research comparing the number of stories your news department produces with the number produced by the rest of the stations in your market. You know your station has been consistently ahead of the others. So you decide to design a station-image survey that will, among other things, examine what the audience thinks of some key on-air personalities.

Your survey will examine what the audience thinks of various on-air personalities, including the person who gives the weather, the sportscaster, and the anchor person. The results of your survey show strong positive attitudes about the sportscaster and weather person. The news, however, is weak, both in programming and the audience perceptions of the anchor. You decide not to make any personnel changes in sports and weather. The sportscaster is not new to the market. She has been anchoring sports on another local station, is well respected, and has a large following. The person giving the weather has been with the station for more than twenty years and is also well respected.

But what do you do about news? You conduct a follow-up survey and find that the public perceives your station as having only a few reporters. Although your station has a full staff of reporters, the audience is not aware of it. Second, you discover that the public thinks the competing stations cover national news much better than your station does. Finally, you find out that when there is fast-breaking news, the audience consistently turns to other stations.

Making Decisions Based on Image Surveys

Working with the news director, you decide to begin using more of your news staff in an on-air capacity; thus, you are able to eliminate the misconception that your station is understaffed. You begin a promotional campaign with billboards and newspaper advertisements promoting the station's newscasts and you remodel the news set to make it more visually appealing. A national news story is inserted at the beginning of each newscast. You also schedule some live reports during the day from the scenes of fast-breaking news events.

Six months later you begin to see the fruits of your labor. Your share of the audience for your local news now equals that for the rest of your programming. Without the station-image survey your station's news might still be in the cellar. A year later another image survey shows all aspects of your programming are highly rated and the station's overall image has improved substantially.

Methods of collecting data for image surveys are as varied as the surveys themselves. Remember, however, that most station-image surveys are designed to dig deeper than the ratings do. Sophisticated computer analysis of this type of questionnaire can tell which programs are popular and how they compare with other programs.[7]

SALES AND MARKETING RESEARCH

The sales manager had worked for months on the account. But the drugstore owner was convinced the newspaper was the only place to advertise and could see no reason for trying the broadcast media. When the drugstore had advertised on the station there was no noticeable increase in sales.

The sales manager went back and researched the ad copy in the drugstore's commercials. The commercials told how long the drugstore had been in the community and how good the service was. Further research into the newspaper ads used by the drugstore showed a single product was featured with a sale price. The sales manager knew that if she could convince the druggist to advertise on the station and feature a single product at a sale price, the druggist would see results. The druggist decided to give the station one more try. The commercial advertised suntan lotion at a sale price and the product sold out the first day the commercials aired.

The sales and marketing research that examined the ad copy of the commercials and the content of the newspaper ads was necessary for the sales manager to convince the druggist to advertise on the station. Much more sophisticated sales and marketing research is possible, including learning about which products are bought by people who tune to a given station and even breaking down such information into Zip-Coded areas and demographic characteristics.

FOCUS GROUPS

Some stations find *focus groups* an efficient way to conduct research on listener and viewer preferences. Focus groups are small assemblages of people brought together to discuss, under the guidance of a trained group leader, topics selected by the station. The station staff records the discussions and analyzes them for trends in audience opinions about the station and its programming.

BEHIND THE SCENES OF A NETWORK ELECTION POLL

"They told me at the doughnut shop we had a reporter from NBC here, and by golly we do!" The pollster looked up from her interviewing sheets to see a weathered farmer in bib overalls smiling as if he had just reaped the biggest corn harvest in three counties. The woman's quick explanation that she was just freelancing for NBC as a precinct pollster for this election day did not seem to bother the farmer in the least. It was obvious that his stories of meeting a real live "network person" would be heard in front of many a crackling fire.

Planning the Poll

Public-opinion polls have become very sophisticated, and in radio and television are used frequently by broadcast journalists, especially at election time.[8] We take for granted the election-night predictions of state and national winners that are made only a few hours after the polls close and before all the votes are counted. We no longer sit in amazement as computers, network commentators, and sample precincts determine the fate of democracy. We expect it.

What is behind these public-opinion polls? It all starts in the network planning rooms, where statisticians and polling experts pore over mounds of computer data, analyzing voting trends in every state. How many persons voted in which county for what candidate? How do these figures correlate with state and national voting trends? Do they correlate often enough to produce a sample precinct, that tiny subdivision of each community used to predict a winner long before the local election officials finish their work? A sample precinct may be found in the heart of Los Angeles or a tiny Maine fishing village. They are chosen by the same random-sampling techniques we learned about when we studied broadcast ratings. Together, they become part of the state-by-state predictions of House and Senate, gubernatorial, and presidential elections.

Selecting the Pollsters

Equally critical to the election poll are the precinct pollsters selected by the networks to interview voters. Some of the pollsters are associated with local broadcasting stations and are invited by the network to represent them at a polling place in their area. At the network's office in New York, other people are in charge of coordinating this national group of volunteers, who will receive a modest fee for their day's work. Coordinating a national core of pollsters and training them by mail are big jobs and must be carefully planned and executed.

Planning Interviews

Imagine yourself as an interviewer participating in a network election-day poll. What will your job be like? It will actually start about a month before the election with a call from a network executive asking you to represent the network as a precinct pollster. About two weeks later you will receive a large packet of instructions. You will be asked to read the instructions carefully and to call a special telephone number if you need advice or if something unexpected happens. The number belongs to a state supervisor hired by the network. The instructions will tell you where you will conduct the polling, give you the exact precinct and voting location, and list the name and telephone number of the local precinct official, along with a suggestion that you contact that person before election day. Because you are just one part of a national systematic effort you will be asked to follow the directions very carefully. You have two responsibilities: to conduct the poll and to report the results to the network.

Each pollster is assigned certain periods in which to conduct the precinct poll. For example, you might be assigned to interview people coming out of the polling place between 6:00 A.M. and 9:00 A.M. and then call in those results to the network at 9:15 A.M. The second interview time might be between 12:30 P.M. and 1:30 P.M., in which case you would call in your second report to the network at 1:45 P.M.

You will also be told which sequence of voters to interview—for instance, every fifth or every third voter. The interval depends upon the precinct's size. Your instructions would also tell you your quota of completed interviews—perhaps thirty within those two time periods.

Conducting the Poll

Finally election day arrives. In the predawn darkness you start your drive to the polling place, leaving yourself plenty of time to get there. You have been assigned to a precinct in a small rural area. The morning sun is just coming up when you roll into the sleepy little town. You pass an old general store, a gas station, and a cafe where a few people are downing their first morning coffee. The precinct voting place is in the town library. You have seen such buildings before, one of the old Carnegie structures with an American flag outside. Inside, the precinct workers all are ready for the first voter. You are greeted by the local precinct official you talked to earlier on the phone. She eyes your official NBC badge and asks, "How about a cup of coffee and some of Mabel's cookies?"

You accept and begin to wonder what the day will bring.

"Who are you going to interview?" the precinct official asks.

"I'm supposed to interview every third person, starting with the first one to vote after 6:00 A.M."

Before the precinct official can answer, Mabel says, "I'll tell you who that's goin' to be." She has a big grin on a face that reveals years of toil on a farm. "You can betcha Harvey Clodfelter will be the first through that door. Twenty years and he ain't missed being first yet."

You sense that election day is one of those special days ranking with the church suppers, the school picnic, and the beginning of the county fair. It seems light years away from the network control center in New York, where computers and statisticians are adding the finishing touches for the marathon telecast to follow that night.

But here is where it all begins. You start, sure enough, with Harvey Clodfelter. Your day will be filled with Harveys. They will ask you who you are, where you live, why you are there, and when you will be back. They will offer you their own brand of election philosophy, tell you why this candidate or that candidate will win, explain why the electoral college should be abolished, and complain how the big-city folks are going to sway the election. To each of them you will give a ballot, similar in some ways to the one they have just completed. The ballot will be labeled SECRET BALLOT, and when they have finished filling it out they will put it in a big envelope. Later you will open that envelope and tally the results.

The ballot requires information that is much more detailed than merely telling for whom the respondents voted. It asks them whether the TV debates affected their choice, whom they voted for in the last presidential election, their ethnic background, the occupation of the head of their household, and such basic demographic information as age and sex. The ballot also asks what issues influenced their decisions, what the candidates' positions on the economy and foreign policy are, and whether those positions affected their decision.

If Harvey Clodfelter, the first person to finish voting after 6:00 A.M., refuses to be interviewed, then you continue counting until you reach the next interval. Remember, your goal is to interview every third person. It is a hard day. Some local people do not trust outsiders, and you get completed interviews from only three of those whom you have asked. Nevertheless, you keep working toward your quota. You make certain you offer a questionnaire only to those individuals who have actually voted. That is one advantage the election-day polls have over preelection polls. You are also careful not to interfere in any way with the actual election process—voting and the checking of voter registration.

So as not to influence the results of your poll, you try to remain polite while refraining from talking about candidates, issues, political beliefs, or anything else that might bias the results. The only information you will supply is on the mechanics of completing the secret ballot; you will not answer any requests for political information. If a person does not understand a question or the directions, a polite suggestion to reread the ballot, not a rephrasing of the question or directions, is necessary. In other words, you will not want to influence, either by word or action, the response of the person being polled.

Phoning Results to the Network

At precisely 9:15 A.M. you walk up the stairs to the library's office and call New York with the results of your first poll. To help make sure every item is understood by the data collector, you use word identifications for each of the possible answers: Alpha for A, Bravo for B, Charlie for C, Delta for D, and so on. A typical conversation might go like this:

NBC OPERATOR: "NBC. Your location code, please."
INTERVIEWER: "This is Ohio, location 400." (The NBC operator will then repeat your code and give you the name of your polling location in order to be sure NBC has the correct information.)
NBC OPERATOR: "I'm ready for the answers."
INTERVIEWER: "Question one, Alpha."
NBC OPERATOR: "One, Alpha, Okay. Go ahead."
INTERVIEWER: "Two, Charlie."
NBC OPERATOR: "Two, Charlie, right."
INTERVIEWER: "Three, Delta.[9]"

You then walk back down the stairs, nibble another cookie, and wait until the specified time of 12:30 P.M., when you will repeat the procedure.

In New York, the data from your poll and those of other pollsters scattered throughout the United States are fed into a computer, which tabulates them and predicts the winners. So why does the network not start predicting at noon on election day? The answer to that is a matter of geography. Because the West, owing to the time-zone difference, votes later than does the East, such predictions could influence western voters. A news report stating that a certain candidate has won by a landslide in New York may convince people in California that they should also vote for the probable winner. On the other hand, that news might make the candidate's California supporters so complacent that they stay home and give the opponent the election. Let us examine this reporting process more closely.

REPORTING PUBLIC-OPINION POLLS

As either a consumer of broadcast communication or someone who will someday report polls, you should be aware that there is more to an opinion poll than just the results. A gullible public and press are often the objects of persuasion campaigns based on public-opinion polls. Broadcast journalists, partly because of the time restrictions placed on them, find it difficult to report all of the pertinent data from an opinion poll. Because of this, the following scenario could very likely happen. A wire service reports that a senatorial candidate is leading his opponent by 20 percent. The radio station reports the story, only to find out later that the poll consisted of twenty-five persons who stopped by for coffee at the candidate's office. The sampling error alone would make the results questionable, to say nothing of the probable bias. We can assume that anyone who takes the time to have coffee at a candidate's office is not vehemently opposed to that person.

Reporting a public-opinion poll can be just as important, perhaps more so, than conducting the poll itself. Both the American Association of Public Opinion Research and the National Council on Public Opinion Polls have adopted guidelines for reporting public-opinion polls. An emphasis on "precision journalism" has also fostered more responsible and intelligent interpretation of polls.[10] Other guidelines have been offered by both journalists and educators. The aim of this attention is to educate the public and the media in the polling process. Even if the press is unable, because of space or time limitations, to report exactly how the poll was set up, it should be able to decide intelligently whether even to use the poll, and if so whether it warrants detailed interpretation.

What are the important considerations in reporting polls?[11] The first is the sample. If it is not a random-probability sample, then it may be questionable. In other words, for the poll to be credible, the probability of one person being selected should equal that of all other persons who might be selected. If the sample covers only your state, then it is not good sense to apply predictions based on it to the entire country.

A second important consideration is who sponsored and conducted the poll. A poll favoring a Democratic governor that is sponsored by the state's Democratic committee and conducted by Democratic precinct workers would be suspect.

Third, it is important to define the sample population. If a poll taken among Maine voters states that the senator from Maine has a 60 percent chance of winning his party's nomination for president, the chances are that the estimate is inflated. Knowing who was sampled permits a better judgment of the poll's results.

A fourth consideration in reporting polls is how the respondents were contacted. A telephone survey, for example, would automatically eliminate all those who could not afford to have a telephone.

Fifth, it is important to know what questions were asked. "Will you be voting for our fine Senator Claghorn?" will elicit answers favoring Senator Claghorn.

Sixth, it is important to know how many people were surveyed. In this way we can determine the margin of error.

Seventh, the poll should distinguish samples and subsamples in interpreting its findings. For instance, a poll may be reported to have a sampling error of ±4 percent, based on a random sample of 600 persons. However, if 50 of those respondents were Indians, it would not be correct to predict how the Indian population would vote and still claim the sampling error was ±4 percent. Because only 50 Indians were sampled, the margin of error would be closer to 14 percent.

Eighth, it is important to report the error margin. Merely knowing what it is but not telling the listening or viewing audience has little purpose. Error margins are one of the best-understood factors in an opinion poll. If you claim that a candidate has an edge of 5 percentage points over his opponent but do not tell the public that the margin for error is 10 percent, then you have misled them. Moreover, if you predict a winner when the margin of error makes a contest too close to call, then you also have misled the public.

A ninth important consideration in reporting polls is how the sample compares with the total population. If the population is 50 percent Hispanic yet only 33 percent of your sample is Hispanic, then you would be asking for trouble by predicting the election without compensating for that variation.

The tenth consideration is the all-important headline. People can be persuaded by headlines. The radio or television reporter who announces, "Opinion poll predicts Senator Doolittle will sweep the state," and then comes back after the commercial to report a poll of the senator's precinct workers has been irresponsible.

Public-opinion polls are not foolproof, and they can be used incorrectly. The key is to understand them and know their pitfalls. The broadcast manager who editorializes in favor of a candidate may be in hot water if he or she predicates that support on a sloppy opinion poll. A TV news director who sends her staff out to conduct an opinion poll on election eve and violates all the rules of random sampling is asking for trouble if she reports the results of the poll as though they were representative of the total population. Take the opportunity to read more about opinion polling. Design a sample poll. Ask yourself if it meets all of the criteria discussed above. Would you be able to draw general conclusions from your results?

ALTERNATIVE FORMS OF DATA COLLECTION AND APPROACHES TO RESEARCH

New technology is accelerating the polling process even beyond the computer analyses so common in today's surveys.[12] With this technology broadcasters will be able to conduct opinion polls faster and with fewer personnel than in the past. One of the newer polling systems uses a series of tape recorders and a computer. The computer dials a telephone number. When the phone is answered, the tape recorder takes over and plays a recorded message to the party answering the phone. It then records the person's response. Of course, the results of such a poll are only as accurate as the information being asked. If the questions are poorly phrased or if there are errors in the sampling procedure, then it is difficult for the poll to reflect the opinions of the population accurately.

Researchers have used an electronic device called an *oculometer* to examine how people watch TV commercials. Utilizing a sensor that monitors the eye movements of viewers, the oculometer has shown, for example, that viewers are attracted immediately to people in TV commercials, that their attention span is rather short, and that certain visual elements in a television commercial can actually distract the viewer from the product being advertised.

In research centers, colleges and universities, advertising-research firms, and almost any other place where broadcast-related research goes on, paper-and-pencil questionnaires are often being replaced by still other electronic data-gathering instruments.

Electronic Response Systems

Electronic data gathering, a process in which a member of an audience punches a button or has his or her pulse read electronically, is not new to broadcasting. In 1937 Frank Stanton of CBS and Paul Lazarsfeld, a well-known social scientist, developed Big Annie, perhaps the first *electronic response indicator (ERI)* used to measure how an audience reacts to radio shows. The listener could push one of three buttons labeled *like, dislike*, and *neutral*. Big Annie had a short life, but it did spark interest in electronic data gathering.

One instrument designed to gather data was invented by a professor of telecommunication at the University of Oregon, Elwood Kretsinger. He called it a *chi-square meter*. It simultaneously records an individual's responses to whatever he or she happens to be listening to or watching. Researchers compare responses among different individuals or groups by using a statistical test called the chi-square. The chi-square meter was not Kretsinger's first venture into response systems. In the late 1940s he invented a device called the *wiggle meter*, which measured how much people fidget. The device had three components. A wire strung on the back of a chair was fed into an amplifier, which in turn measured the amount of wiggling done by the person in the chair. Results were fed onto a long roll of graph paper.[13]

The increased interest in ERIs has resulted in a number of commercial firms that manufacture the systems. They are usually installed in a large auditorium, one to every seat or desk. Each unit has five buttons that can electronically record anything from an answer to a five-part multiple-choice question to one of five attitudinal responses such as "strongly agree," "agree," "neutral," "disagree," and "strongly disagree." The entire auditorium of ERIs can be monitored by a computer and can register continuously the response of the entire room or sections of the room, or calculate statistical comparisons between various groups of people.

Assume you have produced an instructional television program and now want to test its effectiveness. You arrange for a physical-education class to use a room equipped with ERIs, and at periodic intervals in the program you ask the class members to respond to a series of questions with which they will evaluate your program's effectiveness. When the program is over, a computer printout tells you how the class scored on every question. Those questions the students missed may indicate that you need to revise one or more sections of the program. If you had not been using ERIs, you could have waited until the program was over and given the students a paper-and-pencil test to evaluate what they had learned. However, you then would have had to score the tests and calculate the statistical computations yourself or wait until they could be done by computer.

Galvanic Skin Response

Not all of the methods that test audience reaction necessitate pushing a button or wiggling. There are also systems that measure changes in the conductivity of the skin, which are called *galvanic skin responses (GSRs)*. One system uses a computer to analyze the data from about eighty subjects wired for GSR readings. The developers claim that it enables radio stations to figure out how to "recycle" listeners—that is, make them return to the station once they have sampled it.[14] Still other systems have used brain waves to measure audience response. For example, researchers at the Princeton Medical Center used an electroencephalograph to examine the brain waves of people watching television.[15]

Gathering data from either ERIs or GSRs has two advantages over older methods of data collection. First, it is fast, especially when monitored by a computer. What formerly took hours of work and computation can now be calculated almost instantaneously.

Second, it can measure detailed aspects of audience reaction, thus providing much more sophisticated data bases. When decision-makers can monitor audience reaction every few seconds, they have a more useful account of how an audience is reacting to a message.

Role Observation: Producing an Ethnography

Although such data-collection measures as the ERI and the GSR are useful in highly controlled laboratory conditions, the very artificiality of those conditions can sometimes limit the applicability of research results. Recently, researchers have been turning less to laboratory-oriented methods of data collection and more to real-life settings such as living with families and observing how they use radio and television. Such approaches can produce an *ethnography*, an account of how individuals have interacted with their environment and with each other and, in the case of broadcasting, how radio and television have functioned in this interaction.[16]

SUMMARY

This chapter examined research in telecommunication. Five types of research were discussed: historical, descriptive, qualitative, experimental, and developmental. Historical research examines the past and its implications for the future. Descriptive research, such as surveys, examines a current phenomenon. Qualitative research is concerned with the meaning of messages in a cultural context. Experimental research compares two different phenomena. Experimental research manipulates an independent variable, then observes any phenomenon resulting from that manipulation. The phenomenon to be observed is the dependent variable. Developmental research is used in instructional television and helps determine if a program meets its objectives. Of these five types of research, descriptive research is the type most commonly used in the broadcasting industry.

Research extends to all areas of the telecommunications industry. Besides that done by the government and rating services, research is conducted by cable companies, private research firms, local stations, networks, and students and faculty at colleges and universities.

Survey research common in local stations includes station-image surveys, and sales and marketing surveys. Station-image surveys often use more sophisticated methodology. Ranking of programs, questions on station operating procedures, and semantic-differential scales are just some of the many methods researchers use to discover an audience's perceptions of a station and its programming. Sales and marketing surveys examine advertising campaigns looking at such factors as the content of commercials, on-air schedules, and the sale of products in the marketplace. Focus-group research brings together small assemblages of people who discuss topics under the guidance of a trained moderator.

Network opinion polls predict the outcome of elections by sampling voters after they leave the polling place.

Public-opinion polls are usually the domain of the broadcast journalist and are frequently used around election time. But merely conducting a good poll is not enough. Reporters should consider information about the sample, sponsor, population, contact method, questions, sampling error, and headlines in order to communicate responsibly the results of a poll.

New technology is significantly altering the way in which research data are collected. Such methods as electronic response indicators and galvanic skin responses are being used more often to hasten the collection and processing of raw data. Role observation that produces an ethnography is also gaining increased attention.

OPPORTUNITIES FOR FURTHER LEARNING

Carter, K., and Spitzack, C. *Doing Research on Women's Communication*. Norwood, NJ: Ablex, 1988.

COHEN, J., AND GLEASON, T. *Social Research in Communication and Law*. Newbury Park, CA: Sage, 1990.

CONVERSE, J. M., AND PRESSER, S. *Survey Questions: Handcrafting the Standardized Questionnaire*. Newbury Park, CA: Sage, 1986.

HSIA, H. J. *Mass Communication Research Methods: A Step-By-Step Approach*. Hillsdale, NJ: Erlbaum, 1988.

LOWERY, S. A., AND DEFLEUR, M. L. *Milestones in Mass Communication Research*, 2nd ed. New York: Longman, 1988.

ROSENGREN, K. E., WENNER, L. A., AND PALMGREEN, P., eds. *Media Gratifications Research: Current Perspectives*. Beverly Hills, CA: Sage, 1985.

SINGLETARY, M. W. *Communication Theory and Research Applications*. Ames: Iowa State University Press, 1988.

STARTT, J., AND SLOAN, W. D. *Historical Methods in Mass Communication*. Hillsdale, NJ: Erlbaum, 1989.

STEMPEL G., AND WESTLEY, B., eds. *Research Methods in Mass Communication*, 2nd ed. Englewood Cliffs, NJ: Prentice-Hall, 1989.

TARDY, C. H. *A Handbook for the Study of Human Communication: Methods and Instruments for Observing, Measuring, and Assessing Communication Processes*. Norwood, NJ: Ablex, 1988.

WILLIAMS, F., ed. *Measuring the Information Society*. Beverly Hills, CA: Sage, 1988.

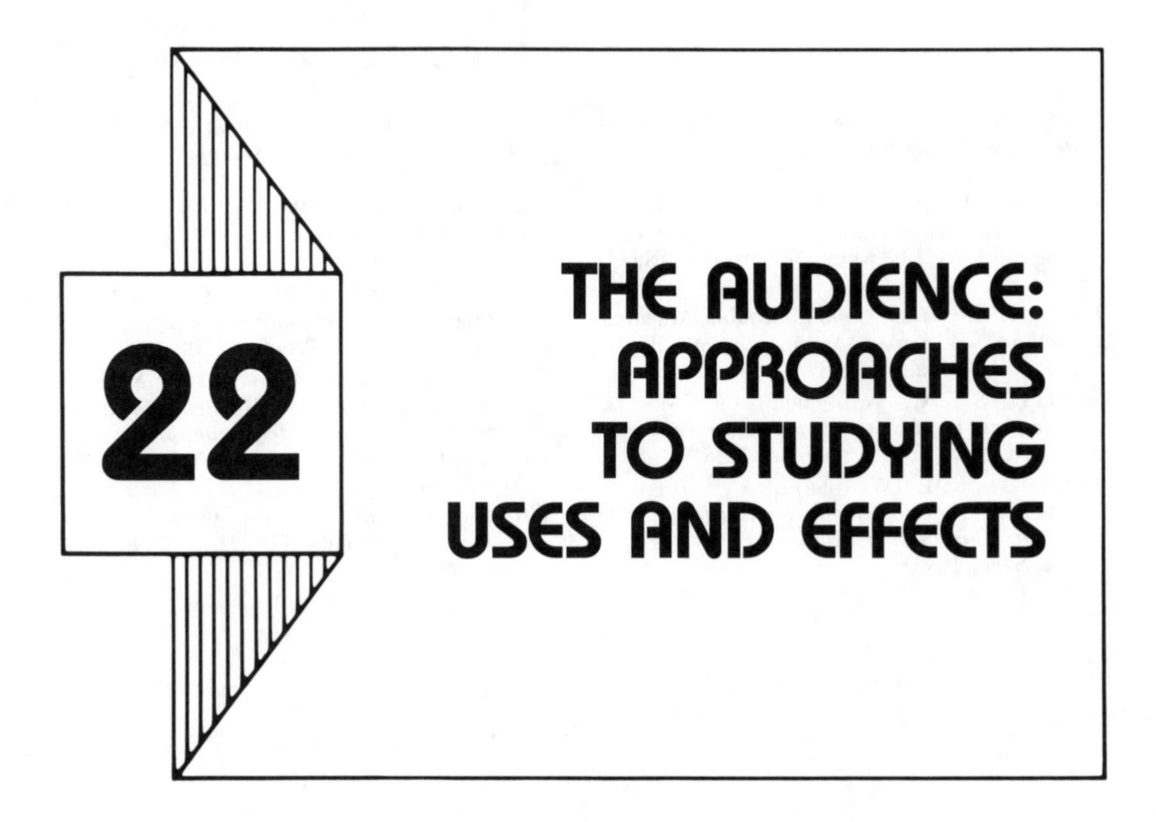

22 THE AUDIENCE: APPROACHES TO STUDYING USES AND EFFECTS

While radio and television impact on our lives, we have not passively accepted this impact. For example, research attempts to determine in what ways we use television programming, and what effect this programming has on our lives. In this chapter we examine how, over the years, researchers have looked at the mass audience, the messages directed toward that audience, and how the audience consumes those messages. We cannot possibly touch on every area, nor is it possible to cite every study beyond some early works that have set the direction for further research. What is possible, however, is to give the reader an awareness of the media audience of which we are a part, and in what way experts have chosen to examine this audience.

Although this book is about broadcasting and telecommunication, much of what we know about the mass audience and the uses and effects of media messages, we have learned from studying many different media. Therefore, we will frequently use the terms "mass" and "media" in place of broadcasting or telecommunication.

To begin to understand the audience for broadcasting and telecommunication, imagine you are media buyer for an advertising agency. Selecting a magazine in which to advertise your client's products requires you to examine the magazine's audience. Naturally you need to know the circulation, but you also want to know something about how the readers think, what their interests are, how old they are, what their income is, and how much education they have had. If the magazine is well established, this information will be readily available to you in a readership survey.

Now assume you need to buy time on a major television network. Here, identifying the audience becomes much more difficult. Although you could choose certain types of programming that would reach certain audiences, such as Saturday-morning programs for children, you are dealing mainly with a large, unidentified audience.

To say only that a broadcast audience is too big and complex to study would be to admit defeat much too soon. Instead, this chapter will examine what we do know about the broadcast audience, realizing all the time that researchers are continually working to learn more about it. We have been able only to scratch the surface of this inquiry, partly because human behavior is a highly complex phenomenon. Our studies of television violence and the effects of broadcast programming overlap with research in many other disciplines. Moreover, cooperative research across international boundaries will be invaluable.

UNDERSTANDING THE AUDIENCE

Electronic media have been studied by scholars, organizations, and the industry itself. Some of the more classic studies remain landmarks in the history of research and have influenced the way we are affected by and interpret the media. Studies such as Harry Field and Paul F. Lazarsfeld's *The People Look at Radio*, completed under the auspices of the National Association of Broadcasters, The National Opinion Research Center at the University of Denver, and the Bureau of Applied Social Research at Columbia University, opened up a new era in systematic evaluation of electronic media. Published in 1946 by The University of North Carolina Press at Chapel Hill, *The People Look at Radio* examined such fundamental questions as "Do you have a radio in working order?" and "If you had to give up either reading the newspaper or listening to the radio, which one would you give up?" Eighty-seven percent of the more than 3,000 people interviewed had a radio in working order and 62 percent said they would rather give up the newspaper.

Since 1946 we have changed the way we measure people's perceptions about the mass media, partially because of more sophisticated means of analysis aided by the power of the computer. We have also altered our views and theories on the way media affect us and the way we use media.

EARLY PERCEPTIONS: THE BULLET THEORY

Theorists used to look at the audience as a disconnected mass of individuals who received communication in much the same way that sitting ducks receive birdshot. This approach to media effects was labeled the *bullet theory*, or sometimes the *hypodermic theory*. Part of this misconception developed during World War I as a result of scare tactics employed in propaganda campaigns. As researchers began to realize that the behavior of humans was more complex than that of sitting ducks, their perception of the mass audience began to change.

The first change was in the concept of the mass. Instead of being viewed as a huge body of isolated persons exhibiting similar reactions to media messages, the audience gradually came to be studied as a group of individuals held together by different social systems and reacting to messages partly on the basis of their interaction with others within these systems. A political commercial may affect different persons in different ways. Those same people may interact with others about the commercial. These other persons may determine, just as much as and perhaps many times more than the media does, how a person reacts to the commercial.

REVISING THE BULLET THEORY: OPINION LEADERS

Interaction among members of the audience is part of the natural process of information dissemination and processing. Imagine for a moment that you are discussing a new TV series with your friend. Your friend tells you how great the series is, says you would enjoy the fast action, and suggests you watch it. Since you value your friend's opinion, you decide you will take a break from studying to watch the show. Why did you make that decision? Undoubtedly because you valued your friend's opinion enough to be influenced to take action—to watch the program. In doing so you demonstrated how other persons, whom we call *opinion leaders*, influence opinions concerning media messages.[1]

Since you had not seen the program, your friend's description was all you knew about it. By first hearing about the program from your friend, you demonstrated how messages flow to and through the mass audience. We call this process the *two-step flow*, even though many more than two steps may be taken before the information reaches all the people who eventually learn about it. Remember that the two-step flow can apply to both acquiring information and being persuaded by it.

TOWARD INDIVIDUAL DIFFERENCES

In revising the bullet theory, researchers have used three approaches to the study of audience reaction to media messages: individual differences, categories, and social relationships.[2] *Individual differences* implies that we react as individuals differently to different media messages. No longer was it possible to look upon the mass audience as one group of people who would react in similar ways to the same message. Such factors as personality differences came into play. Terms such as "attitudes" and "motivation" took on new meaning as researchers attempted to better explain the effects of mass media. Individual differences cause us to react differently to media messages. Researchers began to examine why people selectively expose themselves to certain media messages and avoid others and why different people can perceive the same message differently.

THE AUDIENCE AS CATEGORIES

The *categories approach* originated in the needs of advertisers to reach audiences with specific characteristics. Although the simplest way to group an audience into categories is by demographics—sex, age, and so on—researchers are looking more and more at *psychographics*—values, beliefs, attitudes, and lifestyles. The categories approach to the mass audience can be simple or complex. For the ad buyer wanting to reach eighteen- to twenty-one-year-old females, the application of the theory is mechanical. But for social scientists wanting to know how categories of people think and how they interact with other categories of people, the approach becomes much more complex.

SOCIAL RELATIONSHIPS

If we want to use these interrelationships to understand how people react to broadcast messages, the process becomes even more sophisticated. Buying an ad to reach the homemaker is one thing; buying an ad to reach the homemaker who interacts with another homemaker viewing a competing commercial is something else.[3]

Concentrating on such interaction describes the *social-relations* approach to studying the audience. Interpersonal communication is important to this approach, as is the realization that although the media can help disseminate the initial message, how it is retransmitted, discussed, and rediscussed among audience members will determine the effect of that message.

We can readily see that these three approaches not only overlap but are in some ways all part of the communicative process. Different persons may view that process in different ways. A psychologist concerned with an individual's behavior might feel more comfortable with an individual-differences approach but would be foolish to ignore the other approaches. Similarly, an advertiser wanting to reach a specific audience might be concerned with categories but could not ignore the relationships among people that receive the attention of the social-relations approach.

PROCESSING MESSAGES

In discussing the individual-differences approach it is necessary to talk about exposure, perception, and retention. Each influences our reaction to media messages.

Selective Exposures

Research has taught us that we selectively expose ourselves to certain types of programming, a process called *selective exposure*. If a politician is delivering a televised address, you might tune in the program because you agree with the politician's views. On the other hand, you might tune in the program because you disagree with the politician. In either case, you selectively exposed yourself to the program.

Selective Perception

Second, the perceptions you hold before watching the televised address will affect how you react to it. If you are extremely loyal to the politician, you might agree with everything she says regardless of what she says, so much so that even if her opponent said the very same thing, you might totally disagree with him. You would be guilty of *selective perception*. It is not a serious crime, but one that can distort how you react to messages.

Selective Retention

Third, because of your selective perception you might retain only those portions of the address with which you agree. If you perceive the entire address as favorable, you may remember all of it. If you perceive it as unfavorable, you may wipe it entirely from your mind. If parts of the address affect you positively, you remember those parts while forgetting the negative ones. Or the negative ones may be the very ones you remember. Either way, how you originally perceived the address determines what you retain, a process called *selective retention*.

CREDIBILITY

Two other factors that can affect how we perceive broadcast messages are source credibility and media credibility.

Source Credibility

Source credibility is the credibility of the original source of the communication. In the case of a politician's televised address, it would be the credibility of the politician. If you perceive the politician as highly credible, then the chances are that your reaction to her will be favorable. Research has assigned many subordinate factors to source credibility, among them dynamism, trustworthiness, and competence.

Source credibility lies partially in the source and partially in how the source is perceived by the audience. In other words, how media messages affect us is related to how we perceive the source of those messages. Source credibility and both interpersonal communication and mass communication all are part of the communication process.

Media Credibility

Media credibility is the overall credibility of a medium, such as a local radio station. Two types of media credibility are important. *Intermedia* credibility is the relative credibility of various media. For example, you might determine that television is a more credible medium than radio or newspapers. Intramedia credibility is the relative credibility within the same medium. Here you might decide that one radio station is more credible than another. Over a number of years, studies on media credibility have asked such questions as, "In the face of conflicting news reports, which medium would you be most likely to believe: television, newspapers, radio, magazines, or other people?" Much of the research has found TV to be the most frequent response. Notice that we said "most frequent response," not "most credible medium." Such studies fail to allow for the many possible intervening variables that truly reflect media credibility.

On the basis of such research we might justifiably ask: Is television listed first because it is the most credible medium or because we spend more time with it than with other media? Is there a wide variance among different media in different communities? In some communities would radio or newspapers come out on top because of the credibility of the local press? How many times do we really hear or even recognize conflicting news reports? Is television really the most credible medium, or do we just think so because two of our senses, sight and hearing, can consume the information instead of one, and can do so in color and in motion?

Some of the research that cited television as the most credible medium listed radio as the least credible. If we were to interpret those results, we might have some difficulty explaining some of the effects of radio programming. Back in 1938 many people became hysterical because they believed a report by Orson Welles, acting in the radio play of H . G. Wells's "War of the Worlds," of an invasion from outer space. Those who said it could not happen again were proved wrong in 1977, when Swiss radio aired a program containing mock news bulletins about neutron bombs being dropped in a war between a divided Germany. Mock casualty reports listed 480,000 persons killed. Panicked listeners called the station and received an official apology over Swiss radio.

Source credibility and media credibility interact to produce the persuasive power of the media.Our access to the media, its availability, and our own psychological profile help to determine our reactions to specific media messages.

CATEGORIZING THE BROADCAST AUDIENCE

We have talked about some of the approaches we use to examine the audience. Many of these approaches concentrate on two broad categories of audience classification: demographics and psychographics.

Demographics

Demographics refers to such things as the age, sex, education, income, and race of an audience. Partly because the data are easily obtainable—everything from courthouse records to census figures—demographic characteristics are the most commonly used types of broadcast-audience classification. The rating services, for example, use age and sex as their two principal categories, producing such classifications as women eighteen to twenty-four years of age and men thirty-four to forty-nine.

Arbitron has developed questionnaires approved by its legal counsel that provide demographic breakdowns of ethnic audiences. The questionnaire asked: "How do you describe your family?" It was then validated through personal interviews with respondents who had indicated their race or nationality in the Arbitron diary.

For advertisers, the demographic audience becomes something to reach, something to identify, and something to persuade. Using data from rating services and broadcasting stations, advertisers make time buys based on such considerations as when the highest concentration of women is watching and what the cost-per-thousand is for those women who are reached. Others need data to reach children, males, minorities, or teenagers. The key is to match the product to the audience, and the more specialized the product becomes, the greater the need to reach a specialized audience.

Demographics will most likely continue to be the main identifier of broadcast audiences, for four reasons. First, the information is easily obtained from different sources. Second, the industry is geared to using demographic data. Although the field of psychographics is becoming more important, the average radio-station manager is much more at ease with data on age, sex, and income than with psychological constructs, media involvement scales, or value profiles. Third, advertisers are comfortable with demographic data. An account executive selling a local druggist a radio commercial is on much safer ground talking about the station's high-income audience than attempting to teach a course in psychology while explaining the station's rate card. Fourth, information other than demographics is still subject to conflicting methodologies, not so much in the minds of the researchers collecting the data as in the minds of the industry that uses them. Although a research firm that claims to analyze an audience's value structure may be compared with a firm that analyzes its personality, the broadcast manager may simply groan and say, "Just tell me if they're men or women and how much money they make."

Psychographics

Psychographics refers to such things as *attitudes, values, beliefs,* and *opinions*. Psychographic methodologies range from dividing an audience according to its attitudes about brand preferences, to discovering an audience's subconscious reactions to broadcast programming. Asking consumers why they prefer brand *X* or brand *Y* is one simple way of obtaining psychographic information. Interviewing prison inmates to determine how they react to TV crime shows entails a far more complex psychographic profile. When we talk about the uses and gratifications that people obtain from television, we are talking about psychographic characteristics of the audience.

THE AUDIENCE AS A MASS MARKET

The commercial nature of much of mass media, including television, results in marketing strategists looking at the audience as a potential market for goods and services. The advertising agency media buyer who purchases advertising time on television asks: What group of people in the audience constitutes a market for my client's product or service, and how can I best reach that audience? The buyer considers the mass audience as a large marketplace of potential consumers. Viewing the audience from a demographic or psychographic perspective is not as important as viewing the audience as a marketplace, where both demographic and psychographic characteristics determine if a marketing strategy will succeed.

FUNCTIONAL USE OF MEDIA

We can use psychographics to study the functional uses of mass media. Assume we ask this question: "What function does television play in our lives?" We might answer:"We use television to escape from reality," "We use television to be entertained," or "We use television to learn what's happening in the world." In doing so we have mentioned three functions of broadcasting—escape, entertainment, and information. These different "uses" of media have gained the attention of scholars for decades. Field and Lazarsfeld were asking questions about the use of radio in the 1940s, and over the years inquiries have continued, through many different approaches and research methodologies.[4] Some of these approaches are discussed in the following sections.

PLAY THEORY

Using a data-gathering procedure called *Q-sort*, Professor William Stephenson completed research on how individuals representative of different audience types feel about the media. From this research evolved Stephenson's *play theory*, which, applied to broadcasting, suggests that we use radio and television as a means of escaping into a world of play not accessible at other times.[5]

USES AND GRATIFICATIONS

Stephenson's play theory is part of a wider body of research and theory focusing on what uses we make of media and what gratification we get from exposing ourselves to them. This research has been conducted among populations ranging from farmers in developing countries to American homemakers to children to the elderly.[6] For example, some research suggests children use television as a social activity that permits things seen on TV to be used as topics of discussion with friends and family. Research also suggests that because children develop critical viewing skills they interact, not just react, to television.[7]

Researchers continue to debate not only the different types of uses and gratifications but also the research methodologies that attempt to identify them. Part of the debate arises from a conflict between the individual-differences approach and the categories approach to the study of media effects. Consider a TV program. We could argue that a soap opera provides certain role models for homemakers or college students. We might also suggest that audience reaction to soap operas can be classified not in demographic terms but rather in psychographic terms.

Soap operas have certain uses for people having certain motivations or psychological characteristics. We could argue, however, that even this approach is unsatisfactory, since each individual is different and many different individuals may have many different uses for the same soap opera. How we learn what uses these many different individuals or groups of individuals make of the media is another dilemma. Do we test them individually in tightly controlled laboratories, "wiring" them psychologically in order to get at the depths of their thought processes? Or do we sample a large population of respondents in a survey? These are the questions that confront researchers and theorists working to determine exactly what use we do make of the media affecting our lives.

AGENDA SETTING

Agenda setting means that the media not only inform us but also tell us what we should be informed about.[8] In other words, the media set an agenda for our thought processes; media tell us what is important and what we should know and need to know.

Sophisticated analyses have made it possible to isolate those media that are dominant agenda setters—no small task since many communities have more than fifty different media. Newspapers, radio, television, books, and magazines are all important. By keeping track of which media are important to specific populations and then concentrating on those media, researchers can build a theoretical base for agenda setting.

The *agenda-setting function* becomes more pertinent when we realize that media have suddenly become some of the main determinants of how we perceive our world. The media, in effect, structure our world, and we in turn reinforce this structure. An empirical test of the agenda-setting function took place in 1968, when Maxwell McCombs and Donald Shaw examined the presidential elections. Since then, McCombs and others have continued to research agenda setting.

Research on agenda setting is also winding its way into media decision making—specifically, how and why gatekeepers select the news they do and feed it to the public. The gatekeeper agenda appears to originate in the wire services. Thus, although the local press sets the public's agenda, the wire services set the agenda for the gatekeeper.

SOCIALIZATION

Closely aligned with how we use media is the effect media have on our acquisition of culture and social norms. Although a significant amount of research concentrates on the effects of television on the socialization of children, we know that socialization continues throughout our lives, and broadcasting can affect this socialization at any time.

The Meaning of Messages

As with other approaches to studying the effects of messages on the mass audience, the content of these messages can mean different things to different people. For example, how a TV program containing violence affects a group of male adults can contrast sharply with its effect on a group of small children, whose world and ideas are just being formed and whose socialization is much less developed than that of the adults. The adult might go to bed thinking how great a television character was as the hero. The child may have nightmares about evil forces affecting his or her ability to survive in the world.

Studies Over Time

Because a person is not socialized simply by watching a single program, we must gather data from research across many disciplines in order to begin to theorize exactly how media in general and broadcasting in particular affect our socialization. Moreover, that data must be gathered over time. Few studies examine socialization over time. Most ask a given group of individuals what meaning television or radio has for them and then group the results under the heading of socialization or uses and gratification. Although these methodologies are important, studying the same individuals over a longer period is much more desirable.

Stages in Studying Socialization

Research on the socializing function of broadcasting has three stages. First, it examines the *content* of messages. Such content as the image of women in television commercials, hero figures in prime-time television, and acts of violence are representative of much of what we see or hear on television and radio. The second stage of this research attempts to find whether the people exposed to these messages actually *recognize* them. Were children who saw a given program able to recognize examples of good behavior? The third stage of investigation must determine what *effect* the messages have once they are received.

THE VIOLENCE DEBATE

Of all the effects of broadcasting, none has attracted more attention than the effect of the portrayal of violence on television. It has been the subject of research and debate by academicians, government agencies, schools, and international research organizations alike.

The Kefauver Committee

The issue of televised violence had been raised in 1952 by Senator Estes Kefauver's committee examining juvenile delinquency. But in 1952, television was too new to draw pertinent conclusions, despite testimony by psychiatrists and other recognized authorities. The post-Kefauver attention given to televised violence was the result not only of the presence of violence on television but also of research indicating a possible causal relationship between violence and behavior. Other studies followed.

A Research Agenda: The Surgeon General's Committee

In 1969 Senator John Pastore wrote to the secretary of health, education, and welfare, saying, "I am exceedingly troubled by the lack of any definitive information which would help to resolve the question of whether there is a causal connection between televised crime and violence and antisocial behavior by individuals, especially children."[9] Senator Pastore then called for a blue-ribbon committee of leading scholars to examine the relationship in detail.

Members of the committee, which became known as the Surgeon General's Scientific Advisory Committee on Television and Social Behavior, were chosen by a review of the names of experts on the subject. The final selection process was assigned to the three commercial networks and the National Association of Broadcasters. CBS, seeing its own research director as a possible appointee, withdrew from the selection committee to avoid a conflict of interest. Completed in 1972, the committee's report immediately drew praise, criticism, and varying interpretations. A succinct summary statement said

> There is a convergence of the fairly substantial experimental evidence for short-run causation of aggression among some children by viewing violence on the screen and the much less certain evidence from field studies that extensive violence viewing precedes some long-run manifestations of aggressive behavior. This convergence of the two types of evidence constitutes some preliminary evidence of a causal relationship.[10]

In other words, the report found that television violence may adversely affect some people. The report elicited widespread attention to the violence issue and encouraged researchers to get to the heart of the causal connection.

THEORIES ON THE EFFECTS OF TELEVISED VIOLENCE

The relationship between televised violence and aggressive behavior is best understood from the point of view of the various learning theories, which are summarized below.

Learning Theories

The *catharsis theory* suggests that in our daily lives we build up frustrations that are released vicariously when we watch violent behavior. Therefore, actual benefits accrue from televised violence. This theory is the weakest of the four theories, although some studies have provided limited support for it.[11] The *aggressive-cues theory* suggests that exposure to violence on television will raise the level of excitement in the viewer, and that will trigger violent acts.[12] Closely aligned to this theory is the *reinforcement theory*, which suggests that televised violence will reinforce already existing behavior in an individual.[13] Inherent in such a theory is the assumption that the violent person, because of violent tendencies, perceives televised violence as a real-life experience, whereas the nonviolent person perceives the violent program as entertainment and remains detached from it. The *observational learning theory* suggests that we can learn violent behavior from watching violent programs.[14]

The Relationships of Different Learning Theories

Clearly, all of the theories have merits and none should be discounted. Researchers continue to examine new variations on these four principal approaches, especially as televised violence affects small children. The observational learning theory, for example, could be applied to children in their formative years of growth, when their environment has a significant effect on what they learn. In essence, if television becomes a surrogate parent, it could certainly teach behavior. Later in the child's life, when behavior is more firmly determined, violence learned in the formative years could be reinforced. For the hyperactive or easily excitable child, the aggressive-cues theory might be used to explain emotions easily aroused by exposure to televised violence. The catharsis theory could apply even to the business executive who uses television to unwind and vicariously vent his or her frustrations through the actions of others.

Effects of the Portrayal of Violence on Aggressive Behavior

An early assessment of the relationship between the portrayal of violence and aggressive behavior suggested that:

1. Cartoon as well as live portrayals of violence can lead to aggressive performance on the part of the viewer.
2. Repeated exposure to cartoon and live portrayals of violence does not eliminate the possibility that new exposure will increase the likelihood of aggressive performance.
3. Aggressive performance is not dependent on a typical frustration, although frustration facilitates aggressive performance.
4. Although the "effect" in some experiments may be aggressive but not antisocial play, implications in regard to the contribution of television violence to antisocial aggression remain.
5. In ordinary language, the factors in a portrayal which increase the likelihood of aggressive performance are: the suggestion that aggression is justified, socially acceptable, motivated by malice, or pays off; a realistic depiction; highly exciting material; the presentation of conditions similar to those experienced by the young viewer, including a perpetrator similar to the viewer and circumstances like those of his environment, such as a target, implements, or other cues resembling those of the real-life milieu.
6. Although there is no evidence that prior repeated exposure to violent portrayals totally immunizes the young viewer against any influence of aggressive performance, exposure to television portrayals may desensitize young persons to responding to violence in their environment.[15]

Policy Dilemmas

Although the government has traditionally kept an ear open to complaints about violence, sympathy and rhetoric have been about as much as Congress or the FCC has been willing to offer. To offer more would be to collide head on with the First Amendment to the Constitution and the Communications Act of 1934. Even if the broad "public interest" standard were applied in an effort to curtail violence, court tests would be necessary so that it would not encroach on the Constitution. Although the debate and research will continue, the biggest battle of all may be fought in the political arena, in which no medium yet has been successfully curtailed, except superficially. Nor have the courts been responsive to the violence issue. When a Florida teenager claimed that television violence caused him to commit murder, the jury did not accept the idea.

Then there is the case of Japan, which is receiving more and more attention because of its high level of television violence and its low crime rate. Some possible explanations are that Japanese students are too busy with schoolwork to watch much television, that Japan has strict gun laws, and that Japanese citizens are becoming involved in crime protection.[16] Another possible explanation is Japanese society's emphasis on collective (family, school, company) responsibility, as opposed to American society's emphasis on individual responsibility.

Some programming changes continue to occur from time to time as advertisers begin to place economic pressure on the networks to reduce televised violence. But support among local advertisers for such a movement is minimal, and stations wishing to preempt network programming with locally originated programs containing more violence are finding little to stop them. Cable-TV operators enjoy fewer restrictions than over-the-air television, and over-the-air television claims it must be able to compete with cable-TV.

A DEVELOPMENTAL APPROACH

Studying television viewing over time permits us to learn more about the medium's developmental effects.[17] For example, a child learns words by being exposed to language. By using these words in different contexts a child develops the ability to use language more effectively. As the child gets older the skill with which the child uses language increases substantially.

Now consider that many children are exposed to television at a young age. In some homes, a television set is a surrogate babysitter. In others, parental interaction during television viewing makes the viewing experience active instead of passive. A developmental theory of television viewing over time examines such areas as the effects of parental intervention on a child's television viewing behavior. We know that as children get older their relationships with their parents change. Similarily, we might assume that a youngster's use of television also changes. Just as other experiences in the formative years of life help shape the child's values, habits, beliefs, skills, and view of the world, so does the child's exposure to television.

As the child grows older, such aesthetic elements as plot and portrayals of characters become more important. More serious programming is appealing. How the child in early adolescence reacts to and uses this programming may be determined by what experiences the child had with television during his or her formative years. Moreover, the influence of other media may also play a part. What the child listened to on the radio, saw in the motion picture theatre, or read in a book or magazine can play a part in the youngster's development.

CULTIVATION ANALYSIS

The impact of broadcasting and telecommunication on our lives has resulted in many different ways of examining electronic media. *Cultivation* is one of these areas of inquiry. Others range from attempting to find the answers to simple research questions to major theoretical inquiries.

The Concept

Cultivation analysis seeks to determine what effect television viewing has on the perceptions of the world we inhabit.[18] What view of our world does television cultivate in us and is that view distorted by television? For example, one narrow area of inquiry in the area of cultivation analysis seeks to determine if people who view heavy amounts of television programming that depicts crime and violence are therefore more afraid of the world they live in. This fear could be displayed in a wide range of behaviors from serious paranoia and mental instability, to simply being afraid of going out alone after dark. Television, through heavy viewing and violent programming, has "cultivated" this fear.

The importance of cultivation analysis in international communication research becomes clear when we consider that much of what other countries see on television comes from the United States. For example, if a steady diet of the TV drama "Dallas" cultivates the belief that the average person living in a democratic, capitalist society lives like the stars of "Dallas," then viewers may be much less likely to accept a lesser standard of living in their own country.

Intervening Variables

Although such causal relationships may be valid, research has also sought to isolate variables, other than viewing habits and content, which may prove or disprove cultivation effects. For example, researchers are trying to determine whether having a wide alternative of viewing options, such as is present with cable-TV, may negate some of the effects of heavy television viewing where only a small number of channels are available.

Other variables such as interpersonal communication can also play a part. An individual who lives alone and does not experience close contact with other people may watch more television. The TV programming may show an evil world that the isolated person perceives as real, partly because that individual has no contact with other people, such as friends and family, who may contradict that perception. Similarly, a locale where only a few channels are available for viewing will mean that interpersonal interaction about television viewing may increase because there is more commonality. People see the same thing and therefore have something in common to talk about. This increased interaction may either reinforce or contradict what was viewed on television. On the other hand, when many different viewing options are available, fewer people will see the same programs and there will be less interaction.

Also important is determining whether certain psychological characteristics are present among people who watch a great deal of television and whether these psychological traits, not television, are responsible for how an individual perceives the world. For example, if a person is afraid to go out at night because of a fear of crime on the streets, that person may stay home and watch even more television. Although the television programming being watched may reinforce the person's fears, it may not cause them.

MOOD MANAGEMENT

The theory of *mood management* suggests that people seek to eliminate bad moods and retain good moods. More specifically, mood management is based on the premise that: "(a) individuals strive to rid themselves of bad moods or, at least, seek to diminish the intensity of such moods, and (b) individuals strive to perpetuate good moods and maximize the intensity of these moods."[19] In other words, "to the extent possible, individuals arrange internal and external stimulus conditions so as to minimize bad moods and maximize good moods."[20]

Applied to our choice of television programming, mood management would predict that if we are in a bad mood we therefore want to select a program that will lessen the bad moood. The program might be an action-adventure thriller that will permit us to escape into a fantasy world long enough to forget reality and the conditions that causing the bad mood to exist. For another viewer, that escape might be a romantic comedy; for another a sports program. We may also make other media choices such as reading a good book, or even nonmedia choices, such as going jogging or engaging in some other form of activity.

Our choices may be determined by previous experiences, which have resulted in the past in helping to lessen bad moods. Since watching television successfully eliminated a bad mood in the past, we may, even without realizing why, turn to television again for similar relief.

SUMMARY

Early theories that the audience is a mass of unrelated individuals responding like one homogeneous mass have been greatly altered. Contemporary theorists believe that the audience interacts with the media, permitting the media to be an important part of their lives.

The bullet theory was gradually replaced by theories that began to examine the audience for media messages as having distinct individual differences. The idea of opinion leaders who participated in a two-step flow process between the media and other members of the audience gained importance. From this perspective evolved the concept of the audience as categories of people. Social relationships between people were gradually studied and applied to media theory.

As part of the audience of distinct individuals with individual differences, we process media messages in different ways. For example, we selectively expose ourselves to different media messages. We perceive messages differently and we retain different information from the same messages.

Both the source of the message and the medium through which we receive the information play a part in how we react to media messages. These two factors, called source credibility and media credibility, interact to produce the persuasive power of the message.

Researchers use demographics and psychographics to classify the broadcast audience. Demographics refers to such characteristics as age, sex, income, occupation, and race. Psychographics refers to attitudes, beliefs, values, opinions, and other psychological characteristics of the audience.

We can also view the audience as a marketplace of potential consumers who must be identified, reached, and persuaded to purchase specific goods and services advertised through the mass media.

With these classifications in mind we can begin to consider the functional uses of mass media. Three approaches to how we use media are Stephenson's play theory, the theory of uses and gratification, and agenda setting.

Socialization refers to the way mass media result in our acquisition of cultural and social norms. It is especially applicable to the development of children's perceptions of reality and their uses of media, such as television. Two aspects in the study of socialization are the way an individual perceives media messages and the effects of this perception over time. Content, recognition, and effect are three stages in studying socialization.

The study of television violence has been of interest since the 1950s. Four learning theories impact on whether violent programming results in aggressive behavior. These include the catharsis, aggressive-cues, reinforcement, and observational learning theories. Research on the effects of TV violence has drawn a number of conclusions, among them that both cartoon and live portrayals of violence can lead to aggressive performance on the part of the viewer.

The developmental theory suggests that children learn to use the media in much the same way that they learn other skills, such as language. For each child—depending on such things as parental interaction and intervention—media such as television assume different roles in the child's life. As the child gets older, such things as plot and portrayals of characters take on more meaning.

Cultivation analysis looks at the way the media formulate our perceptions of reality. One narrow focus of cultivation analysis examines whether people who watch large amounts of television—heavy viewers—have a distorted perception of the real world.

Mood-management theory suggests that we work to eliminate bad moods through different activities, such as watching television, and that the choices we make are indicative of choices we have made in the past. Thus, some of our mood-altering behaviors, especially in terms of TV viewing, are the result of learned behavior.

OPPORTUNITIES FOR FURTHER LEARNING

BALL-ROKEACH, S. J., AND CANTOR, M. G., eds. *Media, Audience, and Social Structure*. Newbury Park, CA: Sage, 1990.

BARWISE, P., AND EHRENBERG, A. *Television and Its Audience*. Newbury Park, CA: Sage, 1989.

FERRE, J. P. *Channels of Belief: Religion and American Commercial Television*. Ames: Iowa State University Press, 1990.

GREENBERG, B. S., BURGOON, M., BURGOON, J. K., AND KORZENNY, F. *Mexican Americans and The Mass Media*. Norwood, NJ: Ablex, 1983.

GROSS, L., KATZ, J. S., AND RUBY, J., eds. *Image Ethics: The Moral Rights of Subjects in Photographs, Film, and Television*. New York: Oxford University Press, 1988.

HOOVER, S. M. *Mass Media Religion: The Social Sources of the Electronic Church*. Newbury Park, CA: Sage, 1988.

LIEBERT, R. M., AND SPRAFKIN, J. *Effects of Television on Children and Youth*. New York: Pergamon, 1988.

LIEBES, T. *The Export of Meaning: Cross-Cultural Readings of "Dallas."* New York: Oxford University Press, 1990.

LINDLOF, T. R., ed. *Natural Audiences: Qualitative Research of Media Uses and Effects*. Norwood, NJ: Ablex, 1987.

MEROWITZ, J. *No Sense of Place: The Impact of Electronic Media on Social Behavior*. New York: Oxford University Press, 1985.

MORTENSEN, C. D. *Violence in Communication*. Lanham, MD: University Press of America, 1987.

ROSENGREN, K. E., AND WINDAHL, S. *Media Matter: TV Use in Childhood and Adolescence*. Norwood, NJ: Ablex, 1988.

SIGNORIELLI, N., AND MORGAN, M. *Cultivation Analysis*. Newbury Park, CA: Sage, 1989.

WILLIAMS, T. M. *The Impact of Television: A National Experiment in Three Communities*. Orlando, FL: Academic Press, 1986.

23 CRITICISM AND ETHICS

While working as a TV journalist and later as a radio news director and station manager, the author of this book dealt with many ethical dilemmas, not all of which were resolved easily. I still wonder, whether given the choice again, would I make the same decisions? To have second thoughts is not unusual. Every day working professionals face these same decisions in network newsrooms, in local stations, in policy forums in government.

The fact that radio, television, and telecommunication in all their facets have impacted so tremendously on our lives makes it mandatory that we take the responsibility for understanding, if not challenging, decision-makers and offering constructive criticism. This chapter is designed to sensitize you to the fact that questions about ethics and value judgments do exist in the field of electronic communication and that criticism of media in our society is natural and productive. Within the various topics discussed in this chapter are *questions for critical thinking*, designed to challenge your own decision making. There are no right or wrong answers to these questions, but they do provide a forum for discussion.

THE ROLE OF THE RESPONSIBLE CONSUMER

If this text has accomplished anything, it has made you more knowledgeable about the way the electronic media are organized and operate. This base of knowledge is the first step toward responsible criticism of the field of broadcasting and telecommunication.

A Base of Knowledge

If you read about a congressional hearing examining some aspect of television programming, if your local television station editorializes on an issue of public importance, if your local radio station changes its format because of its ratings, you have a better understanding of these issues and decisions. That in itself is an asset because many people know very little about the way such media as radio and television operate, much less how they affect our lives. These people are passive consumers of the media. You, however, having gained an understanding of the way electronic media operate, are in a position to be much more. You have the opportunity to be active and, above all, responsible consumers of broadcasting and telecommunication. You have the ability to make personal, and most importantly, *responsible* judgments about such things as how much television you watch, what you watch on television, and what your children may watch.

Making Personal Judgments

A radio or TV news director knows that it is impossible to cover all of the events that take place in a given day in a given community. Consequently, judgments must be made on what stories to cover and what stories to avoid. Some of these decisions are based on as simple a formula as time—there are only so many hours in a day and so much time in a newscast. Other judgments are based on such factors as what does the public need to know. Part of being a responsible consumer of media is to understand both of these factors at the personal level. For example, you only have so many hours in the day to watch television. How will you use this time? Will you sit down to watch TV without any thought as to how long you will watch it? Or will you take a more responsible approach and ask yourself, Will I watch television today? Will I watch certain programs? Will I watch them for enjoyment or for informational/educational value? If I do not have the time to watch both, which will I watch, and why? If I listen to radio, will I listen to music for enjoyment or listen to a news and information station to be better informed? Are there short-term and long-term advantages and disadvantages to my decision? By asking these questions you are making personal and responsible judgments about your viewing and listening habits.

Making Cognitive Judgments

At the same time you are making personal judgments, be prepared to make cognitive—mental, thought-based—judgments while you are watching television or listening to radio. For example, when watching a TV newscast, evaluate the pictures that accompany the story. Ask: "Why are certain scenes shown?" Ask: "Why is the person being interviewed and, in my opinion, does that person represent a knowledgeable authority figure on that subject?" Examine the thrust of the story and ask: "Are my own biases determining how I react to the story?" You can make the same assessments about entertainment or dramatic programs based on production, script, or performance characteristics. In other words, we need to be active, not passive consumers about what we hear and see.

The Participatory Process

You can take this active participation responsibility one step further. For example, you may decide to participate in the media decision process by speaking before your city council on the quality and selection of programming on your local cable system whose franchise is up for renewal. You may contribute money to your local public radio or TV station and take advantage of the opportunity to attend meetings of listener or viewer groups. Instead of sitting back and not taking any action, you may choose to write a letter to a member of Congress who influences communication policy, or join a public-interest group that lobbies Congress.

Ask yourself some probing questions: "Have I ever spoken out on an issue of importance to me? Have I ever written a letter to influence decision making on an important issue? Should I speak out? Should I write to an elected official or industry decision-maker? Should I become involved in joining a public-interest group that influences the industry? Should I take an active role in supporting and influencing programming on public television? Can I exert the same influence on commercial radio or television programming by writing to the management of local stations?" Answers to these questions, even thinking about the answers, makes you a more responsible consumer of radio and television.

DIMENSIONS OF CRITICISM

Just as we learned earlier in the book about the support structures for electronic communication, there are different dimensions through which we can critically evaluate this electronic communication and the way it impacts on our lives. For example, the afternoon soap opera has within it certain intrinsic qualities. Some of these qualities are aesthetic-artistic based, such as a camera shot or dull lighting to capture a mood. Other qualities are economic-based, such as scheduling commercials at a time when the demographic characteristics of the audience best support the sponsor's product or the most efficient cost-per-thousand can be achieved.

Policy dimensions are also at work. How explicit the sex scenes or how vulgar the language are to some degree determined by network standards as well as regulatory constraints over indecent programming in over-the-air television.

An examination of the historical aspects of television, for example, must take into consideration the premise that our historical analysis is based on our own selective perceptions, or those of the author who wrote the history book. For example, a history of minorities on American television would be superficial if we were to assume that Afro-Americans began appearing on network television as scripts incorporating Afro-American actors were produced.

Such an assumption would fail to take into account the need for an inquiry into whether some station managers or programmers of network affiliate stations were prejudiced and objected to Afro-Americans on television during the early days of television. Depending on what perspective (or from what prejudice) we approached our historical inquiry would determine how history is written and interpreted.

To better understand a medium such as television, we can therefore approach our critical view—our responsible consumer view—from the different dimensions summarized below.

POLICY DIMENSIONS

Depending in what country they operate, broadcasting and telecommunication are controlled by various laws and regulations that affect what consumers see and hear, and by how revenue is channeled into the company operating the medium of communication.

Forces at Work in the Regulatory Process

Now that we have learned something about the regulatory climate in which broadcasting and telecommunication operate, especially, for example, in reading about international broadcasting, we can use this knowledge to be a more responsible consumer of electronic media. Remember, also, that the regulatory arena of one country is not necessarily better, just different, from that of another country. In some countries the regulatory process is influenced by the regulated industry. The National Association of Broadcasters is a strong lobby organization in the United States and influences FCC decisions and congressional action affecting the commercial broadcasting industry. Stations pay dues to the NAB, which in turn provides many services to the broadcasting industry, lobbying being one of them. Many of the laws and regulations now in force have been shaped by NAB lobbying efforts. The next time you read about an FCC decision or a new law passed by Congress, ask yourself: "What special interests may have been served by this legislation?"

The Impact of Regulation

Other branches of government can also influence laws and regulation. The President of the United States, through the executive branch of government, and the fact the President appoints FCC commissioners, can influence communication policy. For example, much of the buying and selling of radio and television stations during the 1980s occurred because of the deregulated atmosphere in business strongly supported by the executive branch of government. Requirements on news and public-affairs programming have been relaxed, and fewer and fewer radio stations provide significant amounts of news and public-affairs programming. Is the public served by this policy? Are owners of radio stations served by this policy? Are broadcast brokers who deal in the trading of broadcast properties benefiting from this policy? Do politicians benefit from this deregulation?

ECONOMIC DIMENSIONS

Closely related to the policy dimensions are economic dimensions. If you were asked to evaluate critically the economic structure of American system of broadcasting, one aspect you would study would be sources of income.

Income vs. Independence

For example, in the United States, commercial radio and TV stations use advertising as a source of revenue. While your evaluation might start by noting that stations compete for advertising dollars, this would only begin to probe the real impact of this system of broadcasting. Consider the fact that while many small communities have the frequencies available to support more than one local radio station, the economic base is not large enough for more than one station to operate at a profit.

By accepting the above assumption, we confront a number of questions designed to test our level of critical thinking. For example, would a second station, perhaps owned and operated by the government, provide alternative programming to these media-starved communities where the only other medium is a weekly newspaper that publishes one issue per week? Do these small-town radio stations become so dependent on local advertising that offending a local businessperson in a newscast is tantamount to asking for an economic boycott of the station?

We may dig even deeper and ask whether a station, by providing an advertising medium for local businesses, contributes to the economic growth of the community or just engages in redistributing the wealth that already exists. If a local cable system sells advertising, and its franchise is approved by the local city council, is it in the same position as the radio station when it comes to offending the owners of local businesses, which in turn sit on the city council? Does the local newspaper that criticizes the cable service or local radio station in an editorial really have a hidden agenda, namely that it competes for a limited amount of local advertising? Would a noncommercial public radio station supported by listener contributions be an asset to the community? Would a public station gain listeners at the expense of the commercial station, thereby reducing its ability to reach the public and foster economic growth?

Commercial Constraints on Creativity

To what extent can creativity function in any given economic marketplace? By expanding our economic-based inquiry to other countries, we expand even further the ways in which we can critically evaluate the role of media in the marketplace. For example, in Japan a commercial system of broadcasting exists side-by-side with a well-developed public system operated by NHK. Instead of advertising, NHK uses license fees as its primary source of income and the license fees are determined by the government.

As we discovered earlier in this text, NHK prides itself in its independence. But is it really independent? Is the level of independence any broadcasting system exhibits dependent on the freedom with which creative expression can function? Can truly experimental broadcasting exist if it conflicts with what the majority, or at least a large segment of the population, wants to watch? Is artistic expression on the fringe of public acceptance limited by the commercial constraints placed upon a television industry that must capture enough viewers to attract advertisers?

The Global Marketplace

With the increased emphasis on the global economy, a much deeper understanding of the economic issues facing broadcasting and telecommunication is necessary. Moreover, economic policy is many times interrelated with regulatory policy. We touched on one example when discussing high-definition television (HDTV) earlier in the text. Specifically, part of America's HDTV regulatory policy on HDTV concerns the protection of the balance of trade should another country become the market leader in HDTV technology. As responsible consumers, we need to be aware of how the global marketplace affects policy and how different economies affect each other.

THE POLITICAL DIMENSION

Closely related to the economic dimension is the political dimension. One of the most striking uses of the media for political purposes was the effort during World War II to keep American morale high while portraying the enemy as evil as possible. The government hired some of the best directors of Hollywood to produce films that could persuade, influence, and motivate American workers and members of the military. Today, the political consultant designing a campaign commercial or the press secretary engaged in damage control after an embarrassing incident use the media for political purposes.

Other more subtle political dimensions are also present. The broadcast journalist who has spent time covering the governor's office may water down a story about the governor being physically and mentally impaired. When the governor stammers and becomes incoherent in a news conference, the same reporter may report only what the governor said, not the fact that the governor may not be able to perform the duties of the office.

THE HISTORICAL DIMENSION

Assume you decide to write a paper on the history of broadcast ratings and audience research.[1] After researching trade journals and other sources you determine that ratings and audience research have their origins in the early research departments of the broadcasting networks and the early efforts of local stations to learn who was listening to radio.

Your paper would talk about the technology of ratings. At first stations used mostly feedback from the audience in the forms of letters and cards from radio listeners who would report how strong the station's signal was being heard. Station management was as interested in how far the station's signal could be received and how effective the transmitter was working as they were about how many people were listening or to what programs.

Other audience-collection methods were eventually employed. The telephone was used to call a sample of listeners. Questionnaires were developed and mailed to listeners. As the ratings industry itself developed, sophisticated diaries were used and then meters. As we learned elsewhere in this text, today's experimental technology is sophisticated enough not only to tell what is being watched but also to monitor eye movement to see what part of the screen a person is watching. By tracing these developments, you have chosen to let technology determine the direction your paper will take, and in fact your paper is as much a history of ratings technology as it is of broadcast ratings.

SOCIAL DIMENSION

Another approach would be to examine the social aspects associated with ratings. This social dimension may exist side-by-side with the history of ratings technology. If your paper only dealt with the technology of ratings, you would have missed other factors, an understanding of which are necessary to a total understanding of the topic. For example, early research on radio soap operas found, among other things, that people listened for advice and to vicariously live the experiences of the soap opera performers.[2] We can assume that this information was also important and that early studies examining these social dimensions increased the need for more accurate and more detailed audience research. Through such research we began to mold a social perspective about the role of radio in our society.

CULTURAL DIMENSION

Media of all types are products of the culture from which they originate. In the same way, the content of media may reinforce that culture. For example, TV commercials frequently replicate the images of a culture the advertiser wants to reach as a target audience for goods and services. For some consumers that image reinforces their own way of life. For others, the images represent a fantasy world that can be lived vicariously by purchasing the goods or services depicted. True mass media, if it did exist, would reflect a very broad culture. Specialized media reach more narrow cultures. We need only watch an evening of American television to see examples of the youth culture represented in a commercial showing a teenager in brand-name jeans hanging out with other jeans-clad peers on a city street corner. Culture is not, however, necessarily separate and distinct from media. For example, our very use of television as an entertainmment activity is an integral part of many cultures.

SUBDIMENSIONS

Within one dimension can fall subdimensions. Consider, for example, the economic dimension as part of the broader historical dimension. For instance, broadcast ratings began to make radio and TV legitimate commercial media and provided stations, advertisers, and advertising agencies with information that could justify spending ad dollars on broadcasting.

In addition, the ratings were powerful competition against newspaper and advertising circulation figures that further built the economic base of the broadcasting industry. The ability to reach specific target audiences with efficient cost-per-thousand made it possible for broadcasters to weather storms over increased specialization in the magazine industry and even the increased specialization of the broadcast industry, especially where radio was concerned. The broadcast ratings perpetuate themselves by designing their services to meet the needs of the radio and television industry, and they present data in such a way that the stations can best compete in the marketplace.

All of this information is important to a detailed understanding of ratings and audience research. Simply asking questions about the technology of ratings would not have uncovered the total impact the ratings have had on the broadcast industry.

THEORIES OF CRITICISM

The increased attention to critically evaluating the role of media in our society has resulted in the formulation of theories upon which the critical inquiries can operate. The theories encompass the dimensions we have just discussed, but they also offer broader applications of those dimensions over entire media systems.[3] Some of these theories are based in recent history, whereas others are more contemporary. For example, the "classical" or "Marxist" theory is based on the writings of Karl Marx and approaches media as means of production, owned and monopolized by the industrial elite. Media exploit workers, disseminate the ideas of the owners, and stifle ideas that might result in change. The economic and political system helps maintain the media structure.

Growing out of classical Marxist theory was "critical theory," which sought to explain why some of the reforms Marx predicted did not evolve.[4] Within critical theory came an examination of the media's role in creating commercialism, and therefore subverting the economic base upon which some of Marx's predictions for revolutionary change were based.

Another approach is "hegemony" theory, which concentrates on the ideology within which the media function.[5] Economic and political aspects are separated from the ideological aspects. Media, controlled by the ruling class, are seen as helping to create a distorted view of reality through covert but consistent views of reality. A simple interpretation of hegemony theory would be to use the media to convince workers they live in a culturally rich environment when in fact most cannot afford to buy books or recordings.

"Cultural" theory views both the content of media, and our media-use patterns, as reflective of our culture.[6]

ACHIEVING A CRITICAL PERSPECTIVE

Even if we derive from this book a desire to work professionally in broadcasting, telecommunication, or a media-related industry, we know that a critical perspective of broadcasting and telecommunication is necessary both to understand the media and to use it responsibly. The television program director who only looks at ratings to determine how many people are watching—who does not understand the methodology of the ratings and how this methodology is designed to support both the ratings industry and the television station—will not be able to make intelligent programming decisions. Similarly, knowing in what ways the media are part of our culture and understanding how the media reflect, through both content and use, the culture in which we live are important factors in everything from formulating communication policy to reaching target audiences.

AESTHETICS

The term *aesthetics* originates from philosophy and deals with theories of art and beauty. We can use this term in media to put a framework around the decisions that such people as directors and writers make.[7] These decisions determine what we the audience will be exposed to in the finished work of a radio or television production. The decisions may be as simple as deciding what jingle will precede a song on the radio or as complex as what lines, lighting, and camera shots will be used to establish a mood scene in a television drama. Aesthetic judgments are an important part of the decision-communication process.[8]

Handling Dialogue

Consider the aesthetic dimension at the level of an actress and actor in a TV drama. The scriptwriter will develop lines to carry along the plot. Getting the dialogue to communicate the setting and feeling of the drama is important to permit the audience to get from point A to point B. At the same time, however, the scriptwriter may determine that because an actor or actress has a particular strength—say the ability to play romantic roles—then the script will permit those particular talents to be showcased, even though the lines are not critical to the plot. The entire production is enhanced by giving the actress or actor lines that play to the performer's particular strengths. What are your own preferences when watching a television drama?

Directing Decisions

If the romantic lines are written into the script, additional aesthetic decisions must also be made, such as *blocking*—where the actress and actor will stand, sit, or lie down—and *camera shots*, which will determine how the audience will see the romantic scene. For example, if the camera establishes the scene by having the actor and actress walk toward each other across a room, a certain emotion is conveyed. On the other hand, if the actor or actress descends a stairway and the TV camera is positioned at the bottom of the stairway, another emotion is conveyed. If the TV camera is located at the top of the stairway and the audience looks on from above, a different emotion is conveyed.

Establishing Mood

What lighting to use is another aesthetic decision. A dim light will convey one mood whereas a bright light will convey a different mood. The position of lights also comes into play. A horror movie may use lighting that illuminates the underside facial features of the performers, much like placing a flashlight under one's chin.

In conclusion, the decisions a news photographer makes to capture the emotion of a child whose cat is trapped in a well, or a graphic artist makes to enhance the visual elements of a videotex screen, are all aesthetic dimensions of the communication process. Being aware of these different dimensions and others you can read and learn about will increase your awareness of aesthetics and permit you to offer constructive criticism of what you see and hear.

ETHICS AND DECISION MAKING

Our sensitivity to the need for a more critical perspective of broadcasting and telecommunication leads us to the discussion of *ethics*. Decision-makers face ethical decisions every day. Some are easy decisions, such as not distorting the video that accompanies a news story. Others are more complex, such as determining whether to show a particularly grisly accident scene on the evening news.

Ethical decisions also involve more abstract reasoning. For example, assume you acquire the rights to some early TV programs shot in black-and-white and have the technology to color them. Should you change the way they were originally produced to add the dimension of color? What if color was not available when the programs were originally produced? Could you argue that if color had been available, then the programs would have been shot in color and you are just completing what the producer wanted to do all along? What if the production was deliberately shot in black-and-white, even though color was available? Should you add your color tinting to the program now that you own the rights to distribute the program? If you do add color, will you deprive future generations of the ability to see the program in its original art form, and therefore deprive future generations from making their own judgments about the artistic quality of the program? What if you could personally make a very large profit from the distribution of the colorized programs? Would that make a difference in your decision? Are TV programs an artistic national resource that should be legally protected from alteration?

Answers to these critical questions cause us to come to grips with ethical decision making. To some extent, where we place our loyalties will determine our decisions. To understand ethics it is necessary to understand where these loyalties lie.

Loyalty to Oneself

Faced with one's own survival, there is little doubt that an individual will choose oneself. If the broadcast station is on fire the program director will probably choose to leave the equipment and taped programs behind, rather than risk death by smoke inhalation. The choice was simple—loyalty to oneself. Dual loyalties also come into play. For example, while a news director may feel the need to adhere to certain professional standards, faced with a general manager telling the news director to kill a story or get fired, the news director may decide it is better to sacrifice professional ethics than lose one's job. Loyalty to oneself again predominates.

But not all choices are so clear. For instance, if the city where the news director works has a strong journalism-review publication, a magazine published for and about the press, then the news director's decision to kill the story may become public knowledge. The news director would face criticism from peers who would ridicule the decision. Faced with such a decision, how would you decide?

Ask yourself some questions for critical thinking: If you worked for a TV station and the owner of a company that bought a large amount of advertising on the station was arrested on a charge of drug possession would you run the story? Your first answer would probably be a resounding "yes." But now consider some additional dilemmas. What if the owner of the station asked you to kill the story? What if the person arrested was a close friend of the station owner? What if the person was related to the station owner? What if the person arrested was a close relative of yours, such as your cousin or sister? What if you faced being fired for airing the story? Would you make an unpopular decision out of loyalty to your peers?

Some situations may occur where the decision is not so much one of choosing between two courses of action as it is taking an unpopular stand. For example, as a broadcast editorial writer you may have a strong opinion about an issue but know your opinion is not popular among the majority of people in your community. Despite this majority you decide to remain loyal to your own opinion and your editorial reflects this loyalty. Stop and ask yourself: "What issues do I feel strongly about? Would I have the courage to speak out on these issues in an editorial? If my local radio or TV station takes a position in an editorial with which I disagree, would I be willing to appear and air a response?"

Loyalty to the Profession

The news director who worked in the city where the journalism review was published may very well have developed some strong professional loyalties and for a number of reasons. For one, the journalism review may have provided a forum where ethical judgments and decisions were discussed in articles and columns. A sense of right or wrong could be determined. To some extent, the review itself may have helped mold certain professional loyalties. The news director may have also belonged to a professional society where similar "codes" of conduct, if not stated, were implied. These goals could be such things as striving for accuracy, not succumbing to the manipulative pressure of politicians, taking a critical view of government, to name a few. The news director may have won awards and received a certain amount of professional prestige that the news director wanted to protect. Even the presence of a city press club where the news director could go and find a sense of inclusion with others could play a part in developing this loyalty.

Loyalty to the Audience

In the United States, for example, where the Communications Act fosters a sense of serving the "public interest," the loyalty to the audience has been implied, although not always followed. For instance, the executive of a television network, which is traded on the stock exchange, faces the prospect of falling stock prices because of low ratings. The executive may be quick to exchange quality programming for any programming that attracts a large audience and therefore advertising dollars.

Even this decision has room for judgment since we might ask: "Who determines what is quality programming? Should, or does, the audience already make that decision through the broadcast ratings?" To one person, quality programming may be a documentary on wildlife; to another, quality programming may be a sexy and violent, but well-produced, mini-series. Given the choice, which would you watch?

The "public interest" standard is based on the regulation of radio and television because of the limited spectrum space available for over-the-air broadcasting. Do broadcast news programs demand a higher ethical standard than entertainment programs? Should cable TV be held to the same standards as over-the-air broadcasting? Because cable systems have numerous channels from which to choose, frequently in the dozens, should cable be freer to permit programming that some might consider of lesser quality? Should pay-per-view services be held to even less accountable standards? With so many media outlets in the marketplace, should any broadcaster subscribe to audience loyalty beyond simply wanting to capture enough ratings to satisfy advertisers and keep the station or cable system profitable? While such questions may seem crass, they appear often in policy debates over such issues as programming standards, codes of ethics, and regulating fairness in the treatment of important issues.

Mixing Loyalties

While we have categorized different loyalties to better understand decision making, these different loyalties frequently overlap. For example, while preparing a newscast a news director's first loyalty may be to the audience. Stories are selected first on the basis of what the audience needs to know. Even among these stories, the order of presentation in the newscast may be determined on the basis of what the news director's own biases are and what issues the news director may feel are important.

SYSTEM ETHICS

Regardless of how we answer the critical questions posed in this chapter, we are basing our answers on our perceptions of the social system within which we are a part. That system represents an attempt by people to create society and to regulate rules of behavior so the society has some sort of order. In some societies that order is maintained through freedom and various safety valves that permit people to speak out forcefully and to influence public opinion and policy decisions. In other countries the system is maintained through force, which at some point may be resisted by such means as sit-ins, protests, and even revolution.

System Maintenance

Being part of a social system leads us to ask, "What role do broadcasting and telecommunication play in maintaining the system and should the role of media be one of system maintenance or should system maintenance be left to other institutions?" For example, codes of ethics many times are based on what an organization perceives is good for the profession; thus, the code, while supporting specific behavior by media decision-makers, also affirms what the media perceive is right for society.

If the code of ethics calls for fairness, what happens if police and protestors clash in a violent confrontation? Does a sense of fair play demand that the police and the protestors get equal time? Or does the moral aspect of the cause being protested demand coverage, even if police or people with alternative viewpoints are slighted? What if the local TV station is the only available voice for the protestors to air their grievances? What if the TV station is owned by the government and the government is under siege? If the government is replaced, even by means of force, at what point would you consider your station an instrument of the new government? Would you still have your job under the new government? Would you still be alive?

Alienation

Our earlier example of the news director who felt a professional bond with colleagues at the press club points to another issue in system ethics, specifically the problem of media decision-makers becoming alienated from society. For example, many people end their day with a need to know what is happening in the world around them. They may seek out a local television newscast that, they assume, will provide them with important information that may be impacting on their lives. "The Journalist is alienated, not so much from his or her own labor as from what the news product is purported to transform into reality. . . ,"[9] one observer has noted.

The news director at the TV station, however, may approach that same newscast from a somewhat different perspective, not so much from the angle of what the viewer needs to know, but more from what will make a good show. This is not to suggest that TV news directors are only interested in what video is available for the evening news, but every news director, news producer, photographer, editor makes conscious artistic decisions about what will make the news depending on what video is available to accompany the story.

Expanding or Limiting Our Informational Environment

One step beyond the artistic decision is the decision on whether to keep, or eliminate entirely, a news story. While the top news story of the day may not be cut, other less important stories, less important as judged by the people who control the newscast, could be eliminated or repositioned, because they lacked accompanying video. While the news director may be concerned about the audience, that concern may very well be based on how well the newscast will do in the ratings. We might ask: "Do people who work in the newsroom live in a different world from the rest of us? Do the people who work in the newsroom impose their world on us every evening, which in turn distorts the reality of our own world?"

The Interaction of System and Society

While at first we might assume that there is a right and wrong way to approach one's job, and either you work for the station or you work for the audience, that distinction, however, is not made easily and is determined by other factors. For example, we might ask: "Is it wrong for the news director to be concerned first about ratings and second about the audience? If sponsors leave the newscast because of bad ratings, will the news product, regardless of how flawed it may be, suffer even more? If the ratings suffer, could the station cut the newscast from the schedule? If the audience watches the news over a long period of time, does the audience really know its world is being distorted or does the audience perceive what is important by what the newscast says is important?"[10] It may not be so much a question of loyalty as it is the newsroom functioning on a day-to-day basis under standards and rules of behavior that are determined in part by a system of free-enterprise commercial broadcasting and a bureaucracy perpetuating itself for its economic survival.

THE PROBLEM OF ETHICAL EXPEDIENCY

The bureaucracy itself can become a moral issue when we consider the effect it has on the audience. For example, the rules of behavior and standards a broadcast journalist works under appear in many forms and sometimes under deadline pressure. Broadcast journalists write in tight prose and accompany their stories with video. To avoid being manipulated by press agents and others who want to set the news agenda, certain internal safeguards exist. Seasoned practitioners might put it bluntly by calling it "a healthy skepticism for politicians and government." Right or wrong, it exists. Stated another way: "As complex events are reduced to the formula stories to maximize efficiency in reporting, a moral complexity is managed through an *expedient ethics* of action."[11] The journalist must decide how and which stories to cover shaped by the rules of the bureaucracy.

MAKING ETHICAL ASSUMPTIONS

Perhaps you are now, or will become, a practicing professional or responsible consumer of broadcasting and telecommunication. It is hoped that you will use what you have learned from this book, as well as your continued reading and inquiry, to formulate an intelligent understanding of the role of broadcasting and telecommunication in our society. Daily, the media are filled with stories about the way the technology of our information society impacts on our lives. The knowledge you have gained, however, also assumes a responsibility to challenge traditional assumptions about our media world. With the increased emphasis on our global society, these assumptions will become even more important as we begin to formulate policies that must complement the policies of other nations.

Something as powerful as the electronic media demands a cooperative world order if we are to respect the rights of individuals as well as the cultures, societies, and even governments that may be much different from our own. A judgment made in the newsroom in one country can result in a satellite-fed television newscast impacting on dozens of nations and hundreds of millions of people. With that power comes responsibility, but even more important, comes the ability to operate within some framework of one's own ethical judgments, which must constantly be thought out, and often modified.

CRITICISM AND ETHICS IN AN INFORMATION SOCIETY

As we move toward an information society, the challenge to theorists will be to explain these changes and show how we must alter our traditional views of media to encompass our information society. For example, ethical questions about invasion of privacy through the accumulation of massive amounts of data on citizens become more important. Interactive media, which permit instant polling of opinions and programming tastes, make it possible for media suppliers to target programming on the basis of psychographic information. Can such targeting be used to reinforce our programming tastes, or for that matter our political beliefs, even without our knowledge? Will those who possess the power of control over information be able to manipulate others, making subgroups and subcultures conform to the wishes of the power brokers in our information society? These are the questions that will confront consumers, practicing professionals, and policymakers as we approach the twenty-first century.[12]

SUMMARY

Professionals who work in broadcasting and telecommunication must make critical judgments that affect what we see and hear. At the same time, we as viewers and listeners have an affirmative obligation to be responsible consumers of these mediated messages and to use our knowledge to challenge and offer constructive criticism. We must also understand a framework of ethics within which we can better understand the effect these decisions have on individuals and society.

At the heart of responsible criticism is a base of knowledge. This knowledge will continue to be increased as we learn more about broadcasting and telecommunication.

Some of the judgments we make are personal, highly individualized judgments. Others include cognitive judgments such as asking if our biases determine how we react to something we see on television or hear on the radio. We must also take an active, not a passive, role in our evaluation.

Different dimensions of criticism are also present. These include policy, economic, political, historical, social, and cultural dimensions as well as subdimensions.

These different dimensions are also found within various theories of criticism, such as the classical Marxist theory, critical theory, hegemony theory, and cultural theory.

We also need to be aware of the aesthetic elements in media since each aesthetic element was produced through conscious decisions of the writer, producer, and director. By understanding these different dimensions and theories we begin to achieve a critical perspective of media, which helps to make us more responsible consumers of media in our global society.

Determining the ethical basis for our decisions involves determining where our loyalties lie. For example, these loyalties may be to ourselves. On the other hand, our loyalties may be to our profession or to our colleagues. We may have loyalties to our audience and make decisions based on what we feel the audience should see or hear. In still other cases, loyalties may be mixed.

Broadcasting and telecommunication operate as part of a social system. People who work in these industries are participants in this social system and may make decisions based on this association. You may succumb to the social system or be alienated from it, either of which can affect media decision-makers. In broadcast journalism, for example, association or alienation from the system can cause the new product to be distorted.

Still being formulated, however, is the way in which dimensions, theories, and ethical framework presented here will serve to explain our transition to an information society.

OPPORTUNITIES FOR FURTHER LEARNING

ALLEN, R. C. *Channels of Discourse: Television and Contemporary Criticism*. Chapel Hill: University of North Carolina Press, 1987.

BROWN, M. R. *Television and Women's Culture: The Politics of the Popular*. Newbury Park, CA: Sage, 1990.

CARY, J. W. *Media, Myths, and Narratives: Television and the Press*. Newbury Park, CA: Sage, 1987.

CASSATA, M., AND SKILL, T. *Life on Daytime Television: Tuning in American Serial Drama*. Norwood, NJ: Ablex, 1982.

CHRISTIANS, C. G., ROTZOLL, K. B., AND FACKLER, M. *Media Ethics: Cases and Moral Reasoning*. New York: Longman, 1987.

DOWNING, J., AND OTHERS, eds. *Questioning The Media*. Newbury Park, CA: Sage, 1990.

ENTMAN, R. M. *Democracy Without Citizens: Media and the Decay of American Politics*. New York: Oxford University Press, 1989.

LEMERT, J. B. *Criticizing the Media: Empirical Approaches*. Newbury Park, CA: Sage, 1990.

MONTGOMERY, K. C. *Target: Prime Time—Advocacy Groups and the Struggle Over Entertainment Television*. New York: Oxford University Press, 1989.

SAVAGE, R. L., AND NIMMO, D. *Politics in Familiar Contexts: Projecting Politics Through Popular Media*. Norwood, NJ: Ablex, 1988.

SLACK, J. D. *Communication Technologies and Society: Conceptions of Casuality and the Politics of Technical Intervention*. Norwood, NJ: Ablex, 1987.

SLACK, J. D., AND FEJES, F. *The Ideology of the Information Age*. Norwood, NJ: Ablex, 1987.

SMYTHE, D. W. *Dependency Road: Communications, Capitalism, Consciousness, and Canada*. Norwood, NJ: Ablex, 1981.

SUSSMAN, H. S. *High Resolution: Critical Theory and the Problem of Literacy*. New York: Oxford University Press, 1989.

TUROW, J. *Playing Doctor: Television, Storytelling, and Medical Power*. New York: Oxford University Press, 1989.

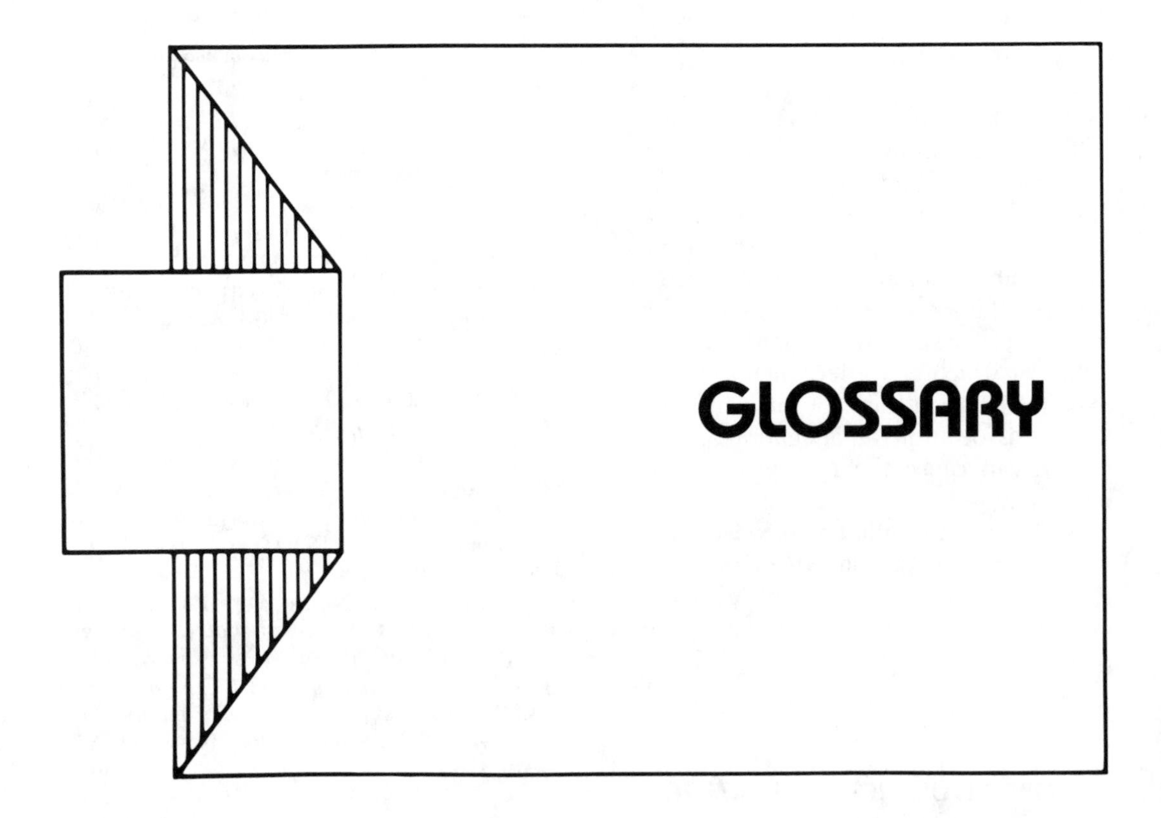

GLOSSARY

AAAA American Association of Advertising Agencies.

ABC (1) African Broadcasting Company; (2) American Broadcasting Company; (3) Australian Broadcasting Company.

Access channels cable-television channels for general public use.

Accountable programming term used in educational television to describe a program meeting a specified set of instructional objectives.

ACT Action for Children's Television.

ADI area of dominant influence; a term used in ratings.

AEJ Association for Education in Journalism.

Affiliate a broadcasting station bound by contract with a particular broadcasting network or wire service.

Agenda-setting function the theory that media set an agenda for our thought processes; they tell us what is important and what we should know and need to know.

Aggressive-cues theory the theory that exposure to violence on television will raise the level of excitement in the viewer.

All-channel receiver capable of receiving AM and FM radio signals.

Alternator developed by Ernst Alexanderson at the General Electric Laboratories. Used to modulate early voice broadcasting.

AM amplitude modulation.

AM stereo dual-channel broadcasting on AM frequencies. A common method is to use one channel as amplitude modulation and the other channel as frequency modulation.

Analog continuously operating, such as a telephone, radio, or television transmission. Analog computers accept and process continuous, or real-world, signals, such as voltage fluctuations.

Anik Canadian satellite system.

Annual billings broadcast station's bill to advertisers for commercials carried over a one-year period.

AP Associated Press, a print and broadcast wire service.

Apogee point at which a satellite is closest to earth.

Application specific use to which a computer can be put.

Application software specific computer instructions for particular applications, such as payroll and purchasing.

AP Radio network of the Associated Press.

ARB Arbitron rating survey.

ARD Federal coordinator of West German radio broadcasting.

Armature revolving iron core of the alternator.

ATS Application Technology Satellite.

Audio actuality the recording of the actual sounds in the news for incorporation into radio newscasts.

Audion three-element vacuum tube invented by Lee de Forest.

Average quarter-hour persons an estimate of the number of people listening to a station during any quarter-hour in a specified time period.

AWRT American Women in Radio and Television.

Banks groups of control switches on a master control console used to program various portions of an audio or video production. Also called "buses."

Bartering an advertising arrangement in which stations receive programming in return for providing advertising time; widespread in syndicated programming.

BBC British Broadcasting Corporation.

BEA Broadcast Education Association.

BRC Broadcast Rating Council.

Broadcast Signals sent via radio or television.

Capital-intensive business a business in which maximum costs occur immediately from the start; exemplified by the cable industry.

Cash-flow statement a statement of the amount of money necessary to keep the station running smoothly and to avoid going into debt.

Catharsis theory the theory that in our daily lives we build up frustrations that are released vicariously when we watch violent behavior.

CATV community-antenna television, or cable TV.

CBC Canadian Broadcasting Corporation.

CCTV closed-circuit television.

Cellular radiotelephone systems a series of small transmitters with limited-coverage cells that make up the core of a mobile telephone system.

Chain broadcasting early term for network broadcasting.

Chart of accounts ledger used to record station finances.

Clearance ratio the number of network affiliate stations that agree to air a network program.

Clear channel one on which dominant stations broadcast over wide areas virtually interference-free within their primary service areas and most of their secondary service areas.

Coaxial cable cable consisting of a wire core surrounded by a layer of plastic, metal-webbed insulation, and a third layer of plastic.

Coherer small glass tube used in Marconi's experiments to create and break an electrical connection.

Columbia a label, developed in the early days of broadcasting, for such companies as the Columbia Broadcasting System (CBS) and the Columbia Phonograph Broadcasting System, Incorporated.

Common carrier a communications channel that handles interstate and international communications by electronic means, and whose services are open to the public for hire.

Communication the movement of messages between senders and receivers.

Communication model a stop-action picture of the communicative process.

Comparative balance sheet a comparison of assets and liabilities between two points in time.

Comparative income statement a comparison of income between two points in time.

Component television a TV set whose monitor, tuner, speakers, video screen, and other attachments are sold separately.

COMSAT Communications Satellite Corporation. Formed by the Communication Satellite Act of 1962.

COMSTAR satellite system launched by COMSAT and leased by AT&T.

Conduction the use of ground or water, both electrical conductors, to replace a second wire in a telegraph hookup.

Construction permit permission granted by the FCC to begin construction of a broadcast facility.

Co-ops (1) broadcast news networks, also called *informal* networks, created by a group of radio or TV news personnel; (2) trade-out advertising agreements between advertisers and the individual advertising outlet.

Co-op advertising a split in the cost of advertising, usually between a retail outlet and the manufacturer.

CPB Corporation for Public Broadcasting.

CPM Cost per thousand; the cost of broadcasting a commercial message to 1,000 people.

CPU central processing unit; the core of a computer, containing the processor and main memory, which reads and interprets the instructions from a computer program.

CRTC Canadian Radio-Television Commission.

CTV abbreviation for Canada's Television Network, Ltd.

Cume persons (cumulative audience) the number of different individuals or households watching or listening to a given station or program at least once during a certain time period.

CWA Communication Workers of America.

Delayed feedback noninstantaneous response to information received; differentiates mass communication from intrapersonal and interpersonal communication.

Demographics a term referring to such things as the age, sex, education, income, and race of an audience.

Descriptive research research that describes a current condition.

Developmental research the process of continually perfecting something, such as an ITV program.

Diary method in a rating survey, a method of data collection utilizing a diary.

Digital a term that describes operations, signals, or transmissions that are broken up into binary code for the computer—a series of on-off pulses (bits of information)—and transmitted virtually noise-free, unlike analog or continuous transmissions.

Direct-broadcast satellites (DBS) high-powered satellites that beam signals directly to small home antennas.

Directional antennas a group of strategically placed broadcast antennas transmitting a signal in a specific direction so as to form an irregular rather than a circular contour.

Directional stations radio stations, primarily in the AM band, that use directional antennas in order to keep their signals from interfering with those of other stations.

Director the person responsible for the entire production of a program.

Direct-wave propagation radio-wave pattern in which signals are transmitted in a direct line of sight.

Dish dish-shaped antenna that receives signals from a satellite.

Dissolve a smooth change from an image produced by one television camera to an image produced by a second television camera, film, slide, or videotape.

DMA Designated market area.

Double billing fraudulent practice of charging advertisers twice.

Drop cable cable from the subtrunk of a cable system to the home terminal.

Earth station a station that transmits microwave signals to a satellite and also receives those signals.

Electromagnetic spectrum the range of levels of electromagnetic energy, or frequencies.

Electromagnetic waves energy traveling through space at the speed of light; used to transmit radio and television signals.

ENG Electronic news gathering.

ERI Electronic response indicator.

ETV educational television; all noncommercial television programming and commercial programming produced especially for educational purposes, whether or not the program is used for direct classroom instruction.

Experimental research sophisticated research that entails a controlled experiment.

Fairness Doctrine FCC rule requiring equal air time for controversial issues.

FCC Federal Communications Commission.

Feedback information received in response to information already imparted.

Fiber optics the use of thin strands of glass to carry as many as a thousand or more cable channels. Also used for data communication.

Field of experience the accumulation of knowledge, experiences, values, beliefs, and other qualities that make up a person's identity.

Fixed position an advertising term referring to a commercial placed at the same time every day.

Floppy disk a single flexible plastic magnetic device that stores computer programs and information.

FM frequency modulation.

Focus groups small groups of people brought together to discuss topics selected by the researcher or station and guided by a trained group leader.

Forfeiture the most common sanction imposed on a station by the FCC; it is based on the violation and on the ability of the station to pay.

Franchise a contractual agreement between a local governmental unit and a cable company.

Frequency (1) broadcast-rating term indicating how often a viewer has tuned to a given station; (2) position on the electromagnetic spectrum.

FR3 French Regional Broadcasting Service.

Gatekeeper the person directly involved in relaying or transferring information from one individual to another through a mass medium.

Geostationary (synchronous) satellite an orbiting satellite traveling at a speed proportional to that of the earth's rotation, and thus appearing to remain stationary over one point on the earth.

GHz gigahertz (1 billion hertz, or cycles per second). (See *kHz* and *MHz*.)

Grid one of the elements in a three-element vacuum tube. The other two are the filament and the plate.

Ground wave wave adhering to the earth's surface.

Ham informal term for amateur radio operators.

HDTV high-definition television; produces large-screen television without a grainy effect.

Head end the combination of humans and hardware responsible for originating, controlling, and processing signals over the cable system.

Hertz (Hz) last name of Heinrich Rudolph Hertz; commonly used as an abbreviation for cycles per second in referring to electromagnetic frequencies.

Heterodyne circuit improved detector of radio waves invented by Reginald A. Fessenden.

Home terminal (1) receiving set for cable TV transmissions, either one-way or two-way; (2) device connecting the drop cable of a cable system to the receiving set.

HUT households using television.

Hyping using promotional efforts to increase the size of an audience during a rating period.

IBA Independent Broadcasting Authority (British Television Network).

IBEW International Brotherhood of Electrical Workers.

ICA International Communication Association.

ILR Independent Local Radio (British Radio Network).

Image orthicon one of the first pickup tubes used in early television broadcasting.

INA Institute Nationale de l'Audiovisuel (France).

Induction process by which a current in one antenna produces a current in a nearby antenna.

Informal networks broadcast news networks created by a professional group of radio or TV news personnel; also called co-ops.

INTELSAT International Telecommunications Satellite Organization.

Interactivity the ratio of user activity to system activity in teletext and videotex systems.

Interface the connection among computer hardware, software, and people.

Intermittent service area an area that receives service from the ground wave but is beyond the primary service area and thus experiences some interference and fading.

Interpersonal communication communication between two or more persons in a face-to-face situation.

Intrapersonal communication communication within ourselves.

Ionosphere the upper level of the atmosphere, which reflects radio waves back to earth.

ITU International Telecommunication Union.

ITV (1) instructional television (programming designed specifically for direct or supplemental teaching); (2) the Independent Television Network.

JCET Joint Committee on Educational Television.

Jingles a set of short musical recordings, all designed around a common musical theme and usually related to a station's call letters.

kHz kilohertz (1 thousand hertz or cycles per second); measure of a position on the electromagnetic spectrum.

Local channels AM channels located at the upper end of the AM band and operating with a power no greater than 250 watts at night and 1,000 watts during the day.

Long lines term used by AT&T to describe long-distance-telephone communication links.

Lowest unit charge minimum charge on a station rate card.

LPTV Low-power television; stations that rebroadcast the signals of TV stations to outlying areas, much like translator stations; and can also originate programming.

Marisat a COMSAT satellite that makes data, telex, and voice communications available interference-free; especially valuable for marine operations.

Mass audience the audience reached by the mass media.

Mass communication the process by which messages are communicated to a large number of people by a mass medium.

Master control console the heart of a TV control room through which both the audio and the video images are fed, joined together, and improved, perhaps by special effects, for the on-air image.

MBS Mutual Broadcasting System.

MDS Multipoint distribution systems; systems that use microwaves to transmit TV signals from a master omnidirectional microwave antenna to smaller microwave antennas.

Media plural of *medium*.

Media credibility the effect of various media on how mass-communication messages are perceived.

Medium channel of communication, such as radio or television; singular of media.

Menu a display of all the options in an interactive computer program that are available to the user at the computer terminal.

Message intensity the value or importance of an event or its potential impact in relation to other events or potential news stories.

Meter method a method of broadcast ratings in which a monitoring device installed on TV sets is connected to a central computer, which then records channel selection at different times of the day.

MGM Metro-Goldwyn-Mayer.

MHz megahertz (1 million hertz or cycles per second). (See *kHz* and *GHz*.)

Microwave a very short wave of higher frequency than that of standard broadcast transmission; usually measured in gigahertz (billions of cycles per second).

Millimeter waves a generic term referring to electromagnetic waves that are between a centimeter and a millimeter long.

Mix to join and separate the pictures from various television cameras for a composite on-air image.

Modem a device that adapts digital transmissions, as from a computer, to analog transmissions, such as a voice or broadcast transmission, and vice versa.

MPATI Midwest Program for Airborne Television Instruction.

MSA metro survey area (rating term).

NAB National Association of Broadcasters.

NARBA North American Regional Broadcasting Agreement.

NASA National Aeronautics and Space Administration.

NBC National Broadcasting Company.

NCTA National Cable Television Association.

Network clipping the fraudulent billing practice whereby a station certifies to a network that a network commercial has been aired when in fact it has not.

NHK Nippon Hoso Kyokai (Japanese broadcasting system).

NPR National Public Radio.

NRBA National Radio Broadcasters Association.

NTIA National Telecommunications and Information Administration.

Opinion leader person interpreting messages originally disseminated by the mass media.

ORTF Office de Radiodiffusion-Television Francaise.

Pay cable a system in which cable subscribers pay a fee in addition to the standard monthly rental fee in order to receive special programming.

Pay-per-view cable a pay-cable arrangement that charges the subscriber on a per program basis.

PBS Public Broadcasting Service.

Perigee the closest point to the earth of a satellite's orbit.

Photophone a device invented by Alexander Graham Bell whereby the voice could be transmitted over light waves; the forerunner of today's fiber-optic light-wave communication.

Physical noise breakdown in communication caused by some physical quality or object interfering with the communicative process.

Plate one of the elements in a three-element vacuum tube. The other two are the filament and the grid. Early tubes used just a plate and filament.

Plumbicon a device that superseded both the image orthicon and the vidicon; it can capture color images with the sensitivity of the human eye. It is a trademarked name of the Amperex Corporation.

Prime time the time of the largest audience, when a station charges the highest price for advertising: 7 to 11 P.M. for TV, 7 to 9 A.M. and 4 to 6 P.M. for radio. (Radio prime time varies with the market and lifestyle trends.)

Production companies businesses that produce broadcasting programs for adoption either by networks or by individual stations through syndication; commonly called *production houses*.

Program directors people responsible for selecting programs for airing, scheduling their air time, and overseeing the production and direction of locally produced programs.

Program 1,2,3 Swedish radio networks.

Projection an estimate of the characteristics of a total universe based on a sample of that universe. Term frequently used in projection polling.

Psychographics a term referring to such things as the attitudes, values, beliefs, and opinions of an audience.

PTA National Congress of Parents and Teachers.

Public broadcasting the operation of the various noncommercial radio and television stations in the United States.

Public service advertising (PSA) advertising designed to support a nonprofit cause or organization. Most of the time or space for this advertising is provided free as a service to the public by the print or broadcast media.

RAB Radio Advertising Bureau.

RAM Random-access memory; that portion of computer memory where information can be transferred in and out. It is not permanent, as is ROM (read-only memory).

Random sampling selection process in which each unit of the larger portion has an equal chance of being selected.

Rating percentage of a given population who are tuned to a radio or television station during a given time period.

RCA Radio Corporation of America.

Record rotation how often records are changed on the playlist plus how frequently they are played on the air.

Regional channels channels assigned where several stations operate, none having a power of more than five thousand watts.

Relay satellite A satellite capable of bouncing messages back to earth.

Repeater satellite A satellite that can both receive signals and retransmit them back to earth.

ROS run of schedule.

RTNDA Radio-Television News Directors Association.

Sales network a group of broadcasting stations linked together by a financial agreement that benefits all member stations by offering advertisers a joint rate.

Sampling the process of examining a small portion of something in order to estimate its characteristics.

Saturation schedule the airing of a heavy load of commercials over a short period of time.

SBC Swedish Broadcasting Corporation.

SCA (1) Speech Communication Association; (2) Subsidiary Communication Authority.

Selective exposure exposing oneself to communication believed to coincide with one's preconceived ideas.

Selective perception perceiving only those things that agree with one's preconceived ideas.

Selective retention remembering only those things that agree with one's preconceived ideas.

Semantic noise breakdown in communication caused by misunderstanding the meaning of words.

Share an estimate of the percentage of listeners to a particular station in comparison with listeners to all other stations or programs during a given time period.

Silicone detector crystal used in early radio receiving sets to detect radio waves.

SIN Spanish International Network.

Skip a section of the earth's surface that neither ground nor sky waves reach.

Sky-wave propagation a radio-wave transmission pattern in which the signals travel up, bounce off the ionosphere to the earth, and rebound from the earth in a continuing process.

SMATV Satellite Master-Antenna Television; a system that receives signals from satellites.

Soft interconnects cooperative associations of cable operators who work together to attract advertisers and air commercial programming, but whose systems are not physically connected.

SPJ, SDX Society of Professional Journalists, Sigma Delta Chi.

STV Subscription television; a system whereby the signal from a TV station is scrambled and the subscriber pays a monthly fee for a decoder that unscrambles it.

Subtrunk cable a secondary cable that branches out from the main trunk in a cable TV system and carries the signal to outlying areas.

Superheterodyne circuit an improvement on Fessenden's heterodyne circuit; developed by Edwin H. Armstrong.

Supering positioning a picture from one TV camera on top of another picture from a second camera; special effect controlled by the master control console.

Superstation a radio or TV station that has expanded its listening and viewing audiences by beaming its signals to satellites, which in turn relay those signals to cable systems and their subscribers.

Sweep the period of a rating survey.

Switcher (technical director) the person responsible for operating the master control console.

Synchronous see Geostationary.

Syndicated programming programming distributed directly to the stations, not through a network, although it may have first appeared on a network.

Syndicator company supplying syndicated programming to local stations.

Talent raid CBS's acquisition of talent from other networks in 1948; sometimes refers to similar actions by ABC in 1976.

TDF Telediffusion de Francaise.

TDRSS Tracking Data Relay Satellite System; a Western Union satellite system that supplements NASA's system for ground-tracking U.S. manned space missions.

Telecommunication electronic communication involving both wired and unwired, one-way and two-way communications systems.

Teleconference a video conference conducted over a long distance.

Teletext one-way transmission of textual information employing the unused scanning lines (the vertical blanking interval) of a television signal.

Television household a broadcast-rating term used for any home merely having a television set, as distinguished from a household actually using television.

Telstar early satellite used for the first transatlantic television broadcast.

TF1 Television Francaise 1.

Tiering a process whereby cable operators lump different channels or services into tiers and charge their subscribers an additional fee to receive the channels offered by a particular tier.

Toll broadcasting an early term for commercial broadcasting; first used by radio station WEAF.

Trade-out an agreement in which a product or service is traded for advertising on a station.

Transaction the process whereby information is sent and received and feedback occurs.

Transfer to send and receive information.

Transistor wafer-thin three-layer crystal used extensively in electronic equipment; performs many of the functions of the three-element vacuum tube.

Translators television transmitting antennas, usually located on high natural terrain.

Transmit to send information.

Transponder a receiver/transmitter in a communications satellite.

Trunk cable the primary cable or main transmission line of a cable-TV system.

TSA total survey area (rating term).

Two-step flow the process by which information disseminated by mass media is (1) received by a direct audience and then (2) relayed to other persons.

Two-way cable cable system capable of both sending and receiving data.

UHF Ultra-high frequency.

Universe the whole from which a sample is being chosen. In broadcast ratings, this can be the sample area, metro area, or rating area.

Untouchables a term used in the cable business for potential subscribers who for some reason choose not to hook up to cable.

UPI United Press International; a print and broadcast wire service.

UPI Audio Radio network of United Press International.

Valve an early two-element vacuum tube.

VBI vertical blanking interval; the thick black bar that appears on a TV screen when the vertical-hold adjustment is manipulated.

VCR Video-cassette recorder.

VHF very-high frequency.

Videodisc a device such as a long-playing record that produces both a TV picture and sound.

Videotex a two-way interactive wired system providing textual information; it can transmit over cable and telephone lines.

Vidicon a sensitive television tube that followed the image orthicon.

VOA Voice of America.

Voltaic pile the first practical energy cell, developed by Alessandro Volta.

VTR videotape recorder.

WARC World Administrative Radio Conferences.

Wavelength the distance between two waves.

Westar Western Union Satellite System.

WICI Women in Communication, Inc.

Wireless term used for early radio.

Wireless telephone term used for an early invention by Nathan B. Stubblefield.

YLE (YLEISRADIO) state-controlled Finnish broadcasting monopoly.

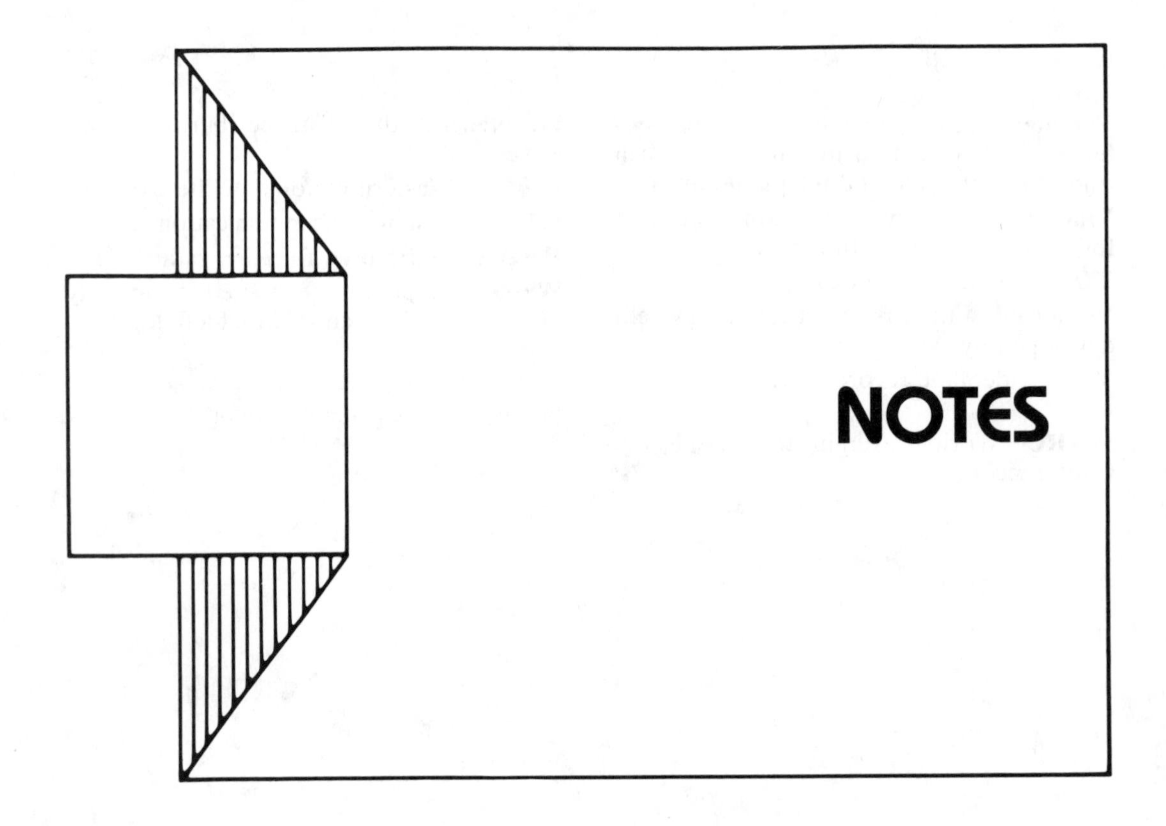

CHAPTER 1

[1]We need to keep in mind that researchers disagree about what transmit, transfer, and transact mean and that definitions change across disciplines. Communication may mean one thing to a computer scientist and something entirely different to a philosopher.

[2]Some of the more well-known models that have developed over the years are found in: C. E. Shannon and W. Weaver, *The Mathematical Theory of Communication* (Urbana: University of Illinois Press, 1964); G. Gerbner, "Toward a General Model of Communication," *Audio Visual Communication Review*, 4 (Summer 1956), 171–99; H. D. Lasswell, "The Structure and Function of Communications in Society," in *The Communication of Ideas*, ed. L. Bryson (New York: Harper & Row, 1948), p. 37; W. Schramm, "How Communication Works," in *The Process and Effects of Communication: An Introduction to Theory and Practice* (New York: Holt, Rinehart & Winston, 1960), p. 72; F. E. X. Dance, "Toward a Theory of Human Communication," in *Human Communication Theory: Original Essays* (New York: Holt, Rinehart & Winston), 1967, p. 296. General works include Berlo, Dance, and others. See also the more recent collection of essays edited by R. W. Budd and B. D. Rubin, *Approaches to Human Communication* (New York: Spartan Books, 1972). A classic work is C. Cherry, *On Human Communication* (New York: John Wiley, 1957).

[3]Components vary considerably, as we can see from the following terms, all of which are found in various communication models or discussions of them: communicator, message, receiver, channel, medium, noise, feedback, delayed feedback, information source, transmitter, signal, noise source, receiver signal, destination, perceptual dimension, percept, event, media, form, content, man, machine, encoder, decoder,

interpreter, communication skills, attitudes, knowledge, social system, culture content, treatment code, sensing, hearing, touching, smelling, tasting, environment, point in time.

[4]As conceptualized as early as 1954 by Wilbur Schramm in the first edition of his classic text *The Process and Effects of Mass Communication* (Urbana: University of Illinois Press).

[5]J. C. McCroskey and L. Wheeless, *Introduction to Human Communication* (Boston: Allyn & Bacon, 1976). The term's conceptual application is credited to P. F. Lazarsfeld and R. K. Merton, "Friendship as a Social Process: A Substantive and Methodological Analysis," in *Freedom and Control in Modern Society*, ed. Monroe Berger and others (New York: Octagon, 1964), p. 23.

[6]K. Lewin, "Channels of Group Life: Social Planning and Action Research," *Human Communication Research* (1947), 143–53.

[7]See, for example, K. Starck and J. Soloski, "Effect of Reporter Predispositions in Covering a Controversial Story," *Journalism Quarterly*, 55 (Spring 1977), 120–25.

[8]Although many studies have examined the various relationships, the concept of opinion leader was first reported and applied to current mass communication in P. F. Lazarsfeld, B. Berelson, and H. Gaudet, *The People's Choice* (New York: Columbia University Press, 1948). Opinion leaders can also act in strictly interpersonal communication. However, it is in reference to mass media that we use the term here.

[9]W. Schramm and J. Alexander, "Broadcasting," in *Handbook of Communication*, eds. I. De Sola Pool and others (Chicago: Rand McNally, 1973), p. 586.

CHAPTER 2

[1]Readers interested in early electrical communication should consult Hugh G. J. Aitken, *Syntony and Spark: The Origins of Radio* (New York: John Wiley, 1976); Silvanus P. Thompson, *Michael Faraday: His Life and Work* (New York: Macmillan, 1898); W. Rupert MacLaurin, *Invention and Innovation in the Radio Industry* (1949); reprint ed., New York: Arno Press, 1971); Gleason L. Archer, *History of Radio to 1926* (New York: American Historical Society, 1938); J. A. Fleming, *The Principles of Electric Wave Telegraphy* (London: Longmans, Green, 1908); Richard T. Glazebrook, *James Clerk Maxwell and Modern Physics* (New York: Macmillan, 1896).

[2]Accounts of Marconi's life may be found in Orrin E. Dunlap, *Marconi: The Man and His Wireless* (1937; reprint ed., New York: Arno Press, 1971); W. P. Jolly, *Marconi* (Briarcliff Manor, N.Y.: Stein & Day, 1972); Degna Marconi, *My Father Marconi* (New York: McGraw-Hill, 1962); Niels H. de V. Heathcote, *Nobel Prize Winners in Physics: 1901–1950* (New York: Henry Schuman, 1953); R. N. Vyvyan, *Marconi and Wireless* (East Ardsley, England: E. P. Publishing, 1974).

[3]Marconi, *My Father Marconi*, p. 27.

[4]Ibid., p. 28.

[5]Ibid., pp. 38–39.

[6]Ibid., p. 36.

[7]Ibid., pp. 100–104.

[8]"Wireless Signals Across the Ocean," *New York Times*, December 15, 1901, pp. 1–2; "Wireless Telegraphy Across the Atlantic," *Times* (London), December 16, 1901, p. 5.

[9]P. T. McGrath, "Marconi and His Transatlantic Signal," *Century Magazine*, 63 (March 1902), 769; George Iles, "Marconi's Triumph," *World's Work* (February 1902), 1784.

[10]McGrath, "Marconi and His Transatlantic Signal," p. 781.

[11]"Signor Marconi's Experiments," *Times* (London), December 19, 1901, p. 5.

[12]Much has been written on the early development of the Marconi companies. Particularly useful to this text were W. J. Baker, *A History of the Marconi Company* (New York: St. Martin's, Press, 1971); Gleason L. Archer, *History of Radio to 1926* (New York: American Historical Society, 1938); L. S. Howeth, *History of Communications: Electronics in the United States Navy* (Washington, D.C.: Bureau of Ships and Office of Naval History, 1963); Hiram L. Jome, *Economics of the Radio Industry* (London: A. W. Shaw, 1925); see Thorn Mayes, "History of the American Marconi Company," *The Old Timer's Bulletin*, 13 (June 1972), 11–18; as cited in Lawrence W. Lichty and Malachi C. Topping, eds., *A Source Book on the History of Radio and Television* (New York: Hastings House, 1975); and Jolly, *Marconi*.

[13]The first proposed title of the company was Marconi's Patent Telegraphs Ltd., to which Guglielmo Marconi himself objected (Baker, *History of the Marconi Company*, p. 35).

[14]*Investors World* (October 7, 1898), 484.

[15]Howeth, *History of Communications*, p. 36.

[16]Archer, *History of Radio to 1926*, pp. 81–82.

[17]See Lee de Forest, *The Father of Radio* (Chicago: Wilcox & Follett, 1950). Fleming's work is discussed in J. A. Fleming, *An Elementary Manual of Radio Telegraphy and Radio Telephony* (London: Longmans, Green, 1908), pp. 204–11; and George G. Blake, *History of Radio Telegraphy and Telephony* (1928; reprint ed., New York: Arno Press, 1974), pp. 238–60.

[18]Harlow, *Old Wires and New Waves*, pp. 462–63, as cited in Archer, *History of Radio to 1926*, p. 92.

[19]"October Meeting of the American Institute of Electrical Engineers," *Electrical World*, 43 (November 3, 1906), 836–37; also published by de Forest as "The Audion: A New Receiver for Wireless Telegraphy," *Scientific American Supplement* (November 30, 1907), 348–56.

[20]J. A. Fleming, "Wireless Telegraph Receiver," *Electrical World*, 43 (December 8, 1906), 1117.

[21]Lee de Forest, "Wireless Telegraph Receiver," *Electrical World*, 43 (December 22, 1906), 1206. See also note 40.

[22]Jome, *Economics of the Radio Industry*, p. 208; *Marconi Wireless Telegraphy Company of America* v. *De Forest Radio Telephone and Telegraph Company*, 236 Fd. 942, affirmed by the Circuit Court of Appeals in 243 Fd. 560.

[23]De Forest, *Father of Radio*, pp. 325–26.

[24]Ibid., p. 457.

[25]Charles Susskind, "de Forest, Lee," in *Dictionary of Scientific Biography*, Vol. 3, ed. Charles Coulston Gillespie (New York: Scribner's, 1975), pp. 6–7.

[26]See Elliot N. Sivowitch, "A Technological Survey of Broadcasting's Prehistory," *Journal of Broadcasting*, 5 (Winter 1970–71), 1–20. For the most part, research into the role of induction and conduction in the development of the radio has been overlooked by historians, partly because such approaches became scientifically obsolete in later years. Research focusing on the political, economic, and social implications is still needed. It would provide some important "micro" insights into how new scientific knowledge is applied to the invention process. Sources for such research studies are still available and more easily accessible than those of inventors in the early nineteenth century.

[27]For accounts of Stubblefield, see Sivowitch, "Technological Survey," pp. 20–22; Thomas W. Hoffer, "Nathan B. Stubblefield and His Wireless Telephone," *Journal of Broadcasting*, 15 (Summer 1971), 317–29; and Harvey Geller, "The Man History Overheard," *Circular-Warner/Reprise*, 7 (December 8, 1975), 1–4.

[28]Geller, "The Man History Overheard," p. 2.

[29]*Washington Post*, August 10, 1940, cited in Hoffer, "Stubblefield," p. 322.

[30]Hoffer, "Stubblefield," p. 322.

[31]Quoted in H. M. Fessenden, *Builder of Tomorrow* (New York: Coward-McCann, 1940), p. 77. The remainder of the letter reads:

> "The Government cannot legally pay for the patents issued to you, but by the proposition herein made, you are allowed a salary that will enable you to easily bear such expense and thus own the patents, the Bureau reserving the right to make use of such patents or of such devices as you may invent for its use in receiving meteorological reports and transmitting Weather Bureau information.
>
> "I am of the opinion that you would have a better opportunity here not only to test your present devices, but also while enjoying a remunerative salary and having your traveling expenses paid be able to devise new apparatus that would inure both to the profit of the Government and to your own individual benefit.
>
> "Congress recently gave us $25,000 which we expended in making additions to our buildings in Washington. We therefore have plenty of room and can easily fit up such laboratory as you may need. You will also find instrument makers, blacksmiths, metal workers and artisans of many classes, the services of whom will be freely placed at your disposal.

"If this general proposition meets with your approval, I would thank you to sign the enclosed agreement and return same to this office" (p. 77).

[32]Ibid., p. 81.

[33]Ibid., p. 80, Quoting *Popular Radio*, 1923.

[34]Fessenden, *Builder of Tomorrow*, p. 94.

[35]Ibid., p. 95.

[36]Ibid., p. 98. The specific terms were outlined in Fessenden's letter to Queen & Company dated June 12, 1902:

"About the contract, I note that you have made a mistake. I made a memorandum of our conversation, so that I should not forget it, within a few moments of its occurrence as my memory is rather poor sometimes. 'Mem. of agreement with Mr. Grey. 50 mile transmission to sell for $5,000 or $100 per mile. Allowing $500 for the cost of apparatus, and allowing Queen and Co. an additional $1,000 for manufacturers profit and office expenses, leaves $3,500 for division. Mr. Grey asked what portion of this I thought should be my share. Told him $2,000, leaving him $1,500 in addition to other $1,000. Said he thought this satisfactory.'

"I see that the contract was drawn up hurriedly, so possibly the amount allotted to me, i.e., $1,250, is a mistake, as I do not think that you will say that 'a profit of $2,500 on apparatus which can be bought in the open market for $500 is too small.' According to the agreement sent, your profit would be $3,250 and mine $1,250.

"I have assumed that there has been a slip, and have had another contract drawn up, in agreement with my memorandum of the agreement made in our conversation and forward it for your signature, if agreeable. I think that if you get, as you will according to this agreement, $3,000 for each 50 mile transmission on outside work and $3,600 on each pair for the navy, your company will make a good thing. 30 pairs at this rate would make over $100,000 and there is little doubt but that you will get at least this.

"My object was to give you such liberal terms that would be an object for you to push matters, and that you should be enabled to make a good round sum before any company was formed. Even in case the company was formed it was my intention to see that Queen and Co. retained a profitable connection, if my influence could manage it, and in case the company were formed sooner than anticipated, it is my intention to see that the contract is continued, if I can possibly arrange this (and my influence would naturally be strong to this end) until you have made an amount of profit which should be entirely satisfactory to you."

[37]Ibid., p. 105. See the chapter in Fessenden's book entitled "National Electric Signalling Company." See also the accounts of Sivowitch, "Technological Survey"; Howeth, *History of Communications*; Archer, *History of Radio to 1926*; and Blake, *History of Radio Telegraphy and Telephony*.

[38]Details of many of Fessenden's experiments can also be found in Sivowitch, "Technological Survey"; Howeth, *History of Communications*; Archer, *History of Radio to 1926*; and Blake, *History of Radio Telegraphy and Telephony* as well as other works cited.

[39]Archer, *History of Radio to 1926*, pp. 102–3.

[40]De Forest, *Father of Radio*, p. 268.

[41]Ibid., p. 260.

CHAPTER 3

[1]R. Franklin Smith, "Oldest Station in the Nation," in *American Broadcasting: A Source Book on the History of Radio and Television*, ed. Lawrence W. Lichty and Malachi C. Topping (New York: Hastings House, 1975), pp. 114–16 (originally published in *Journal of Broadcasting* (Winter 1959–60), 40–55.

[2]Gordon R. Greb, "The Golden Anniversary of Broadcasting," in *American Broadcasting*, ed. Lichty and Topping, pp. 95–96 (originally published in *Journal of Broadcasting* (Winter 1958–59), 3–13.

[3]Ibid., p. 98.

[4]Ibid., p. 102.

[5]Two accounts of WHA's early history are Werner J. Severin, "WHA-Madison 'Oldest Sta-

tion in the Nation' and the Wisconsin State Broadcasting Service" (paper presented at the meeting of the Association for Education in Journalism, Madison, Wisconsin, 1977); and *The First 50 Years of University of Wisconsin Broadcasting: WHA 1919–1969, and a Look Ahead to the Next 50 Years* (Madison: University of Wisconsin, 1970.)

[6]Our account of WWJ's early history is drawn from R. J. McLauchlin, "What the Detroit *News* Has Done in Broadcasting," in *American Broadcasting*, ed. Lichty and Topping, pp. 110–13 (originally published in *Radio Broadcast*, June 1922, pp. 136–41); WWJ Detroit: WWJ, 1936; and the brochures "WWJ Broadcasting Firsts" and "WWJ Radio One," published by WWJ in 1970 in commemoration of the station's sixtieth anniversary.

[7]The account of KDKA's history is from *The History of KDKA Radio and Broadcasting* (Pittsburgh: KDKA, n.d.). See also *American Broadcasting*, ed. Lichty and Topping, pp. 110–13.

[8]*The History of KDKA*, p. 10.

[9]Besides WHA, WWJ, and KDKA there were many different licensees—including municipalities—of early radio stations. The first city station was WPG in Atlantic City, New Jersey. It went on the air on January 3, 1925, with 500 watts of power, broadcasting from a station in the rear of the Atlantic City Senior High School building. *The Atlantic City Press* of May 27, 1927, reported a *Radio Fan* magazine survey (whose reliability has been questioned) showing that WPG was the seventh most popular radio station in the United States. The ten most popular stations, according to the survey, were (1) WJZ, New York; (2) WEAF, New York; (3) KDKA, Pittsburgh; (4) WLS, Chicago; (5) WGY, Schenectady, New York; (6) WBBM, Chicago; (7) WPG; (8) KFL, Los Angeles; (9) WBZ, Springfield, Massachusetts; and (10) WOC, Davenport, Iowa. (M. M. Anapol, "WPG Atlantic City: A Forgotten Chapter in the History of Broadcasting," paper presented at the annual meeting of the Speech Communication Association, San Antonio, Texas, November 1979.)

[10]Gleason L. Archer, *History of Radio to 1926* (New York: American Historical Society, 1938), p. 164.

[11]Ibid., p. 157.

[12]Ibid., pp. 162–63.

[13]Ibid., pp. 112–13.

[14]There are many accounts of the formation of RCA. These include Archer's *History of Radio to 1926*; his *Big Business and Radio* (1939; reprint ed., New York: Arno Press, 1971), pp. 3–22; and Eric Barnouw, *A Tower in Babel* (New York: Oxford University Press, 1966), pp. 52–61.

[15]Barnouw, *Tower in Babel*, pp. 44–45.

[16]Ibid., p. 49.

[17]W. R. MacLaurin, *Invention and Innovation in the Radio Industry* (1949; reprint ed., New York: Arno Press, 1971), p. 123.

[18]Ibid., p. 106.

[19]*Report on Chain Broadcasting* (Washington, D.C.: Federal Communications Commission, 1941), p. 10.

[20]Barnouw, *Tower in Babel*, p. 181.

[21]Archer, *History of Radio*, p. 276.

[22]*American Radio Journal*, 1 (June 15, 1922), 4; cited in William Peck Banning, *Commercial Broadcasting Pioneer: The WEAF Experiment* (Cambridge, Mass.: Harvard University Press, 1946), p. 94.

[23]Banning, *Commercial Broadcasting Pioneer*, p. 93.

[24]Ibid., pp. 231–36.

[25]See Archer, *Big Business and Radio*, pp. 133–65.

[26]Ibid., p. 169.

[27]Quoted in Ibid., p. 173.

[28]FCC *Report on Chain Broadcasting*, p. 17.

[29]Ibid., p. 92.

[30]Manuel Rosenberg, *Advertiser*, 14 (August 1943), 1–2, 24.

[31]FCC release (71159), October 12, 1943.

[32]Ibid.

[33]"Blues Sales Record an Outstanding One," Press release, Blue Network, 1942.

[34]Press release, American Broadcasting Company, March 29, 1945.

[35]Press release, American Broadcasting Company, June 15, 1945.

[36]FCC *Report on Chain Broadcasting*, p. 23.

[37]"The Way We've Been . . . and Are," *Columbine*, 2 (April–May 1974), 1. (*Columbine* is a corporate publication of CBS.)

[38]FCC *Report on Chain Broadcasting*, pp. 26–28.

[39]Eric Barnouw, *The Golden Web: The History of Broadcasting in the United States* (New York: Oxford University Press, 1968), p. 40.

[40]McLauchlin, "What the Detroit *News* Has Done," p. 186.

[41]Archer, *Big Business and Radio*, p. 424.

[42]Barnouw, *The Golden Web*, p. 242.

[43]The perspective on FM growth is from a Cox Broadcasting report, undated, but published in the 1970s. *Cox Looks at FM Radio: Past, Present and Future* (Atlanta: Cox Broadcasting Corporation), pp. 81–82.

[44]Stephen F. Hoffer, "Philo Farnsworth: Television's Pioneer," *Journal of Broadcasting*, Spring 1979, p. 157. Scholars who explore some of the earlier historical accounts of the telegraph will find that books written around 1850 elevate Bain's work above that of Samuel F. B. Morse. For his information on Bain, Hoffer cites Y. K. Zworykin, E. G. Ramberg, and L. E. Flory, *Television in Science and Industry* (New York: John Wiley, 1958), pp. 3–4.

[45]*From Semaphore to Satellite* (Geneva: International Telecommunication Union, 1965), p. 61.

[46]Hoffer, "Philo Farnsworth," p. 157; Zworykin, Ramberg, and Flory, *Television in Science and Industry*, p. 4.

[47]Important sources for the material on Philo Farnsworth are Romaine Galey Hon, ed., *Headlines Idaho Remembers* (Boise: Friends of the Bishops' House, 1977), p. 39; *Idaho Statesman*, July 13, 1953; Hoffer, "Philo Farnsworth" pp. 153–65. Another valuable source is George Everson, *The Story of Television* (New York: W.W. Norton & Co., Inc., 1949), p. 266. I am grateful to the staff of the Duke University library for helping me locate many of the documents I needed in order to check facts and obtain early perspectives on the development of radio and television.

[48]Hoffer, "Philo Farnsworth," p. 154. Everson places it as near Bever City, Utah (*Story of Television*, p. 15).

[49]Hoffer, "Philo Farnsworth," p. 156.

[50]Ibid.

[51]Ibid., p. 160.

[52]Ibid., pp. 160–61, citing the following patents:

United States Patent 1,773,980, Philo T. Farnsworth of Berkeley, California, Assignor, by Mesne Assignments to Television Laboratories, Inc., of San Francisco, California, A Corporation of California, Television System, August 26, 1930, p. 1.

United States Patent 1,773,981, Philo T. Farnsworth of Berkeley, California, Assignor, by Mesne Assignments to Television Laboratories, Inc., of San Francisco, California, A Corporation of California, Television Receiving System, August 26, 1930, p. 1.

[53]See K. B. Benson, "A Brief History of Television Camera Tubes," *Journal of SMPTE*, 90 (August 1981), 708–12.

[54]Hoffer, "Philo Farnsworth," p. 158.

[55]Ibid., p. 159, in reference to the Farnsworth-Zworykin dispute.

[56]Ibid.

[57]Ibid., pp. 161–62.

[58]*Columbine*, 2 (April/May 1974), 8.

[59]See Benson, "Television Camera Tubes," p. 708.

[60]Ibid.

[61]Ibid.

[62]The tube has been continually improved upon, with special attention given to its optical and scanning systems. The Plumbicon has the ability to capture color images with the sensitivity of the human eye. Two other tubes achieving acceptance since the Plumbicon are the Saticon—a registered trademark of NHK, the Japanese Broadcasting Corporation—and the Newvicon—a registered trademark of Matsushita Electronics Corporation. The circuitry and design of the latter two are derived from the vidicon family.

[63]Ron Whittaker, "Super 8 in Broadcasting, CATV and CCTV: Current Technology and Applications" (unpublished paper, University of Florida, 1975). The specific advantages Whittaker saw for Super-8 broadcast use were: (1) electronic image enhancers can increase image sharpness and provide clarity on television comparable with 16mm film; (2) advances in the film-emulsion process have reduced graininess; (3) Super-8 works well in low light conditions; (4) Super-8 equipment is still more portable and lighter than ENG equipment; (5) at low light levels, a picture "lag" or "smear" can occur with many electronic cameras, whereas film can han-

dle the greater brightness range; (6) film can be processed in as little as fifteen minutes; (7) the Super-8 camera is small and inconspicuous compared with most ENG equipment, which can be especially important when news teams cover such things as civil unrest, in which the presence of TV cameras can trigger crowd reaction; and (8) stringers can use Super-8 cameras easily without much training and at less cost than an ENG setup.

[64]Sources on the history of television recording include Albert Abramson, "A Short History of Television Recording," *Journal of SMPTE*, 64 (February 1955), 72–76, and "A Short History of Television Recording: Part II," *Journal of SMPTE*, 82 (March 1973), 188–98; and Joseph Roizen, "Video-tape Recorders: A Never-Ending Revolution," *Broadcast Engineering* (April 1976), 26–30, and "The Videotape Recorder Revolution," *Broadcast Engineering* (May 1976), 50, 52–53.

[65]See Roizen, "Video-tape Recorder Revolution."

[66]"Radio: State of the Art 1989," *Broadcasting* (July 24, 1989), 37–69; "Growth Depends on Finer Tuning," *Advertising Age* (November 28, 1988), 5–13; "AM: Band on the Run," *Broadcasting* (November 11, 1985), 35, 46, 48, 50, 52.

CHAPTER 4

[1]William Chauncy Langdon, *Bell Telephone Quarterly* (July, 1923), 5. (Page references are to the 1979 reprint.)

[2]Ibid.

[3]Ibid.

[4]Ibid., p. 6.

[5]Ibid.

[6]Of interest is the account that:

> Bell had been carrying on studies with a laboratory device called the phonautograph, which made mechanical tracings of sound vibrations. He made an improved phonautograph, using the bones and drum of a human ear procured for him by a friend, Dr. Clarence Blake. Bell explained how his experiments with this improved instrument led him to his conception of the telephone as follows: "I was much struck by the disproportion in weight between the membrane and the bones that were moved by it; and it occurred to me that if such a thin and delicate membrane could move bones that were, relatively to it, very massive indeed, why should not a larger and stouter membrane be able to move a piece of steel in the manner I desired? At once the conception of the membrane speaking telephone became complete in my mind. . . . The arrangement thus conceived in the summer of 1874 was substantially similar to that shown . . . in my patent of March 7, 1876." Bell described to his father, while on a visit to the family home at Brantford, Ontario, on this date (1874) his conception of his electric speaking telephone using his undulating current principle. (*Events in Telecommunications History* (New York: AT&T, 1979), pp. 2–3. Date (1874) added.

[7]Prior to the formation of the company, on August 10, 1876,

> the world's first long-distance telephone call (one-way) was received at Paris, Ontario, by Bell from his father and from his uncle at Brantford, Ontario, over telegraph lines 8 miles and 68 miles away, respectively. Referring to this call, Bell said: "This Brantford experiment is of historical interest . . . because it led to the discovery of the proper combination of parts in a telephone to enable it to become operative upon a long line; and because upon this occasion occurred the first transmission of the human voice over a telegraph line in which the transmitting and receiving telephone was miles apart."
>
> Permission to use this telegraph line was granted by Lewis B. McFarlane, a telegraph manager, who entered the telephone business in 1879, was president of The Bell Telephone Company of Canada from 1915 to 1925 and chairman of the board from 1925 to 1930. (*Events in Telecommunications History*, p. 3.)

[8]Hubbard offered to sell the telephone invention to Western Union for $100,000, but the offer was refused. (*Events in Telecommunications History*, p. 3.)

[9]Chauncy, *Bell Telephone Quarterly*, p. 8.

[10]Ibid., p. 10.

[11]Ibid.

[12]Ibid., p. 13.

[13]Ibid., p. 14.

[14]The District Telephone Company of New Haven, organized by licensee George W. Coy along with Walter Lewis and H. P. Frost, had twenty-one subscribers. (*Events in Telecommunications History*, p. 5.) "First" exchanges the following year (1879) included:

February	Louisville, Ky.
February 15	Minneapolis, Minn.
February 24	Denver, Colo.
March	Indianapolis, Ind.
March 15	New Orleans, La.
April 1	Richmond, Va.
April 2	Providence, R.I
April 26	Toronto, Canada
May 1	Montreal, Canada
June	Omaha, Neb.
June 1	Burlington, Vt.
June 4	Topeka, Kans.
June 11	Dubuque, Ia.
August 1	Augusta, Ga.
August 15	Camden, N.J.
August 21	Galveston, Tex.
August 26	Charleston, S.C.
September 1	Portland, Me.
September 1	Brantford, Canada
September 20	Raleigh, N.C.
September 22	London, Canada
November 1	Little Rock, Ark.
November 15	Mobile, Ala.
December	Nashville, Tenn.

(*Events in Telecommunications History*, p. 7.)

[15][Elisha Gray] worked his way through Oberlin College as a carpenter. His work in physics at Oberlin quickly narrowed to electrical applications. He invented a self-adjusting telegraph relay, a telegraphic switch, and a repeater. He continued in the firm of Gray and Barton only two years, after which he devoted himself entirely to electrical research. Gray was working on a harmonic telegraph at the same time as Bell, and perfected what he called a telephone to transmit musical sounds. The idea of transmitting vocal sounds occurred to him, and on Feb. 14, 1876, he filed a caveat (a confidential report of an invention which is not fully perfected) in the U.S. Patent Office. His caveat indicated that he was on the same track as Bell, but had not worked out his transmitter as fully. And on that same day, but a few hours earlier, Alexander Graham Bell had filed a patent application for his telephone, thus anticipating Gray. Gray's most important invention thereafter probably was the telautograph which transmits facsimile handwriting and drawings. At the time of his sudden death in 1901, he was experimenting on underwater signaling to vessels at sea. (*Events in Telecommunications History*, p. 2)

[16]Competition in the telephone business developed from the Western Union Telegraph Company through its newly established subsidiaries, the American Speaking Telephone Company and the Gold & Stock Telephone Company. These companies used Thomas A. Edison transmitters and Elisha Gray receivers.

> The Handset Telephone—Robert G. Brown, who became chief operator at the Western Union (Gold & Stock) exchange in New York City, devised in 1878 (possibly in April but more probably in May) the first handset, mounting an Edison transmitter and a Gray receiver on a bar of metal. Brown was sent to France in 1879 to open an exchange at Paris. There, his handset was lightened to make it more easily handled by women operators. It became popular in France and Europe despite limitations that made it unacceptable in the United States, and for years was known as the French or Continental telephone. Brown died at St. Petersburg, Florida, October 2, 1947, at the age of 93. (*Events in Telecommunications History*, p. 5.)

[17]Chauncy, *Bell Telephone Quarterly*, pp. 23–24.

[18]Ibid., p. 24.

[19]Ibid.

[20]Ibid., p. 25

[21]Two of the companies, Southern New England Telephone and Cincinnati Bell, were unaffected by the divestiture since AT&T owned a minority interest in them.

[22] Formerly called American Bell, in 1983 it was renamed as a result of further negotiations in which AT&T agreed to cease using the name Bell except for Bell Labs.

[23]For two perspectives on cellular radiotelephone services, see "Cellular Radio," *Broadcasting* (June 7, 1982), 38–42; and W. Ginsberg, "New Cellular Radio Systems Promise Bright Future for Entrants and Users," *Broadcasting* (March 1982), 36–37. Control over

cellular radio telephone systems was initially a controversial regulatory issue, which centered on the stake AT&T would have in developing cellular radio in competition with other companies. The Justice Department said AT&T's involvement would be "blatantly anti-competitive." (United Press International, January 3, 1982.)

[24]N. Sweet, "Airplane Phoning Gets Off the Ground," *Crain's Chicago Business* (January 17, 1983), T10. This article describes one of the first plane-to-ground telephone services.

[25]Ibid. The same concept of cell-to-cell switching that is embodied in cellular radiotelephone systems is also utilized in plane-to-ground systems.

[26]Two reports from the trade press are E. Gold, "Trends in Teleconferencing Today Indicate Increasing Corporate Use," *Communications News*, 19 (October 1982), 48–49; and "Teleconferencing Today," *Communications News*, 19 (February 1982), 43–66.

[27]See Waldo T. Boyd, *Fiber Optics* (Indianapolis: Howard W. Sams & Co., Inc., 1982), p. 28–44.

[28]Julie Amparano Lopex, "Broadcasters Urged to Try Fiber Optics As Alternative to Satellite Transmission," *Wall Street Journal* (February 27, 1990), p. B4.; Also, A.C. Deichmiller, "Advances in Fiber Optics Are Spurring Broadband Services to Home and Office," *Communications News* (January 1981), 52–53; "First Network TV Program Via a Lightwave Link, *Communications News*, 17(January 1980), 54.

[29]Early optic cables employed three types of fibers. The most common, multi-mode graded-index fiber, is used in some cable systems. It has broad bandwidths and a diameter of from fifty to sixty-three micrometers—less than the thickness of human hair.

Single-mode fibers are five micrometers thick (much less than the thickness of multi-mode graded-index fibers), and are mostly experimental, since their small diameter creates problems in installation and maintenance.

Multi-mode step-index fibers have less bandwidth and carry less information. Source: A. C. Deichmiller, "Advances in Fiber Optics Spurring Broadband Services to Home and Office," *Communications News*, 18 (January 1981), 52–53.

CHAPTER 5

[1]With this in mind, we can determine the wavelength of radio waves by simple division. For example, given that electromagnetic waves travel at a speed of 186,000 miles per second, if 10,000 complete waves pass a given point in one second, the wavelength of each wave would be 18.6 miles (186,000 ÷ 10,000). Now compute the wavelength of a higher frequency, 535 kilocycles (535,000 cycles per second). We divide 186,000 by 535,000. The answer is 0.3477 miles. Since there are 5,280 feet in one mile, we can convert our wavelength to feet by multiplying 5,280 by 0.3477. The answer is a wavelength of 1,836 feet.

[2]The video portion of the signal is sent over AM; the audio signal by FM. "The effective radiated power of the aural transmitter shall not be less than 10 percent nor more than 20 percent of the peak radiated power of the visual transmitter" (FCC Rules 73.682 (a)(15)). FCC Rules 73.881 define the television broadcast band as "the frequencies . . . extending from 54–890 megahertz which are assignable to television stations. These frequencies are 54 to 72 megahertz. (channels 2 and 4), 76 to 88 megahertz (channels 5 and 6), 74 to 216 megahertz (channels 7 through 13), and 470 to 890 megahertz (channels 14 through 83)." Because channel 6 is part of the FM broadcast band, it can be heard on the lower end of most FM radios. Approximately 4 MHz of the frequency range allocated to television stations are used for video transmission (FCC Rules 73.699, figure 5).

[3]Not to be confused with the three primary pigment colors: red, yellow, and blue.

CHAPTER 6

[1]See Richard Witkin, "Live Images Transmitted Across Ocean First Time," *New York Times*, July 11, 1962, p. 16, and "Europeans Beam First Television to Screens in U.S.," *New York Times*, July 12, 1962, pp. 1, 12. The specific agreement, "Cooperative Agreement between the National Aeronautics and Space Administration and the American Telephone and Telegraph Company for the Development and Experimental Testing of Active Communications

Satellites" provided backup launching systems and stipulated that all data resulting from the experiments were to be made available to NASA.

[2]That bit of political rhetoric was meant to pull down the fence that FCC commissioner Newton Minow had put up a year earlier between the administration and commercial broadcasters with his famous "vast wasteland" speech before the National Association of Broadcasters convention.

[3]Jack Gould, "TV: Telstar and World Broadcasting," *New York Times*, July 11, 1962, p. 71.

[4]Leonard H. Marks summarized these events in an article in the *Journal of Broadcasting* entitled, "Communication Satellites: New Horizons for Broadcasters," 9 (Spring 1965), 97–101. The article also summarized issues and asked probing questions.

[5]This account of the launch of Syncom II and Hughes Aircraft Company's part in it draws upon "Mr. Watson I Want You," *Vectors*, 15 (Summer/Fall 1973), 7–9.

[6]Anthony Lewis, "Sarnoff Suggests Industry Merger," *New York Times*, August 8, 1962, pp. 1, 14.

[7]Western Union eventually established its own satellite system.

[8]"Interactive Satellite ATS-6 Brings People Together," *Broadcast Management Engineering*, 10 (November 1974), 30–44.

[9]*Broadcasting*, October 10, 1988. Ads were not the only mark of Alpha Lyracom's presence. In a filing before the Federal Communications Commission, Pan American Satellite questioned the intentions of who will be served by the INTELSAT VII series satellites. PAS objected to Comsat's participation in the INTELSAT VII series satellites calling it "speculative investment" that would be underwritten by United States rate payers who would end up paying for a satellite system dedicated to foreign domestic application. "Bird Deal," *Broadcasting*, October 10, 1988, p. 49.

[10]Information on the Westar series was obtained from "*Westar* Satellite Backgrounder," a publication of Western Union, July 1981.

[11]"Satellites Flying Toward the Future," *Broadcasting* (July 17, 1989), 44.

[12]"GE Americom and Alascom to Launch Two Satellites," *Broadcasting* (October 10, 1988), 49.

[13]"The Uncertain Future of DBS," *Broadcasting* (May 13, 1989), 42.

[14]Kevin Goldman and Laura Landro, "Four Media Giants Enter Venture for Direct-Broadcast TV Service," *Wall Street Journal* (February 22, 1990), B4.

[15]"European DBS: Off to an Unsteady but Enthusiastic Start," *Broadcasting* (July 17, 1989), 55–56.

CHAPTER 7

[1]"Cable: The First Forty Years," *Broadcasting* (November 21, 1988), 38.

[2]An assessment of the medium can be found in T. Streeter, "The Cable Fable Revisited: Discourse, Policy, and the Making of Cable Television," *Critical Studies in Mass Communication*, 4 (1987), 174–200.

[3]T. L. Childers and D. M. Krugman, "The Competitive Environment of Pay Per View," *Journal of Broadcasting and Electronic Media*, 31 (Summer 1987), 335–342.

[4]See "The Growing Radio-Cable Connection," *Broadcasting* (May 22, 1989), 58–60.

[5]W. Walley, "Victim of Its Own Success," *Advertising Age* (November 28, 1988), S-7, S-22.

[6]"The Cable Network Programming Universe," *Broadcasting* (August 22, 1988), 32–33.

[7]J. R. Finnegan and K. Viswanath, "Community Ties and Use of Cable TV and Newspapers in a Midwest Suburb," *Journalism Quarterly*, 65 (Summer 1988), 456–63, 473.

[8]K. Goldman, "Broadcasters, Cable Enter 'Era of Blur'," *Wall Street Journal* (September 28, 1989), p. B1.

[9]P. H. Feinberg, "Cable television—getting ready for the next decade," *Broadcast Financial Journal* (May–June 1988), 12–14.

[10]For a perspective on the problem of disconnects, see S. Spillman, "CATV Operators Churning 34 Over Disconnects," *Advertising Age* (November 15, 1982), 52.

CHAPTER 8

[1]For example, land mobile services in the UHF spectrum.

[2]Background on multipoint distribution service can be found in *New Technologies Affecting Radio and Television Broadcasting* (Washington, D.C.: National Association of Broadcasters, 1981), pp. 6–7.

[3]See, for example, "Small Earth Stations Blossom into Big Businesses," *Broadcasting* (December 22, 1980), 32, 34, 35, 38. See also *SMATV: Strategic Opportunities in Private Cable* (Washington, D.C.: National Association of Broadcasters, 1982).

[4]*SMATV*, pp. 28–29.

[5]See the National Association of Broadcasters' research report "Subscription Television" (Washington, D.C., 1980).

[6]See, for example, R. Anderson, "Satellite-Aided Land Mobile Radio System Could Prove Cost-Effective," *Communications News*, 18 (March 1981), 56–59.

[7]This system is based on the Dow Jones Alert News Service. See "Dow Alert Service Spawns New Generation for Radio," *Electronic Media* (January 13, 1983), 5, 20.

[8]R. A. Shaffer, "Promising Uses Are Emerging for Millimeter Radio Signals," *Wall Street Journal* (June 26, 1981), p. 29.

[9]Ibid.

[10]"Banishing Ghastly Music from Your Car Radio," *Business Week* (February 12, 1990), 95.

[11]A National Association of Broadcasters' research report on radio services quoted Emil L. Torick, Director of Audio Development for the CBS Technology Center, as predicting the change to digital radio. See M. L. DeSonne, *Radio, New Technology, and You* (Washington, D.C.: National Association of Broadcasters, 1982), p. 24. A perspective on digital audio can be found in this same report on page 18.

[12]Mark R. Levy, "Home Video Recorders and Time Shifting," *Journalism Quarterly* (Autumn 1981), 404–5.

[13]Ibid.

[14]"Zap! Zap!, Video Games Are Back," *Newsweek* (March 14, 1988), p. 39.

[15]B. Loveless, "The Broadcasting of Teletext," *Educational and Industrial Television*, 16 (June 1979), 34–35.

[16]Ibid., p.35. See also "U.S. in Need of Videotex Primer," *Advertising Age Electronic Media Edition* (June 10, 1982), 23.

[17]W. Paisley, "Computerizing Information Lessons of a Videotex Trial," *Journal of Communication* (Winter 1983), 155.

CHAPTER 9

[1]W. Walley, "Network TV pressure point," *Advertising Age* (June 27, 1988), 102.

[2]The author is especially grateful to the staff of the public-information department at ABC, which provided many original documents dealing with the history of that network.

[3]Ibid., pp. 68–69.

[4]Ibid.

[5]"Capcities + ABC," *Broadcasting* (March 25, 1985), 31–33.

[6]Source for the CBS–Olympic contract: Patrick McGeehan, "CBS Grabs Olympics," *Advertising Age* (May 30, 1988), 3.

[7]"CBS Sports Spending Spree," *TV Technology* (October 1989), 3.

[8]For an analysis of the strategies behind network diversification see: John Dimmick and Mikel Wallschlaeger, "Measuring Corporate Diversification: A Case Study of New Media Ventures by Television Network Parent Companies," *Journal of Broadcasting & Electronic Media,* 30 (Winter 1986), 1–14.

[9]"Is Westwood One Doomed to Be No. 2?" *Business Week* (December 12, 1988), 94.

[10]"Fox Broadcasting Co.: The birth of a network," *Broadcasting* (April 6, 1987), 88.

[11]"In the Race for Viewers, the Networks Fall Further Behind," *Business Week* (January 9, 1989), 81.

CHAPTER 10

[1]The account of this broadcast is drawn from C. C. Clark, "Television in Education," *School and Society*, 48 (October 1, 1938), 431–32.

[2]"Metropolitan Art Is to Be Televised," *New York Times*, May 26, 1941, p. 21.

[3]The account of this series is based on "NBC's Educational Television Series," *School and Society*, 63 (February 16, 1947), 110.

[4]William M. Dennis, "Transition to Visual Education," *National Education Association Journal*, 35 (October 1960), 424.

[5]Amo DeBernardis and James W. Brown, "A Study of Teacher Skills and Knowledge Necessary for the Use of Audio-Visual Aids," *Elementary School Journal*, 46 (June 1946), 550–56.

[6]"A Research Fellowship in Television Education," *School and Society*, 69 (April 16, 1949), 278.

[7]"The U.S. Commissioner of Education on the Television," *School and Society*, 72 (December 23, 1950), 427.

[8]"Colleges and Universities Prepare Television Programs," *School and Society*, 72 (September 2, 1950), 155–56.

[9]For a history of the MPATI, see Norman Felsenthal, "MPATI: A History 1959–1971," *Educational Broadcasting Review*, 5 (December 1971), 36–44.

[10]See Richard J. Stonesifer, "The Separation Needed Between ETV and ITV," *A Communication Review*, 14 (Winter 1966), 489–97.

[11]Ibid., p. 490, quoting Doris Willens, "ETV: An Uncertain Trumpet," *Television Magazine*, (February 21, 1964).

[12]*Public Television: A Program for Action*. Report of the Carnegie Commission on Educational Television (New York: Bantam, 1967), Preface.

[13]Corporation for Public Broadcasting, *1981 Annual Report* (Washington, D.C.: Corporation for Public Broadcasting, 1981), p. 11.

[14]Ibid., p. 18.

[15]"New Public TV Service Names Board," *Broadcasting* (September 18, 1989), 65.

[16]"CPB Narrows Its Budgetary Focus," *Broadcasting* (May 23, 1988), 42.

[17]"PBS Ends 'Sponsor' Taboos," *Advertising Age* (July 5, 1982), 4.

[18]Ibid., supplement sheet.

[19]Paul B. Brown and Maria Fisher, "Big Bird Cashes In," *Forbes* (November 5, 1984), 176–78, 182.

[20]New York Times News Service, "Use of Video Spreads Among Corporations," *The Chapel Hill Newspaper* (August 28, 1988), p. 16D, citing research of D/J Brush Associates.

[21]Material provided the author by Federal Express Employee Communications, June 1, 1989.

CHAPTER 11

[1]The author is indebted to Will Davis of Capitol Broadcasting for his many insights that helped formulate this chapter.

CHAPTER 12

[1]National Public Radio, "Chinese Gardens," Weekend Edition No. 891104, 1989.

[2]David Traub, "Video-Computing's Third-Wave," *Videography* (October 1989), 46.

CHAPTER 13

[1]Canadian Broadcasting Act of 1968.

[2]"The Educational and Cultural Television," Radio Quebec brochure, 1988.

[3]Ibid. French derivations in spelling have been changed to standard English usage; for example, "program" from "programme."

[4]"This Is Independent Broadcasting," IBA publication 1988; "IBA Factfile," 1988–1989.

[5]"Local Oracle," *Airwaves* (Spring 1988), 23.

[6]B. Toman, "Britain's Commercial-TV License Sale to Be Open to Foreign-Media Bidders," *Wall Street Journal* (March 12, 1990), p. B6A.

[7]Especially valuable in preparing the information on the Nordic countries, including Finland was: "Nordic Television in Transition," booklet published by The Nordic Council of Ministers, (1989).

[8]A. G. Sennitt, ed. *World Radio TV Handbook* (New York: Billboard Publications, Inc., 1989). Sennitt's book was especially valu-

able in preparing material for both chapters 13 and 14 and is hereby gratefully acknowledged.

[9]The source for G.D.R. Television is *Fernsehen der DDR*, 1989.

[10]Examples of these themes can be seen in G.D.R. Television's promotional brochure issued in 1989 prior to the reunification efforts in Germany, which stated:

> "The series entitled "Race Against Time," which now stretches to some sixty programmes, provides innumerable examples of the use of microelectronics and biochemistry in G.D.R. industry and agriculture, illustrating the ways in which hi-tech helps to improve conditions for people at work. Basic knowledge on how to operate computers and put them to good use is provided in the 'Computer Time' series. 'Review' looks at recent research and international trends in science and technology, while issues of concern in the field of medicine, including AIDS, the environment and raw materials supply are examined in the presence of leading scientists in the URANIA debate series." (*Source*: Fernshen der DDR.)

[11]See David E. Powell, "Television in the U.S.S.R.," *Public Opinion Quarterly*, 39 (Fall 1975), 287–300.

[12]Announced in the United States as part of a series of news stories by the rock singer Frank Zappa on the Financial News Network in February 1990. *Source*: Associated Press story by Jay Sharbutt, "Rock Star Frank Zappa Begins Stint on Financial News Network," *The News and Observer* (February 26, 1990), p. 10D.

[13]Pyotr Smirnov, "The Breakfast Show," *Soviet Life* (May 1988), 2.

[14]See Milton Hollstein, "French Broadcasting After the Split," *Public Telecommunication Review*, 6 (January/February 1978), 15–19.

[15]Information for the section on the People's Republic of China is either quoted directly or paraphrased from *Radio, Film and Television in the People's Republic of China* (Beijing: Ministry of Radio, Film and TV, n.d.) citing through 1986. Provided the author by the Foreign Affairs Bureau of the Ministry of Radio, Film and TV. Translation corrections into standard English have been inserted when quoting directly.

[16] Ibid. p. 8.

[17]Sources on early broadcasting in China and the Republic of China is from material provided the author by the Broadcasting Corporation of China and the Voice of Free China of the BCC, 1989.

[18]The Central Broadcasting System is an independent broadcasting division under the BCC and is responsible for broadcasts to Mainland China.

[19]Quoting from the information material supplied by the BCC:

> "Each day and all day, the station reports of traffic conditions on the highway to help motorists arrive at their destinations both quickly and in a safe manner. Another example is the station at Chung-hsing New Village in Nantou County of Central Taiwan. This station is the Agricultural Special Station. The station broadcasts in both Mandarin and Amoy in order to provide farmers and fisherman with the newest information and technology."

[20]David W. Chen, "Tuning in to Air Power," *Free China Review*, 39 (August 1989), 50–53. In a youth feature in the *Free China Review* a writer noted:

> "The BCC, which has been influenced by ICRT's style, has tried a somewhat different approach—hiring disc jockeys whose English is as good, if not better, than their Chinese. In fact, one of their most popular disc jockeys is an Overseas Chinese who just graduated from the University of California at Berkeley. As a result, it is not uncommon to hear someone of the BCC speak occasionally in English, then translate it into Chinese. Radio personalities now seem more approachable, which helps raise station ratings and draw in more advertising as well." (*Source*: "Tuning in to Air Power," p. 52.)

[21]*Television Yearbook of the Republic of China*, 1986–1987, p. 34–35.

[22]Ibid.

[23]Ibid.

[24]"Outline of NHK," NHK publication, 1988.

[25]"Sound Broadcasting," NHK publication, 1988.

[26]NTT Annual Report, 1989. The company lists 283,000 employees.

[27]Ibid. Nippon Directory Development Co. Ltd. is the joint ITT-WD and NTT venture; Battelle Memorial Institute is the joint U.S. research firm, and the battery is being developed with Moli Energy Ltd. of Canada. NTT was involved in more than 130 different companies by the early 1990s.

[28]This section was prepared from wire services; *The Wall Street Journal*; Sennitt's *World Radio TV Handbook*; CNN; various editions of the *Encyclopedia Americana*; B. Rubin, "Reshaping the Middle East," *Foreign Affairs*, 69 (Summer, 1990), 131–146; E. Sciolino, "Iraq Yearns for Greatness and an Identity," *New York Times*, (August 4, 1990), p. 1, 3, Sec. 4.

CHAPTER 14

[1]"RCI: An Overview," Radio Canada International information sheet.

[2]*BBC Handbook*. This publication is updated annually.

[3]Radio Orange used the facilities of the BBC.

[4]"40th Anniversary of Radio Nederland, The International Service," booklet describing the jubilee year of Radio Nederlands, 1987, p. 6.

[5]From the 1989–1990 Program Schedule of Radio Moscow's North American Service.

[6]From the 1989 Program Schedule of Radio Moscow's World Service.

[7]"Soviets Bar Episode of Popular TV Program," *Raleigh News and Observer*, December 31, 1989, p. 22a.

[8]*Radio, Film and Television in The People's Republic of China* (Beijing: Ministry of Radio, Film and TV, Foreign Affairs Bureau), p. 9. (English-language section.) Provided to the author in correspondence of August 24, 1989.

[9]Ibid.

[10]*Radio, Film and Television in The People's Republic of China* (Beijing: Ministry of Radio, Film and TV, n.d., citing through 1986), p. 8.

[11]Information on the Voice of Free China and the broadcasting system of the Republic of China is from material provided the author by "VOFC, the Voice of Free China Broadcasting Monthly," No. 76, and the VOFC listener's pamphlet of June 1989.

[12]An interesting comparison appears in No. 76 of the "VOFC Broadcasting Monthly," which states:

> "In terms of cost-effect, the VOFC compares favorably with other major international broadcasting services. For example, the VOFC has an annual budget only 1/24 that of the VOA and 1/19 that of the BBC. In the size of the work forces, it is only 1/10 of the VOA. However, for each of the letters from its listeners, it spends only US$93 while the figures for the VOA and the BBC are US$767 and US$252 respectively." (p. 30)

[13]Ibid.

[14]ABC Annual Report, July 1987–June 1988.

[15]Information for this section is from material provided the author by British Satellite Broadcasting.

[16]"Woman with a Mission," *Airport* (April 1989), 31–32. Information for this section of the text provided the author by Super Channel. Countries served in the early 1990s include Germany, Austria, Switzerland, Belgium, Netherlands, Luxembourg, Sweden, Norway, Finland, Denark, UK, Ireland, Hungary, Yugoslavia, Spain, and France.

[17]Information on TeleClub is from the Office of the Director and correspondence with the Assistant to the Director, Manuela Meier.

[18]The license for the TeleClub experiment was reviewed in 1990 and originally granted as part of an experiment to test demand for new media technologies. The legal basis for the experiment was based on:

> the article on radio and television in the Swiss Constitution
>
> the regulation on local radio experiments (RVO)
>
> the report of the Swiss Committee of experts in a General Media License (EK MGK)
>
> "The license granted to the Swiss association for subscription television (STA), forming the basis of the experiment carried out by TeleClub, is intended to provide some indication as to the demand for pay TV at the end of this experimental phase. Furthermore, it is to be ascertained what are the effects of

pay TV on the other media, especially the film industry." (*Source*: TeleClub public-information materials.)

[19]An early analysis of the topic can be found in "New Kids on the Bloc: Commercial Broadcasting Struggling for Toehold in Eastern Europe," *Broadcasting* (January 8, 1990), 101–102.

[20]Ibid.

[21]Ibid.

[22]C. Sims, "Baby Bells Scramble for Europe," New York Times News Service, *Raleigh News and Observer* (December 17, 1989), pp. 1G–3G.

CHAPTER 15

[1]The Wireless Ship Act of 1910, Public Law 262, 61st Congress, June 24, 1910.

[2]The Radio Act of 1912, Public Law 264, 62nd Congress, August 13, 1912, Sec. 1.

[3]See Edward F. Sarno, Jr., "The National Radio Conferences," *Journal of Broadcasting*, 13 (Spring 1969), 189–202.

[4]Ibid. For a summary of Department of Commerce action during this period, see Marvin R. Bensman,"Regulation of Broadcasting by the Department of Commerce, 1921–1927," in *American Broadcasting: Source Book on the History of Radio and Television*, ed. Lawrence W. Lichty and Malachi C. Topping (New York: Hastings House, 1975), p. 5.

[5]*Hoover* v. *Intercity Radio Co., Inc.* 286 F. 1003 (D. C. Cir), February 25, 1923.

[6]*United States* v. *Zenith Radio Corporation et al.* 12F. 2d 614 (N. D. Ill.), April 16, 1926.

[7]See Eric Barnouw, *A Tower in Babel: A History of Broadcasting in the United States* (New York: Oxford University Press, 1966), p. 175.

[8]Attorney General's Opinion, 35 Ops. Att'y Gen. 126, July 8, 1926.

[9]See H. Doc. 483, 69th Congress, 2nd Session.

[10]The Radio Act of 1927, Public Law 632, 69th Congress, February 23, 1927, Sec. 3.

[11]S. Doc. 164, 73d Congress, 2d Session, February 26, 1934.

[12]Communications Act of 1934, Sec. 326.

CHAPTER 16

[1]FCC Annual Report.

[2]The organization of the FCC is from the 1988 FCC Annual Report.

[3]Ibid.

[4]Ibid.

[5]The FCC's authority to issue short-term renewals was granted by the same statute permitting it to issue forfeitures.

[6]Information about the FTC is from *A Guide to the Federal Trade Commission*, 1989.

[7]See "The FTC Advertising Review Process," *Advertising Age*, 48 (July 11, 1977), 142.

[8]Your FTC, p. 26. (Publication of the FTC), n.d.

[9]Discussions of the history, organization, and function of the ITU and of issues it faces can be found in David M. Leive, *International Telecommunications and International Law: The Regulation of the Radio Spectrum* (Dobbs Ferry, N.Y.: Oceana, 1971); and *Global Communications in the Space Age: Toward a New ITU* (New York: John and Mary R. Markle Foundation and the Twentieth Century Fund, 1972).

[10]Harold K. Jacobson, "The International Telecommunication Union: ITU's Structures and Functions," in *Global Communications in the Space Age*, p. 60.

[11]Final Protocol, Documents of the Berlin Preliminary Conference (1903), cited in Leive, *International Telecommunications and International Law*, pp. 83–85. The thrust of the protocol agreement survives in contemporary broadcast regulations.

[12]International Telecommunication Convention (Montreux, 1965). See also Jacobson, "The International Telecommunication Union," pp. 40–41.

CHAPTER 17

[1]Lee Loevinger, "The Role of Law in Broadcasting," *Journal of Broadcasting*, 8 (Spring 1964), 115–17.

[2]Ibid.

[3]See, for example, V. Blasi, "The Newsman's Privilege: An Empirical Study," *Michigan Law Review*, 233 (December 1971).

[4]Loevinger, "Role of Law in Broadcasting," pp. 115–17.

[5]Section 73.120. In particular, state laws vary in the ways candidates for local office can be placed on the ballot.

[6]Gillmor and Barron, *Mass Communication Law*, (St. Paul: West, 1979), p. 282. The case is *Farmers Educational and Cooperative Union of America, North Dakota Division* v. *WDAY Inc.*, 89 N. W. 2d 102, 109 (N. D. 1958).

[7]*Farmers Educational and Cooperative Union of America* v. *WDAY Inc.*, 360 U.S. 525, 79 S. Ct. 1302, 3 L. Ed. 2d 1407 (1959).

[8]*In the Matter of Editorializing by Broadcast Licensees*, 13 FCC 1246, June 1, 1949.

[9]*In the Matter of the Mayflower Broadcasting Corporation and The Yankee Network, Inc. (WAAB)*, 8 FCC 333, 338, January 16, 1941.

[10]*In reference to United Broadcasting Co. (WHKC)*, 10 FCC 515 June 26, 1945.

[11]*In reference to Petition of Robert Harold Scott for Revocation of Licenses of Radio Stations KQW, KPO and KFRC*, 11 FCC 372, July 19, 1946.

[12]Ibid.

[13]*In the Matter of Editorializing by Broadcast Licensees,* 13 FCC 1246, June 1, 1949.

[14]*NAB Legal Guide to Broadcast Law and Regulation* (Washington, D.C.: NAB, 1988), p. 42.

[15]For the early development of legal precedent in the area of regulating obscene, indecent, and profane programming, see James Walter Wesolowski, "Obscene, Indecent, or Profane Broadcast Language as Construed by the Federal Courts," *Journal of Broadcasting*, 13 (Spring 1699), 203–19.

[16]Title 18, United States Code (Codified June 25, 1948, Ch. 645, 62 Stat. 769).

[17]*Pacifica Foundation*, 56 FCC 2d 94 (1975).

[18]The argument was based on four considerations:

(1) children have access to radio and in some cases are unsupervised by parents; (2) radio receivers are in the home, a place where people's privacy interest is entitled to extra deference; (3) unconsenting adults may tune in a station without any warning that offensive language is being or will be broadcast; and (4) there is a scarcity of spectrum space, the use of which the government must therefore license in the public interest.

[19]*FCC* v. *Pacifica Foundation*, 438 U.S. 726(1978).

[20]*Sonderling Broadcasting Corporation, WGLD-FM*, 27 Radio Reg. 2d 285 (FCC, 1973). The appeals case affirming the FCC ruling is *Illinois Citizens Committee for Broadcasting* v. *Federal Communications Commission*, 515 F. 2d 397 (D. C. Cir. 1975). See also Charles Feldman and Stanley Tickton, "Obscene/Indecent Programming: Regulation of Ambiguity," *Journal of Broadcasting*, 20 (Spring 1976), 273–82.

[21]*Pacifica Foundation*, 56 FCC 2d 94 (1975).

[22]Public Notice, FCC 87-153, April 29, 1987.

[23]*Miller* v. *California*, 413 U.S. 15 (1973). For a perspective on the cable television aspects of indecency, see Howard M. Kleiman, "Indecent Programming on Cable Television: Legal and Social Dimensions," *Journal of Broadcasting & Electronic Media*, 3 (Summer 1986), 275–294.

[24]The exception is Saturday nights.

[25]Robert P. Sadowski, "Broadcasting and State Statutory Laws," *Journal of Broadcasting*, 18 (Fall 1974), 435.

[26]Earl W. Kinter, *Michigan Law Review*, 64 (May 1966), 1280–81. Reprinted by permission.

[27]FCC Docket No. 20550, FCC 76-426, adopted June 22,1976. The FCC has revised its EEO requirements over the years; the most recent EEO reporting forms apply into the 1990s.

[28]The plan is a compilation of FCC guidelines and reporting requirements.

[29]*EEOC Sexual Harassment Guidelines*. Additional perspectives on sexual harassment in the workplace can be found in K. A. Thurston, "Sexual Harassment: An Organizational Perspective," *Personnel Administrator* (December 1980), 59–64; and C. R. Klasoon, D. E. Thompson, and G. L. Luben, "How Defensible Is Your Performance Appraisal System?" *Personnel Administrator* (December 1980), 77–83. A summary advisory and procedural statement—NAB Counsel L-026—has been authored by communications attorney Wade H. Hargrove of the firm of Tharrington, Smith & Hargrove, Raleigh, North Carolina.

CHAPTER 18

[1]Source of quote and information for the sections on common carriers was provided by the Federal Communications Commission and included the FCC publication *Common Carrier Services* (1977) and updated information on important court and FCC decisions.

[2]*Common Carrier Services*, p. 2.

[3]Ibid.

[4]*Hush-a-Phone* v. *United States*, 238 F. 2d 266 (D.C. Cir. 1956).

[5]*Carterfone Device*, 13 FCC 2d 420, 423 (1968).

[6]*International Tel. & Tel. Co.* v. *General Tel. & Electronics Corp.*, 351 F. Supp. 1153 (D. Hawaii 1972). Refer also to *United States* v. *Western Elec. Co.*, 1956 Trade Cas. 68,246 (D. N.J. 1956) (consent decree).

[7]Information for this section on cable regulation was derived from FCC rules and publications. Especially helpful was the FCC Information Bulletin *Cable Television* (March 1982).

[8]*United States* v. *Southwestern Cable Co.*, 392 U.S. 157 (1968).

[9]A discussion of the compromise made in the 1972 rules and an example of the issues that can confront a local change of service can be found, respectively, in Harvey Jassem, "The Selling of the Cable TV Compromise," *Journal of Broadcasting*, 17 (Fall 1973), 427–36; and Norman Felsenthal, "Cherry-Picking, Cable, and the FCC," *Journal of Broadcasting*, 19 (Winter 1975), 43–53. The 1972 rules were issued in *Cable Television Report and Order*, 36 FCC 2d 143 (1972). Discussion of specific cable rules can be found in the definitions cited on pp. 7–8 of *Regulatory Developments in Cable Television* (Washington, D.C.: Federal Communications Commission, 1977); see also FCC Information Bulletin 13632, *Cable Television* (March 1979) and the most recent editions of *Cable Sourcebook*.

[10]Vernone Sparkes, "Local Regulatory Agencies for Cable Television," *Journal of Broadcasting*, 19 (Spring 1975), 228–29.

[11]*Regulatory Developments in Cable Television*. These standards are also discussed in *Cable Television* (March 1982).

[12]Keep in mind that franchises do vary from city to city.

[13]See Frederick W. Ford and Lee G. Lovett, "State Regulation of Cable Television, Part 1: Current Statutes," *Broadcast Management/Engineering*, 10 (June 1974), 18, 21, 50; "Part II: States with No CATV Statutes: Short-Term and Long-Term Trends," *Broadcast Management/Engineering*, 30 (June 1974), 20, 21, 22.

[14]The discussion of these requirements is adapted from *Statement of Account for Secondary Transmission by Cable Systems*, Forms CS/SA-1-3 (Washington, D.C.: Licensing Division, United States Copyright Office), p. ii.

[15]Information on compulsory licensing is from the Copyright Office's *Statement of Account SA3* (Long Form) for period beginning January 1, 1989.

[16]Ibid.

[17]Ibid., p. 1.

[18]Ibid.

[19]Ibid.

[20]*Statement of Account*, Short Form (CS/SA-1), p. iii.

[21]Ibid.

[22]Ibid.

CHAPTER 19

[1]From NAB's *Radioactive*.

[2]Dick Stein, "Co-Op Q & A, Part I," *Radioactive*, 3 (May 1977), 16.

[3]See, for example, "Barter Taking $450 million Bite Out of Spot," *Broadcasting* (June 17, 1985), 27–29.

[4]"Perils and Pleasures of Selling Combos," *Broadcasting* (September 23, 1985), 63–64.

[5]"Radio's Promotional Efforts on the Rise," *Broadcasting* (October 3, 1988), 53–55.

[6]See John Coughlan, *Accounting Manual for Radio Stations* (Washington, D.C.: National Association of Broadcasters, 1975), p. 21.

[7]Interview with the late Richard A. Shaheen.

[8]For example, see Barry J. Dickstein, "True Station Value Is Key Ingredient in Broadcast Financing," *Broadcast Management/Engineering* (September 1976), 41–42; and Harold Poole, "What's Your Station's Worth?" *Radioactive* (July 1977), 16–17.

[9]Two sources were particularily useful in completing this section of the text: A checklist on selling broadcast properties provided by the late Richard A. Shaheen, who for many years operated a successful brokerage business. Also, E.G. Krasnow, R.B. Martin, and D. R. Brenner, "The Business Aspects of Selling a Broadcast Station," *Broadcast Management Financial Journal* (July–August 1988), 4–6.

CHAPTER 20

[1]See Donald Hurwitz, "Broadcast Ratings: The Missing Dimension," *Critical Studies in Mass Communication*, 1 (June 1984), 205–15.

[2]"The System for Success: Nielsen," brochure published by Nielsen Media Research (November 1988), p. 8.

[3]Harry Field and Paul F. Lazarsfeld, *The People Look at Radio* (Chapel Hill: The University of North Carolina Press, 1946).

[4]Level of confidence is at the 95 percent level. Detailed discussions of sampling can be found in various statistics books. A general description of the sampling process is Maxwell McCombs, "Sampling Opinions and Behaviors," in *Handbook of Reporting Methods*, eds, Maxwell McCombs, Donald Lewis Shaw, and David Grey (Boston: Houghton Mifflin, 1976), pp. 123–38.

[5]For example, Nielsen uses and intreprets data using the Block/Group Enumeration District (BG/ED) of the U.S. Bureau of the Census. These small geographic areas, when compared with each other, can yield significantly different viewing patterns, even within the same Zip Code areas. *Source*: "Nielsen TV Conquest: The Bg/ED Advantage," brochure published by Nielsen Media Research, February 1989.

[6]See, for example, John Colombotos, "Personal Versus Telephone Interviews: Effect of Responses," *Public Health Reports*, 84 (September 1969), 773–82; and T. F. Rogers, "Interviews by Telephone and in Person: Quality of Responses and Field Performance," *Public Opinion Quarterly*, 40 (Spring 1976), 51–65. The actual questions asked in surveys can also affect the results: Bradley S. Greenberg, Brenda Dervin, and Joseph Dominick, "Do People Watch 'Television' as 'Programs'?" *Journal of Broadcasting*, 12 (Fall 1968), 367.

[7]Various materials published by the rating services are helpful in understanding the rating process. See "Nielsen TV Conquest: The BG/ED Advantage," from Nielsen Media Research, 1989; "Nielsen: The System for Success," from Nielsen Media Research, 1989; *Standard Definitions of Broadcast Research Terms* (Washington, D.C.: National Association of Broadcasters, 1973).

[8]*Understanding and Using Radio Audience Estimates* (Beltsville, Md.: American Research Bureau, 1976), p. 4.

[9]Adapted from Ibid., p. 6.

[10]See V. Gay, "Nielsen Gives in, Holds up People Meters," *Advertising Age* (May 26, 1986), 12.

[11]L. Egerter, "Rating Arbitron's New Diary," *Radio Only*, 8 (April 1989), 19–24.

[12]I. Teinowitz, "People Meters Miss Kida: JWT," *Advertising Age* (July 18, 1988), 35.

[13]"Spanish-Language Net Still Battling Ratings," *Advertising Age* (February 26, 1973). See also, "Language Problem," *Broadcasting*, 45 (April 26, 1976), 5.

CHAPTER 21

[1]Qualitative studies have different interpretations and are looked upon differently by people with different methodological perepectives. See Klaus Bruhn Jensen, "Qualitative Audience Research: Toward an Integrative Approach to Reception," *Critical Studies in Mass Communication*, 4 (1987), 21–36. Robert S. Fortner and Clifford G. Christians, in "Separating Wheat from Chaff in Qualitative Studies," *Research Methods in Mass Communication*, Guido H. Stempel III, and Bruce H. Westley, eds. (Englewood Cliffs, N.J.: Prentice-Hall, 1981), pp. 363–74.

[2]An interesting perspective on network research operations can be found in a paper presented by William S. Rubens to the Advertising Research Foundation's 10th Annual Midyear Conference in Chicago, September 1984, and titled, "Program Research at NBC, Or Type I Error as a Way of Life." Reprinted and distributed by the National Association of Broadcasters. See also H. Stipp, K. Hill-Scott, and A. Dorr, "Using Social Science to Improve Chil-

dren's Television: An NBC Case Study," *Journal of Broadcasting and Electronic Media*, 31 (Fall 1987), 461–73.

[3]Thomas E. Coffin, "Progress to Date in Radio and Television Research" (paper presented at the 1960 conference of the American Association for Public Opinion Research).

[4]Lyn Garson of the PBS Research Office to the author, October 31, 1977.

[5]*A Broadcast Research Primer* (Washington, D.C.: National Association of Broadcasters, 1974).

[6]An example from one type of format can be seen in L. Loro, "Radio Downplays Ratings in New Age," *Advertising Age* (June 20, 1988), 100.

[7]Mark I. Kassof, "Using Multi-Dimensional Scaling to Find Your Station's Niche," National Association of Broadcasters, Research and Planning memorandum, October 1985.

[8]B. Bogart, "Pollsters Given Vote of Confidence," in a special report section of *Advertising Age* (November 2, 1987), S-1 thru S-4.

[9]Adapted from guidelines for NBC pollsters and prepared through the assistance of Denise A. Bittner, who served as an NBC pollster.

[10]See D. Charles Whitney, "The Poll Is Suspect," *Quill*, 65 (July-August 1976), 23; G. Cleveland Wilhoit and Maxwell McCombs, "Reporting Surveys and Polls," in *Handbook of Reporting Methods*, Maxwell McCombs, Donald Lewis Shaw, and David Grey, eds. (Boston: Houghton Mifflin, 1976), pp. 81–95; and Philip Meyer, *Precision Journalism* (Bloomington: Indiana University Press), p. 197.

[11]The ten considerations that follow are adapted from Whitney, "The Poll Is Suspect," p. 25.

[12]William S. Rubens, "High-Tech Audience Measurement for New-Tech Audiences, *Critical Studies in Mass Communication* (June 1984), 204–205.

[13]See "Tell if Speaker Is Boring," *Science Newsletter* (May 24, 1952), 325; and Elwood A. Kretsiner, "Gross Bodily Movement as an Index of Audience Interest," *Speech Monographs*, 19 (1952), 244–48.

[14]See Claude Hall, "Doomsday Machine Will Evaluate," *Billboard* (December 24, 1974), 1, 25–28.

[15]See Ralph Schoenstein, "Watching Howard Cosell for the Love of Science," *TV Guide*, 24 (February 21, 1976), 18–21.

[16]James Lull, "Ethnographic Studies of Broadcast Media Audiences," in *Broadcasting Research Methods*, Joseph R. Dominick and James E. Fletcher, eds. (Boston: Allyn & Bacon, 1985), pp. 80–88.

CHAPTER 22

[1]Paul Lazarsfeld, Bernard Berelson, and H. Gaudet, *The People's Choice* (New York: Columbia University Press, 1948).

[2]See M. L. DeFleur and S. Ball-Rokeach, *Theories of Mass Communication*, 4th ed. (New York: Longman, 1982), pp. 185–93.

[3]A discussion of how the categories approach evolved from the bullet theory and how it fits into current communication theory is found in Wilbur Schramm and Donald Roberts, *The Process and Effects of Mass Communication* (Urbana: University of Illinois Press, 1971), pp. 4–53.

[4]Lazarsfeld, Berelson, and Gaudet, *The People's Choice*.

[5]William Stephenson, *The Play Theory of Mass Communication* (Chicago: University of Chicago Press, 1967).

[6]See, for example, J. D. Robinson, "Mass Media and the Elderly: A Uses and Dependency Interaction," in *Life-Span Communication: Normative Processes*, J.F. Nussbaum, ed. (Hillsdale, N.J.: Lawrence Erlbaum, 1989), pp. 319–37. Also, A. Johnston Wadsworth, "The Uses and Effects of Mass Communication During Childhood," in *Life-Span Communication: Normative Processes*, pp. 93–116.

[7]Wadsworth, "The Uses and Effects of Mass Communication During Childhood," pp. 93–116. An additional perspective can be seen in T. H. A. van der Voort and M. W. Vooijs, "Validity of Children's Direct Estimates of Time Spent Television Viewing," *Journal of Broadcasting and Electronic Media*, 34 (1990), 93–99.

[8]The theory developed through the early work of Maxwell B. McCombs and Donald Shaw, "The Agenda-Setting Function of Mass Media," *Public Opinion Quarterly*, 36 (1972), 176–87.

[9]*Surgeon General's Scientific Advisory Committee on Television and Social Behavior*, 1972.

[10]Ibid.

[11]See, for example, Seymour Feshbach, "The Stimulating vs. Cathartic Effects of a Vicarious Aggressive Experience," *Journal of Abnormal and Social Psychology*, 63 (1961), 381–85.

[12]See Leonard Berkowitz, *Aggression: A Social Psychological Analysis* (New York: McGraw-Hill, 1962).

[13]See Joseph Klapper, *The Effects of Mass Communication* (New York: Free Press, 1960).

[14]Albert Bandura and Richard Walters, *Social Learning and Personality Development* (New York: Holt, Rinehart & Winston, 1963).

[15]George Comstock, "Types of Portrayal and Aggressive Behavior," *Journal of Communication*, 27 (Summer 1977), 189–98.

[16]Charles N. Barnard, "An Oriental Mystery," *TV Guide*, 26 (January 28, 1978), 2–4, 6, 8.

[17]See B. Watkins, "Television Viewing as a Dominant Activity of Childhood: A Developmental Theory of Television Effects," *Critical Studies in Mass Communication*, 2 (1985), 323–27.

[18]See N. Signorielli and M. Morgan, eds., *Cultivation Analysis: New Directions in Media Effects Research*, (Newbury Park, Calif.: Sage, 1990). Also, J. Bryant, "The Road Most Traveled: Yet Another Cultivation Critique," *Journal of Broadcasting and Electronic Media*, 30 (Spring 1986), 231–44.

[19]See D. Zillman, "Mood Management Through Communication Choices," *American Behavioral Scientist*, 31 (January/February 1988), 327–40 (citing p. 328). The theory is discussed in a number of previous research studies. See J. Bryant and D. Zillman, "Using Television to Alleviate Boredom and Stress: Selective Exposure as a Function of Induced Excitational States," *Journal of Broadcasting*, 28 (1984), 1–20; also D. Zillman, "Mood Management: Using Entertainment to Full Advantage," in L. Donohew and others, eds. *Communication, Social Cognition, and Affect* (Hillsdale, N.J.: Lawrence Erlbaum, 1988), pp. 141–71.

[20]Ibid.

CHAPTER 23

[1]See, for example, Donald Hurwitz, "Broadcast Ratings: The Missing Dimension," *Critical Studies in Mass Communication* (June 1984), 205–25. The Hurwitz article provided the author of this book with rich material for examples and is gratefully acknowledged. For other historical-critical perspectives, see Eileen R. Meehan, "Critical Theorizing on Broadcast History," *Journal of Broadcasting and Electronic Media* (Fall 1986), pp. 393–411; George Lipsitz, "This Ain't No Sideshow: Historians and Media Studies," *Critical Studies in Mass Communication* (1988), 147–61.

[2]Herta Herzog, "What Do We Really Know about Daytime Serial Listeners?" in *Radio Research, 1942–1943*, Paul F. Lazarsfeld and Frank Stanton, eds. (New York: Duell, Sloan and Pearce, 1944).

[3]An overview of these and other theories can be found in D. McQuail, *Mass Communication Theory: An Introduction* (London: Sage, 1987). Sources are cited by McQuail.

[4]T. Adorno and M. Horkheimer, "The Culture Industry: Enlightment as Mass Deception," in *Dialectics of Enlightment* (New York: Herder and Herder, 1972).

[5]See A. Gramsci, *Selections from the Prison Notebooks* (London: Lawrence and Wishart, 1971).

[6]Reflected in the writings of, among others, Stuart Hall. See, for example, S. Hall, "The Rediscovery of Ideology: Return of the Repressed Media Studies," in M. Gurevitch and others, eds., *Culture Society and the Media* (London: Methuen, 1972), pp. 56–90. Also, J. Fiske, *Introduction to Communication Studies*, (London: Methuen, 1982).

[7]See David Thornburn, "Television as an Aesthetic Medium," *Critical Studies in Mass Communication* 4 (1987), 161–73.

[8]See David Barker, "Television Production Techniques as Communication," *Critical Studies in Mass Communication*, 2 (1985), 234–46.

[9]For a perspective on ethics in journalism see Douglas Birkhead, "An Ethics of Vision for Journalism," *Critical Studies in Mass Communication*, 6 (1989), 283–94. The quotation is

Birkhead's interpretation of Kenneth Burke, which is cited in Birkhead's article. See Kenneth Burke, *Attitudes Toward History*. (Berkeley: University of California Press), 1959.

[10]Imbeded in the concept of agenda setting.

[11]Birkhead interpreting Burke, p. 285. Emphasis added.

[12]O.H. Gandy, Jr., "The Surveillance Society: Information Technology and Bureaucratic Social Control," *Journal of Communication*, 39 (Summer 1989), 61–76; A. Gillespie and K. Robins, "Geographical Inequalities: The Spacial Bias of the New Communications Technologies," *Journal of Communication*, 39 (Summer 1989), 7–18; J. Brunet and S. Proulx, "Formal versus Grass-Roots Training: Women, Work, and Computers," *Journal of Communication* (Summer 1989), 77–84; G. Murdock and P. Golding, "Information Poverty and Political Inequality: Citizenship in the Age of Privatized Communication," *Journal of Communication*, 39 (Summer 1989), 180–95. The Summer 1989 issue of *Journal of Communication* is devoted to the subject of the information society.

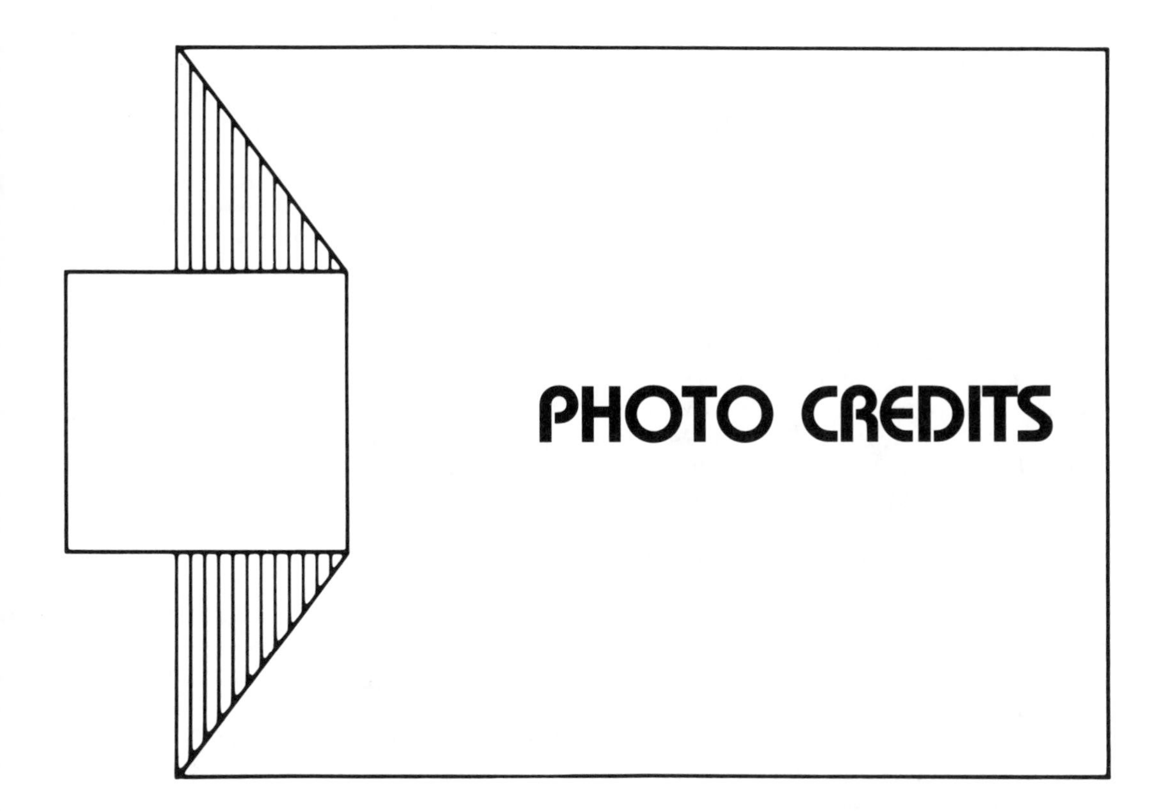

PHOTO CREDITS

Chapter 1

Fig. 1–1, Prentice Hall; Fig. 1–2, CNN and George Bennett; Fig. 1–3, © 1989 The Arbitron Company; Fig. 1–4, WTVD and Capital Cities/ABC, Inc.; Fig. 1–5, WNET/13; Fig. 1–6, Schramm et al. (Editors), HANDBOOK OF COMMUNICATION. Copyright © 1973 by Houghton Mifflin Company. Adapted with permission; Fig. 1–7, Courtesy Canon U.S.A., Inc.; Fig. 1–8, by Alonzo Pond for Wisconsin Public Radio; Fig. 1–9, Western Union and the TRW Defense and Space Systems Group; Fig. 1–10, photo by John R. Bittner, all rights reserved.

Chapter 2

Fig. 2–1, Science Museum, South Kensington, London; Fig. 2–2, Western Union; Fig. 2–3, The Marconi Company Limited, Marconi House, Chelmsford, Essex; Fig. 2–4, The Marconi Company Limited, Marconi House, Chelmsford, Essex; Fig. 2–5, RCA; Fig. 2–6, on display at the Science Museum, South Kensington, London. Photo by John R. Bittner, all rights reserved; Fig. 2–7, certificate photo by John R. Bittner, all rights reserved; Fig. 2–8, The Marconi Company Limited, Marconi House, Chelmsford, Essex; Fig. 2–9, Murray, Kentucky,

Chamber of Commerce; Fig. 2–10, RCA; General Electric Research and Development Center; Fig. 2–11, RCA.

Chapter 3

Fig. 3–1, *Radio News,* July 1923; Fig. 3–2, WWJ; Fig. 3–3, KDKA; Fig. 3–4, RCA; Fig. 3–5, AT&T, Fig. 3–6, © 1949, George Everson. From G. Everson, *The Story of Television: The Life of Philo T. Farnsworth* (New York: W. W. Norton & Co.); Fig. 3–7, RCA; Fig. 3–8, WNET/13 and CBS, Inc.; Fig. 3–9, courtesy, ABC; Wolper Productions, Inc.; Warner Bros.; Phil Gersh Agency; Bresler, Wolff, Cota, and Livingston; Fig. 3–10, © 1989 *EMMY* magazine, all rights reserved; used with permission.

Chapter 4

Fig. 4–1, AT&T; Fig. 4–2, H.M. Boettinger, *The Telephone Book* (New York: Riverwood Publishers Ltd., 1977); Fig. 4–3, photo courtesy Western Electric Company; Fig. 4–4, AT&T; Fig. 4–5, courtesy the Norwegian Telecommunications Administration, from the *Annual Report 1988,* NTA; Fig. 4–6, photo by John R. Bittner, all rights reserved; Fig. 4–7, Western Electric; Fig. 4–8, courtesy of Siecor Corporation, Hickory, North Carolina; Fig. 4–9, *Communications News;* Fig. 4–10, *Communications News;* Fig. 4–11, Times Fiber Communications, Inc.; Fig. 4–12, courtesy of Siecor Corporation, Hickory, North Carolina; Fig. 4–13, courtesy of Siecor Corporation, Hickory, North Carolina.

Chapter 5

Fig. 5–1, *FCC Broadcast Operators Handbook;* Fig. 5–2, *FCC Broadcast Operators Handbook;* Fig. 5–3, *FCC Broadcast Operators Handbook;* Fig. 5–4, *FCC Broadcast Operators Handbook;* Fig. 5–5, *FCC Broadcast Operators Handbook;* Fig. 5–6, Credits: *Left,* photo by John R. Bittner, all rights reserved. *Right,* the Hughes HS-376 satellite, courtesy of British Satellite Broadcasting; Fig. 5–7, photo by John R. Bittner, all rights reserved.

Chapter 6

Fig. 6–1, courtesy Aeronautic Ford and Ford Aerospace & Communications Corporation; Fig. 6–2, NASA; Fig. 6–3, Western Union; Fig. 6–4, British Satellite Broadcasting.

Chapter 7

Fig. 7–1, NCTA; Fig. 7–2, NCTA; Fig. 7–3, NCTA; Fig. 7–4, *Cable Television Business*; Fig. 7–5, NCTA; Fig. 7–6, NCTA.

Chapter 8

Fig. 8–1, Panasonic; Fig. 8–2, SONY; Fig. 8–3, Canon Inc.; Fig. 8–4, HDTV 1125/60 Group; Fig. 8–5, Source: Donald G. Fink, "The Future of High Definition Television," *SMPTE Journal,* February, 1980, vol. 89; Fig. 8–6, National Association of Broadcasters; Fig. 8–7, courtesy of Apple Computer; Fig. 8–8, Sigma Design's LaserView Display System; Fig. 8–9, KCET/Los Angeles, photo by M. Trumbo; Fig. 8–10, Official Airlines Guides, Inc.

Chapter 9

Fig. 9–1, WTVD and Capital Cities/ABC, Inc.; Fig. 9–2, WTVD and Capital Cities/ABC, Inc.; Fig. 9–3, Wall Street Journal Radio Network; Fig. 9–4, The Telemundo Network.

Chapter 10

Fig. 10–1, Office of Information of the University of Houston; Fig. 10–2, courtesy of the Children's Television Workshop; Fig. 10–3, © Family Communications, Inc.; Fig. 10–4, reprinted from *The Saturday Evening Post* © 1986; Fig. 10–5, Photographic Media Center, University of Wisconsin Extension, Madison, Wisconsin; Fig. 10–6, South Carolina Educational Television and Radio Network and the South Carolina State Department of Education; Fig. 10–7, photo by John R. Bittner, all rights reserved; Fig. 10–8, Inland Steel Company.

Chapter 11

Fig. 11–1, Turner Broadcasting Systems, Inc.; Fig. 11–2, © The Arbitron Ratings Company.

Chapter 12

Fig. 12–1, Grass Valley Group.

Chapter 13

Fig. 13–1, TVOntario; Fig. 13–2, IBA; Fig. 13–3, courtesy MTV, from the 1988 *MTV Annual Report*; Fig. 13–4, Norwegian Broadcasting Corporation; Fig. 13–5, from the East German government publication *Fernsehen der DDR*; Fig. 13–6, courtesy, Radio Moscow; Fig. 13–7, photo by Roger Picard for Radio France; Fig. 13–8, M6, by Jacques Le Goff; Fig. 13–9, NHK; Fig. 13–10, CNN.

Chapter 14

Fig. 14–1, courtesy, Radio Nederland; Fig. 14–2, Radio Moscow; Fig. 14–3, TV5 Europe; Fig. 14–4, The Children's Channel.

Chapter 19

Fig. 19–1, WAZY; Fig. 19–2, WAZY; Fig. 19–3, courtesy WBTV, Charlotte, North Carolina; Fig. 19–4, National Association of Broadcasters; Fig. 19–5, National Association of Broadcasters; Fig. 19–6, National Association of Broadcasters.

Chapter 20

Fig. 20–1, © 1989 The Arbitron Company; Fig. 20–2, Arbitron Radio.

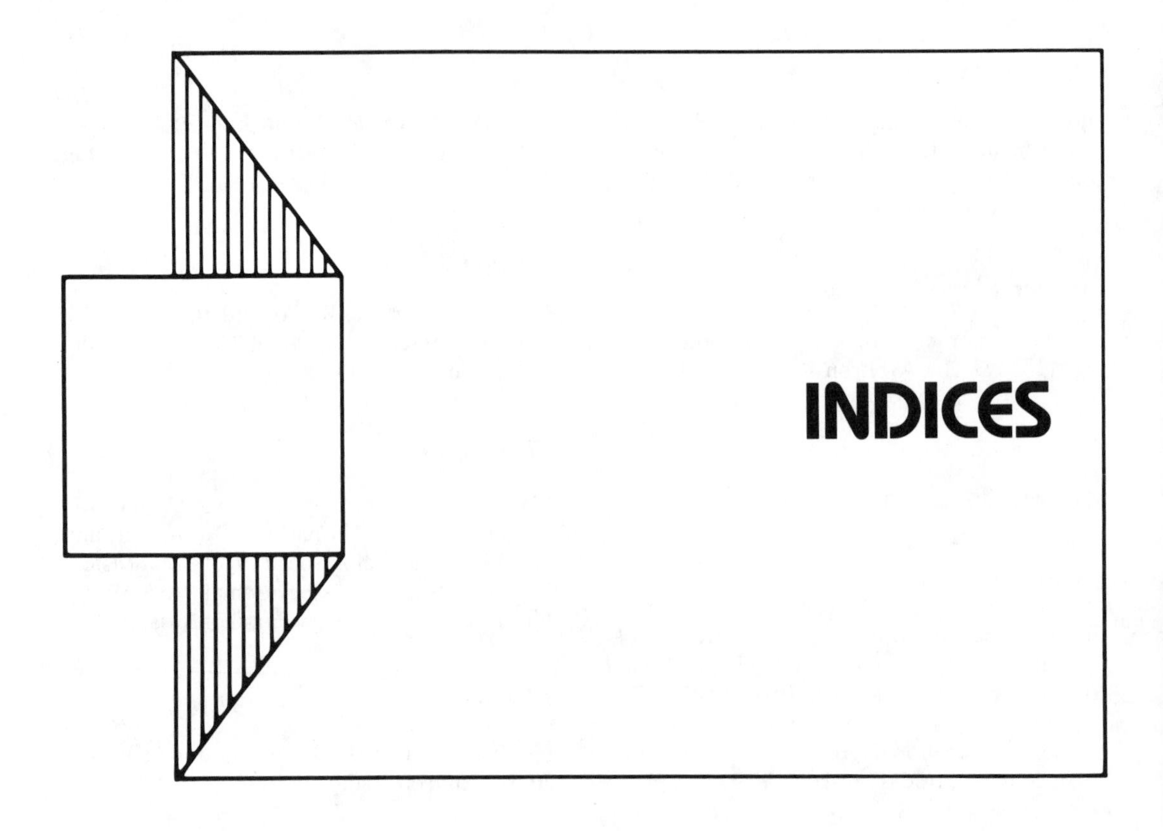

NAME

SUBJECT

1935 1940 1945 1950 1955 1960

1941
Mayflower decision discourages editorializing.

1943
Edward G. Noble buys NBC Blue, becomes ABC in 1945.

1935
Dr. Frank Stanton leaves position at The Ohio State University to join CBS's research efforts.

1938
First over-the-air ETV programming.

1941
FCC issues *Report on Chain Broadcasting.*

1939
G.R. Stibitz and S.B. Williams build Complex Number Calculator.

1937
Howard Aiken develops electromechanical computer.

1939
Television is introduced at the World's Fair.

1938
Orson Welles makes famous "War of the Worlds" broadcast.

1954
Edward R. Murrow confronts Senator Joseph McCarthy's "Red Scare" tactics on *See It Now.*

1953
KUHT signs on as first educational television station under new FCC allocations for ETV.

1956
Ampex engineers demonstrate videotape recording.

1953
ABC merges with United Paramount Theatres, Inc.

1948–1953
FCC freezes television allocations.

1956
United Press International launches audio service, first for a wire service.

1950
First successful recording of color television.

1946
J.P. Eckert and J. Mauchley develop ENIAC computer.

1955
Association for Professional Broadcasting Education is founded. Later becomes Broadcast Education Association (B.E.A.).

1948
CBS conducts the first of various "talent raids" the networks participate in during the late 1940s.

1947
Transistor is invented at Bell Laboratories.

1948
Cable systems begin in Oregon and Pennsylvania.

1945
Blue network changes name to American Broadcasting Company.

1955
Radio programming begins to specialize. Rock and roll formats develop.

1949
Fairness Doctrine is issued.

1961–1968
Midwest Program for Airborne Television Instruction (MPATI).

1958
J.S. Sibley of Texas Instruments develops integrated circuit.

1957
RCA demonstrates color videotape recording.

1960
First televised Presidential debates between John F. Kennedy and Richard M. Nixon.

1961
Spanish International Network (SIN) begins operation.

1935 1940 1945 1950 1955 1960